nature

The Living Record of Science

《自然》百年科学经典

英汉对照版（平装本）

第一卷（下）

总顾问：李政道（Tsung-Dao Lee）

英方主编：Sir John Maddox
Sir Philip Campbell
中方主编：路甬祥

1869-1930

外语教学与研究出版社 · 麦克米伦教育 · 自然科研

FOREIGN LANGUAGE TEACHING AND RESEARCH PRESS · MACMILLAN EDUCATION · NATURE RESEARCH

北京 BEIJING

图书在版编目（CIP）数据

《自然》百年科学经典．第一卷．下，1869-1930 ：英汉对照 ／（英）约翰·马多克斯（Sir John Maddox），（英）菲利普·坎贝尔（Philip Campbell），路甬祥主编．—— 北京 ：外语教学与研究出版社，2016.9（2018.12 重印）

ISBN 978-7-5135-8070-0

Ⅰ．①自… Ⅱ．①约… ②菲… ③路… Ⅲ．①自然科学－文集－英、汉 Ⅳ．①N53

中国版本图书馆 CIP 数据核字（2016）第 225008 号

出 版 人　蔡剑峰
项目统筹　章思英　Charlotte Liu（加拿大）
项目负责　刘晓楠　黄小斌　Chris Balderston（美）
责任编辑　黄小斌
责任校对　王丽霞
装帧设计　孙莉明　天泽润
出版发行　外语教学与研究出版社
社　　址　北京市西三环北路 19 号（100089）
网　　址　http://www.fltrp.com
印　　刷　北京华联印刷有限公司
开　　本　787×1092　1/16
印　　张　30.25
版　　次　2016 年 9 月第 1 版　2018 年 12 月第 2 次印刷
书　　号　ISBN 978-7-5135-8070-0
定　　价　168.00 元

购书咨询：（010）88819926　电子邮箱：club@fltrp.com
外研书店：https://waiyants.tmall.com
凡印刷、装订质量问题，请联系我社印制部
联系电话：（010）61207896　电子邮箱：zhijian@fltrp.com
凡侵权、盗版书籍线索，请联系我社法律事务部
举报电话：（010）88817519　电子邮箱：banquan@fltrp.com
物料号：280700001

《自然》百年科学经典（英汉对照版）

编译委员会

Contents
目录

Volume I

(1869-1930)

Molecules*

J. C. Maxwell

Editor's Note

James Clerk Maxwell was sometimes accused of being a poor lecturer, but this talk delivered to the British Association offers a lucid, engaging picture of the current understanding of the molecular nature of matter. Maxwell made a decisive contribution himself with his kinetic theory of gases, which explained how the macroscopic properties of gases, such as the laws relating pressure, volume and temperature, could be explained from the microscopic motions of the constituent particles. Maxwell's estimate of the size of a hydrogen molecule is only slightly bigger than the modern view. And his discussion of molecular diffusion anticipates the work of Albert Einstein and Jean Perrin on Brownian motion that provided the first real evidence for molecules as physical entities.

AN atom is a body which cannot be cut in two. A molecule is the smallest possible portion of a particular substance. No one has ever seen or handled a single molecule. Molecular science, therefore, is one of those branches of study which deal with things invisible and imperceptible by our senses, and which cannot be subjected to direct experiment.

The mind of man has perplexed itself with many hard questions. Is space infinite, and if so in what sense? Is the material world infinite in extent, and are all places within that extent equally full of matter? Do atoms exist, or is matter infinitely divisible?

The discussion of questions of this kind has been going on ever since men began to reason, and to each of us, as soon as we obtain the use of our faculties, the same old questions arise as fresh as ever. They form as essential a part of the science of the nineteenth century of our era, as of that of the fifth century before it.

We do not know much about the science organisation of Thrace twenty-two centuries ago, or of the machinery then employed for diffusing an interest in physical research. There were men, however, in those days, who devoted their lives to the pursuit of knowledge with an ardour worthy of the most distinguished members of the British Association; and the lectures in which Democritus explained the atomic theory to his fellow-citizens of Abdera realised, not in golden opinions only, but in golden talents, a sum hardly equalled even in America.

---

* Lecture delivered before the British Association at Bradford, by Prof. Clerk Maxwell, F. R. S.

分 子*

麦克斯韦

编者按

有人认为詹姆斯·克拉克·麦克斯韦缺乏演讲天赋，但这篇在英国科学促进会所作的报告中，麦克斯韦对物质分子的那些得到普遍接受的性质描述得非常透彻，给人留下了深刻的印象。麦克斯韦在分子学方面作出了有决定意义的贡献，他提出的气体动力学理论能够说明如何用组成粒子的微观运动来解释气体的宏观性质，如与压力、体积和温度相关的定律。麦克斯韦对氢分子大小的估计只比现在的公认值略大一些。他对分子扩散现象的讨论促使阿尔伯特·爱因斯坦和让·佩兰开始了关于布朗运动的研究，这使人们第一次认识到分子是一种物理实体。

原子是不能被一分为二的实体。分子是组成物质的最小单位。没有人看见或者摆弄过单个分子。因此，分子科学是研究不可见也不可感觉的事物的一门学问，我们无法对它进行直接实验。

人类经常思索很多难以回答的问题。空间是无限的吗？如果是，是从什么意义上讲的？物质世界的范围是无限的吗？在这个范围内是不是每个地方都同等地充满了物质？原子存在吗？或物质是否无限可分？

自从人类开始理性思考以来，关于这类问题的讨论就一直没有停止过。对于我们每个人来说，一旦开始用心智思考，那些古老的问题就会像从前一样令人觉得新奇。不论是在我们所处的19世纪，还是在公元前5世纪，这些问题都构成了科学的基本部分。

我们对2,200年前位于色雷斯的科学组织所知甚少，也不知道他们用何种方式来传播对自然研究的兴趣。不过那时候确实有人毕生追求知识，热情不亚于英国科学促进会中最杰出的成员。当德谟克利特向他的阿布德拉市民开设讲座讲解自己的原子理论时，他获得的高度评价和丰厚报酬即使在今天的美国也很少有人能比得上。

* 皇家学会会员克拉克·麦克斯韦教授在布拉德福德对英国科学促进会作的报告。

To another very eminent philosopher, Anaxagoras, best known in the world as the teacher of Socrates, we are indebted for the most important service to the atomic theory, which, after its statement by Democritus, remained to be done. Anaxagoras, in fact, stated a theory which so exactly contradicts the atomic theory of Democritus that the truth or falsehood of the one theory implies the falsehood or truth of the other. The question of the existence or non-existence of atoms cannot be presented to us this evening with greater clearness than in the alternative theories of these two philosophers.

Take any portion of matter, say a drop of water, and observe its properties. Like every other portion of matter we have ever seen, it is divisible. Divide it in two, each portion appears to retain all the properties of the original drop, and among others that of being divisible. The parts are similar to the whole in every aspect except in absolute size.

Now go on repeating the process of division till the separate portions of water are so small that we can no longer perceive or handle them. Still we have no doubt that the sub-division night be carried further, if our senses were more acute and our instruments more delicate. Thus far all are agreed, but now question arises. Can this sub-division be repeated for ever?

According to Democritus and the atomic school, we must answer in the negative. After a certain number of sub-divisions, the drop would be divided into a number of parts each of which is incapable of further sub-division. We should thus, in imagination, arrive at the atom, which, as its name literally signifies, cannot be cut in two. This is the atomic doctrine of Democritus, Epicurus, and Lucretius, and, I may add, of your lecturer.

According to Anaxagoras, on the other hand, the parts into which the drop is divided, are in all respects similar to the whole drop, the mere size of a body counting for nothing as regards the nature of its substance. Hence if the whole drop is divisible, so are its parts down to the minutest sub-divisions, and that without end.

The essence of the doctrine of Anaxagoras is that the parts of a body are in all respects similar to the whole. It was therefore called the doctrine of Homoiomereia. Anaxagoras did not of course assert this of the parts of organised bodies such as men and animals, but he maintained that those inorganic substances which appear to us homogeneous are really so, and that the universal experience of mankind testifies that every material body, without exception, is divisible.

The doctrine of atoms and that of homogeneity are thus in direct contradiction.

But we must now go on to molecules. Molecule is a modern word. It does not occur in *Johnson's Dictionary*. The ideas it embodies are those belonging to modern chemistry.

另一位杰出的哲学家，即以身为苏格拉底的老师而闻名于世的阿那克萨哥拉在德谟克利特之后对原子学说作出了最重要的贡献。实际上，阿那克萨哥拉和德谟克利特两人的原子学说是如此地针锋相对，一方正确则另一方必错。今晚我们对原子到底是否存在这一问题的讨论,用这两位哲学家的对立理论来表达,是最清楚不过了。

随便取一份物质,比如一滴水,来观察它的性质。就像我们看到的其他物质一样，它是可以分割的。把它分成两份，每一份都保持原来那滴水的所有性质，其他可以分割的物质也是一样。每一个部分除了尺寸比整体小些，其他各方面都和整体相似。

就这么一直分下去，直到分出来的水滴小到我们再也看不见，也无法对它们进行操作。但是大家都明白，如果我们的感官更敏锐，我们的设备更精密，细分过程还是可以接着进行下去的。到此为止不会有什么异议，但是现在问题就来了：这样的细分过程可以永远继续下去吗?

德谟克利特和原子学派的回答是否定的。经过一定次数的分割后，水滴就被分成很多很小的部分，每一部分都不能进一步细分了。也就是我们已经细分到了想象中的原子，原子这个名字的字面含义就是不可分的意思。这就是德谟克利特、伊壁鸠鲁、卢克莱修还有我，你们的演讲者，所赞成的原子论。

另一方面，阿那克萨哥拉认为，水滴被分割成的各个部分，除了和物质性质无关的物体尺寸发生了变化以外，其他一切方面都和整个的水滴类似。因此，如果原来的水滴是可分的，那么分割得到的各个部分也应该是可分的，哪怕分到极小，永无止境。

阿那克萨哥拉学说的根本点是：一个物体的部分和它的整体在所有方面都是相似的，所以被称为同质性学说。阿那克萨哥拉当然没有把它应用到人和动物这样的有机体的身上，但是他认为那些看上去是均质的无机物的确是同质的，并且人类的普遍经验也能证实每一种物质实体毫无例外都是可分的。

原子学说和同质性学说就是这样地针锋相对。

但是现在我们必须转入对分子这个现代名词的讨论，它在《约翰逊词典》里是没有的。分子所包含的概念属于现代化学的范畴。

A drop of water, to return to our former example, may be divided into a certain number, and no more, of portions similar to each other. Each of these the modern chemist calls a molecule of water. But it is by no means an atom, for it contains two different substances, oxygen and hydrogen, and by a certain process the molecule may be actually divided into two parts, one consisting of oxygen and the other of hydrogen. According to the received doctrine, in each molecule of water there are two molecules of hydrogen and one of oxygen. Whether these are or are not ultimate atoms I shall not attempt to decide.

We now see what a molecule is, as distinguished from an atom.

A molecule of a substance is a small body such that if, on the one hand, a number of similar molecules were assembled together they would form a mass of that substance, while on the other hand, if any portion of this molecule were removed, it would no longer be able, along with an assemblage of other molecules similarly treated, to make up a mass of the original substance.

Every substance, simple or compound, has its own molecule. If this molecule be divided, its parts are molecules of a different substance or substances from that of which the whole is a molecule. An atom, if there is such a thing, must be a molecule of an elementary substance. Since, therefore, every molecule is not an atom, but every atom is a molecule, I shall use the word molecule as the more general term.

I have no intention of taking up your time by expounding the doctrines of modern chemistry with respect to the molecules of different substances. It is not the special but the universal interest of molecular science which encourages me to address you. It is not because we happen to be chemists or physicists or specialists of any kind that we are attracted towards this centre of all material existence, but because we all belong to a race endowed with faculties which urge us on to search deep and ever deeper into the nature of things.

We find that now, as in the days of the earliest physical speculations, all physical researches appear to converge towards the same point, and every inquirer, as he looks forward into the dim region towards which the path of discovery is leading him, sees, each according to his sight, the vision of the same quest.

One may see the atom as a material point, invested and surrounded by potential forces. Another sees no garment of force, but only the bare and utter hardness of mere impenetrability.

But though many a speculator, as he has seen the vision recede before him into the innermost sanctuary of the inconceivably little, has had to confess that the quest was not for him, and though philosophers in every age have been exhorting each other to direct their minds to some more useful and attainable aim, each generation, from the earliest dawn of science to the present time, has contributed a due proportion of its ablest intellects to the quest of the ultimate atom.

回到我们原来的例子，一滴水能够最大限度地被分割成一定数量的彼此相似的部分。每一个这样的部分都是现代化学家称作的水分子。水分子包含两种不同的物质，氧和氢，所以它绝不是一个原子。通过一定的处理方式，确实可以把水分子分解成两部分，一部分含氢，另一部分含氧。根据公认的理论，每一个水分子都包含两个氢分子和一个氧分子。至于这些氢分子和氧分子是不是不能再分解的原子，我这里先不去确定。

现在我们明白了分子是什么，它和原子有什么不同。

一种物质的分子是很小的，一方面，如果许多同样的分子聚合在一起，就会形成大量这种物质；另一方面，如果这些分子的某个部分缺失，它们与其他经过相同处理的分子聚合在一起也不能形成原来的物质。

任何物质，无论是简单的还是复合的，都有自己的分子。如果这个分子再被分割，形成的部分就是其他物质的分子。一个原子，如果确实存在的话，应该是一种基本物质的分子。不是每个分子都是原子，但是每个原子都是分子，所以我将使用含义更广的分子这个术语。

我不想浪费时间去详细解释现代化学关于各种物质分子的理论。我作这个报告的目的是讲述分子科学中普遍的而不是具体的问题。不是因为我们恰好是化学家、物理学家或者其他某个领域的专家才对这个与所有物质息息相关的中心问题感兴趣，而是因为我们都属于人类这个物种，其所具备的资质促使我们不断地深入研究事物的本质。

现在我们发现，就像早期物理猜想时代一样，所有的物理研究似乎都汇集到了同一点上；每一个探求者，在眺望发现之途指向的茫茫区域时，虽然目力各有不同，看到的却都是同一件宝物的幻象。

有些人眼中的原子是一个物质点，被有势力场包围着。另一些人则看不到力的存在，只看到裸露而坚硬的不可穿透的实体。

很多人看到幻象在眼前消退而去，躲进那不可思议的渺小之物最隐秘的庇护所之后，不得不承认宝物非他所属；各个时代的哲人们互相劝诫对方去追求更实际、更容易达到的目标。虽然如此，自从科学的启蒙时期直到今天，每一代都不乏最富才智的人投身于对最终原子的探求。

Our business this evening is to describe some researches in molecular science, and in particular to place before you any definite information which has been obtained respecting the molecules themselves. The old atomic theory, as described by Lucretius and revived in modern times, asserts that the molecules of all bodies are in motion, even when the body itself appears to be at rest. These motions of molecules are in the case of solid bodies confined within so narrow a range that even with our best microscopes we cannot detect that they alter their places at all. In liquids and gases, however, the molecules are not confined within any definite limits, but work their way through the whole mass, even when that mass is not disturbed by any visible motion.

This process of diffusion, as it is called, which goes on in gases and liquids and even in some solids, can be subjected to experiment, and forms one of the most convincing proofs of the motion of molecules.

Now the recent progress of molecular science began with the study of the mechanical effect of the impact of these moving molecules when they strike against any solid body. Of course these flying molecules must beat against whatever is placed among them, and the constant succession of these strokes is, according to our theory, the sole cause of what is called the pressure of air and other gases.

This appears to have been first suspected by Daniel Bernoulli, but he had not the means which we now have of verifying the theory. The same theory was afterwards brought forward independently by Lesage, of Geneva, who, however, devoted most of his labour to the explanation of gravitation by the impact of atoms. Then Herapath, in his "Mathematical Physics", published in 1847, made a much more extensive application of the theory to gases, and Dr. Joule, whose absence from our meeting we must all regret, calculated the actual velocity of the molecules of hydrogen.

The further development of the theory is generally supposed to have been begun with a paper by Krönig, which does not, however, so far as I can see, contain any improvement on what had gone before. It seems, however, to have drawn the attention of Prof. Clausius to the subject, and to him we owe a very large part of what has been since accomplished.

We all know that air or any other gas placed in a vessel presses against the sides of the vessel, and against the surface of any body placed within it. On the kinetic theory this pressure is entirely due to the molecules striking against these surfaces, and thereby communicating to them a series of impulses which follow each other in such rapid succession that they produce an effect which cannot be distinguished from that of a continuous pressure.

If the velocity of the molecules is given, and the number varied, then since each molecule, on an average, strikes the side of the vessel the same number of times, and with an impulse of the same magnitude, each will contribute an equal share to the whole pressure. The pressure in a vessel of given size is therefore proportional to the number of molecules in it, that is to the quantity of gas in it.

我们今天晚上的任务是介绍分子科学中的一些研究成果，特别是向你们展示那些关于分子本身的现在已经比较确定的认识。卢克莱修所描述的旧原子论到了现代重获新生。他认为所有物体中的分子都在不停地运动，即便当物体本身处于静止状态时也不例外。在固体中，分子的运动只局限于一个很小的范围，哪怕是利用当前最好的显微镜我们也察觉不到它们的移动。但是对于液体和气体的情况，分子的运动没有受到确切的范围限制，可以在整个物质中移动，哪怕这个整体没有受到任何可见的运动的干扰。

这个过程被称为扩散。它在气体、液体甚至一些固体中持续进行着，可以由实验验证，同时它也是分子运动论最有力的证明之一。

现在分子科学的最新进展，是从研究这些运动着的分子碰撞固体表面的机械效应开始的。飞行中的分子一定会撞击所有置于其中的物质。根据我们的理论，这种持续不断的撞击，就是产生所谓的空气和其他气体压力的唯一原因。

丹尼尔·伯努利可能是第一个想到这一点的人，但是他当时没有我们今天的实验手段来验证他的理论。后来日内瓦的勒萨热也独立提出过这个理论，但是他的工作主要是用原子碰撞来解释重力现象。赫拉帕斯在他 1847 年出版的《数学物理学》一书中，将这个理论更广泛地应用于各种气体。焦耳博士计算了氢分子的实际速度，今天他没有在场实在让人感到遗憾。

大家普遍认为这一理论的进一步发展是从克勒尼希的一篇论文开始的。但在我个人看来，这篇文章本身并没有在前人工作的基础上作出什么改进。不过它引起了克劳修斯教授对这个问题的关注，而后者对以后的理论发展起了很大作用。

我们都知道放置在一个容器中的空气或其他气体会对容器壁以及放置在其中的其他物体表面产生压力。根据气体动力学理论，这个压力完全是由分子碰撞这些表面产生的。这种碰撞带给表面的一系列冲击之间的间隔非常小，产生的效应和连续的压力没有什么两样。

假定分子的速率一定，但数量不同。那么平均来说，每个分子碰撞器壁的次数相同，产生的冲击强度也相同，因此对总压力的贡献也相同。这样一来，一个固定容积的容器承受的压强和其中分子的总数量，也就是其中的气体总量，成正比。

This is the complete dynamical explanation of the fact discovered by Robert Boyle, that the pressure of air is proportional to its density. It shows also that of different portions of gas forced into a vessel, each produces its own part of the pressure independently of the rest, and this whether these portions be of the same gas or not.

Let us next suppose that the velocity of the molecules is increased. Each molecule will now strike the sides of the vessel a greater number of times in a second, but besides this, the impulse of each blow will be increased in the same proportion, so that the part of the pressure due to each molecule will vary as the *square* of the velocity. Now the increase of the square of velocity corresponds, in our theory, to a rise of temperature, and in this way we can explain the effect of warming the gas, and also the law discovered by Charles that the proportional expansion of all gases between given temperatures is the same.

The dynamical theory also tells us what will happen if molecules of different masses are allowed to knock about together. The greater masses will go slower than the smaller ones, so that, on an average, every molecule, great or small, will have the same energy of motion.

The proof of this dynamical theorem, in which I claim the priority, has recently been greatly developed and improved by Dr. Ludwig Boltzmann. The most important consequence which flows from it is that a cubic centimetre of every gas at standard temperature and pressure contains the same number of molecules. This is the dynamical explanation of Gay Lussac's law of the equivalent volumes of gases. But we must now descend to particulars, and calculate the actual velocity of a molecule of hydrogen.

A cubic centimetre of hydrogen, at the temperature of melting ice and at a pressure of one atmosphere, weighs 0.00008954 grammes. We have to find at what rate this small mass must move (whether altogether or in separate molecules makes no difference) so as to produce the observed pressure on the sides of the cubic centimetre. This is the calculation which was first made by Dr. Joule, and the result is 1,859 metres per second. This is what we are accustomed to call a great velocity. It is greater than any velocity obtained in artillery practice. The velocity of other gases is less, as you will see by the table, but in all cases it is very great as compared with that of bullets.

We have now to conceive the molecules of the air in this hall flying about in all directions, at a rate of about seventeen miles in a minute.

If all these molecules were flying in the same direction, they would constitute a wind blowing at the rate of seventeen miles a minute, and the only wind which approaches this velocity is that which proceeds from the mouth of a cannon. How, then, are you and I able to stand here? Only because the molecules happen to be flying in different directions, so that those which strike against our backs enable us to support the storm which is beating against our faces. Indeed, if this molecular bombardment were to cease, even for an instant, our veins would swell, our breath would leave us, and we should, literally,

这就是罗伯特·玻意耳发现的空气压强正比于其密度这一事实的完整动力学解释。它还表明，如果我们将不同批次的气体加入容器，则无论它们的种类是否相同，每个部分都将独立地产生自己的分压强。

下一步让我们假定分子速率增加的情形。因为每秒钟内，每个分子碰撞器壁的次数相应增加，同时每次碰撞的冲击强度也成比例增加，所以每个分子对压强的贡献和它的速度的**平方**成正比。在我们现在的理论中，分子速率平方的增加和温度的增加相对应。这样我们就能解释加热气体导致压强增加的效应，以及随温度增加各种气体体积同比增大的查理定律。

动力论还告诉我们不同质量的气体分子互相碰撞会发生什么结果。质量大的分子比质量小的分子运动速率要小一些，所以平均来说，每个分子，不论质量大小，其动能都相同。

对这一动力学定律的证明是我最先提出的，近来被路德维希·玻尔兹曼博士加以改进和发展。它的一个重要推论就是在标准温度和压强下，1 立方厘米的任何气体都含有同等数量的分子，这就是盖·吕萨克定律关于相同体积气体的动力学解释。现在我们举一个具体的例子来计算一个氢分子的实际速率。

在温度为冰的熔点，压力为一个大气压时，1 立方厘米氢气的质量是 0.00008954 克。这么小的质量的运动（是合在一起还是分散到各个分子，对结果没有影响）到底需要多大速率，才能在 1 立方厘米的容器壁上产生测量到的压强呢？焦耳博士首先对此进行了计算，他的结果是每秒 1,859 米。通常对我们来说，这是一个很大的速率,比任何炮弹的速率都要大。从附表中大家可以看到,其他气体的速率要小一些，但是无论如何都远大于子弹的速率。

现在我们要设想一下这个大厅里的空气分子以每分钟 17 英里的速率向各个方向飞行。

如果所有分子都向同一个方向飞行，它们就会形成速率为每分钟 17 英里的强风，只有从加农炮炮口出膛的风速能够接近这个速率。那么，你我怎么能够在这里保持站立？这只是因为这些分子飞行的方向各不相同，在前面和后面撞击我们的分子冲击作用互相抵消。事实上，如果这种分子碰撞哪怕停止一刻，我们也会静脉肿胀，不能呼吸，一命呜呼。这些分子不只是撞击我们和房间四周的墙壁。考虑到它

expire. But it is not only against us or against the walls of the room that the molecules are striking. Consider the immense number of them, and the fact that they are flying in every possible direction, and you will see that they cannot avoid striking each other. Every time that two molecules come into collision, the paths of both are changed, and they go off in new directions. Thus each molecule is continually getting its course altered, so that in spite of its great velocity it may be a long time before it reaches any great distance from the point at which it set out.

I have here a bottle containing ammonia. Ammonia is a gas which you can recognise by its smell. Its molecules have a velocity of six hundred metres per second, so that if their course had not been interrupted by striking against the molecules of air in the hall, everyone in the most distant gallery would have smelt ammonia before I was able to pronounce the name of the gas. But instead of this, each molecule of ammonia is so jostled about by the molecules of air, that it is sometimes going one way and sometimes another. It is like a hare which is always doubling, and though it goes a great pace, it makes very little progress. Nevertheless, the smell of ammonia is now beginning to be perceptible at some distance from the bottle. The gas does diffuse itself through the air, though the process is a slow one, and if we could close up every opening of this hall so as to make it air-tight, and leave everything to itself for some weeks, the ammonia would become uniformly mixed through every part of the air in the hall.

This property of gases, that they diffuse through each other, was first remarked by Priestley. Dalton showed that it takes place quite independently of any chemical action between the inter-diffusing gases. Graham, whose researches were especially directed towards those phenomena which seem to throw light on molecular motions, made a careful study of diffusion, and obtained the first results from which the rate of diffusion can be calculated.

Still more recently the rates of diffusion of gases into each other have been measured with great precision by Prof. Loschmidt of Vienna.

He placed the two gases in two similar vertical tubes, the lighter gas being placed above the heavier, so as to avoid the formation of currents. He then opened a sliding valve, so as to make the two tubes into one, and after leaving the gases to themselves for an hour or so, he shut the valve, and determined how much of each gas had diffused into the other.

As most gases are invisible, I shall exhibit gaseous diffusion to you by means of two gases, ammonia and hydrochloric acid, which, when they meet, form a solid product. The ammonia, being the lighter gas, is placed above the hydrochloric acid, with a stratum of air between, but you will soon see that the gases can diffuse through this stratum of air, and produce a cloud of white smoke when they meet. During the whole of this process no currents or any other visible motion can be detected. Every part of the vessel appears as calm as a jar of undisturbed air.

们数量巨大，正在四处乱飞，你会发现它们不可能不互相碰撞。每当两个分子撞到一起，其轨道就都会改变，而它们会飞向新的方向。这样每个分子都在不断地改变轨道，因而虽然其速率很快，但要从出发点移开一定距离也需要不少时间。

我这里有一个里面装着氨气的瓶子。氨气的味道大家可以闻得出来。它的分子运动速率是每秒 600 米，因而如果它们的运动轨迹没有因为和大厅里的空气分子碰撞而改变，就算坐得最远的听众都会在我说出氨气这个名称之前闻到它。但是实际上并非如此，每个氨气分子都被空气分子撞来撞去，一会儿向东一会儿向西，就像一只野兔，老是改变方向，虽然步子很快，但是跑不了多远。虽然如此，在离瓶子一定距离以内的听众还是开始闻到氨气的味道了。氨气确实在空气中不断地扩散着，只是速率较慢而已，如果我们把大厅的所有出口都封起来，不让空气流走，并保持几个星期不动它，那么氨气就会均匀地散布在大厅的每个角落。

气体的这个相互扩散的性质是由普里斯特利最先指出的。道尔顿指出这种扩散的发生与相互扩散的气体间发生的具体化学反应无关。格雷姆的研究工作集中在那些能为分子运动提供线索的现象上。他仔细研究了扩散现象，最早得到了可以用来计算扩散速率的结果。

再晚些时候，维也纳的洛施密特教授精确地测量了气体分子相互扩散的速率。

他把两种不同气体分别装入两个相似的竖直试管中，轻的气体放在重的气体上面以防止对流产生。随后打开滑动阀门让两个试管相通。将它们静置大约一个小时以后，他关上阀门，然后测量每种气体中有多少已经扩散到了另一种气体中。

因为大多数气体都是不可见的，所以我得用氨气和盐酸这两种相遇后能生成固体反应物的气体来展示气体的扩散过程。氨气比较轻，所以放在盐酸上面，中间隔着一层空气。但是你们马上就可以看到，这两种气体能通过扩散穿过空气层而相遇，并产生一股白烟。整个过程看不到气体流动或者其他任何可见的运动。容器的每个部分都像一罐未受扰动的空气一样平静。

But, according to our theory, the same kind of motion is going on in calm air as in the inter-diffusing gases, the only difference being that we can trace the molecules from one place to another more easily when they are of a different nature from those through which they are diffusing.

If we wish to form a mental representation of what is going on among the molecules in calm air, we cannot do better than observe a swarm of bees, when every individual bee is flying furiously, first in one direction, and then in another, while the swarm, as a whole, either remains at rest, or sails slowly through the air.

In certain seasons, swarms of bees are apt to fly off to a great distance, and the owners, in order to identify their property when they find them on other people's ground, sometimes throw handfuls of flour at the swarm. Now let us suppose that the flour thrown at the flying swarm has whitened those bees only which happened to be in the lower half of the swarm, leaving those in the upper half free from flour.

If the bees still go on flying hither and thither in an irregular manner, the floury bees will be found in continually increasing proportions in the upper part of the swarm, till they have become equally diffused through every part of it. But the reason of this diffusion is not because the bees were marked with flour, but because they are flying about. The only effect of the marking is to enable us to identify certain bees.

We have no means of marking a select number of molecules of air, so as to trace them after they have become diffused among others, but we may communicate to them some property by which we may obtain evidence of their diffusion.

For instance, if a horizontal stratum of air is moving horizontally, molecules diffusing out of this stratum into those above and below will carry their horizontal motion with them, and so tend to communicate motion to the neighbouring strata, while molecules diffusing out of the neighbouring strata into the moving one will tend to bring it to rest. The action between the strata is somewhat like that of two rough surfaces, one of which slides over the other, rubbing on it. Friction is the name given to this action between solid bodies; in the case of fluids it is called internal friction or viscosity.

It is in fact only another kind of diffusion—a lateral diffusion of momentum, and its amount can be calculated from data derived from observations of the first kind of diffusion, that of matter. The comparative values of the viscosity of different gases were determined by Graham in his researches on the transpiration of gases through long narrow tubes, and their absolute values have been deduced from experiments on the oscillation of discs by Oscar Meyer and myself.

Another way of tracing the diffusion of molecules through calm air is to heat the upper stratum of the air in a vessel, and so observe the rate at which this heat is communicated

但是根据我们的理论，这种气体间相互扩散的运动进程，同样在平静的空气中发生着。区别只是当扩散发生在不同气体间时，追踪分子从一处到另一处的移动要容易一些。

如果我们想要在头脑中构思出一幅表现分子在平静空气中运动的图像，最好去观察一群蜜蜂，每只蜜蜂都拼命地飞来飞去，先朝一个方向飞，然后再朝另一个方向飞，但是整个蜂群不是停着不动，就是在空中缓慢地移动。

有的季节蜂群可以飞得很远，养蜂人为了能在别人的地盘上也能认出自己的蜂群，有时会向蜂群洒一把面粉。现在我们假定面粉正好把蜂群下面一半的蜜蜂染白，而上面一半没有沾上面粉。

如果这群蜜蜂继续散乱地飞来飞去，蜂群上半部就会有越来越多的沾上面粉的蜜蜂，直到它们均匀地分布于上下两部分。但是这种扩散的原因不是因为蜜蜂沾上了面粉,而是因为它们到处乱飞。沾面粉标记的目的只是为了帮助我们识别特定的蜜蜂。

我们没有办法标记一定数目的空气分子，使得它们在扩散到其他分子中之后还能被追踪到。但是我们可以用某些参数的传递来证明它们的扩散。

比如说，如果一个水平空气层向水平方向移动，那么从该层扩散到上下两个相邻层的分子将携带水平动量，并把这个动量传递给相邻的上下空气层。而从相邻空气层中扩散进入这个水平移动的水平层的分子，会减慢水平层的移动。相邻水平层之间的作用，就像两个粗糙固体表面之间的滑动和摩擦。固体之间的这种作用叫做摩擦，在流体的情况下则称为内摩擦或者黏滞力。

实际上这只不过是另一种类型的扩散——动量的横向扩散。其大小可以通过观察第一种扩散，也就是物质扩散的数据计算出来。研究气体通过长细管的蒸腾过程，格雷姆确定了不同气体的相对黏滞系数。黏滞系数的绝对值由奥斯卡·迈耶和我通过圆盘振动的实验结果推导得到。

另一种跟踪分子在平静空气中扩散的方法是加热容器顶层的空气，然后观察热量向下层传递的速率。这实际上是第三种类型的扩散——能量扩散。在直接进行热

to the lower strata. This, in fact, is a third kind of diffusion—that of energy, and the rate at which it must take place was calculated from data derived from experiments on viscosity before any direct experiments on the conduction of heat had been made. Prof. Stefan, of Vienna, has recently, by a very delicate method, succeeded in determining the conductivity of air, and he finds it, as he tells us, in striking agreement with the value predicted by the theory.

All these three kinds of diffusion—the diffusion of matter, of momentum, and of energy—are carried on by the motion of the molecules. The greater the velocity of the molecules and the farther they travel before their paths are altered by collision with other molecules, the more rapid will be the diffusion. Now we know already the velocity of the molecules, and therefore by experiments on diffusion we can determine how far, on an average, a molecule travels without striking another. Prof. Clausius, of Bonn, who first gave us precise ideas about the motion of agitation of molecules, calls this distance the mean path of a molecule. I have calculated, from Prof. Loschmidt's diffusion experiments, the mean path of the molecules of four well-know gases. The average distance travelled by a molecule between one collision and another is given in the table. It is a very small distance, quite imperceptible to us even with our best microscopes. Roughly speaking, it is about the tenth part of the length of a wave of light, which you know is a very small quantity. Of course the time spent on so short a path by such swift molecules must be very small. I have calculated the number of collisions which each must undergo in a second. They are given in the table and are reckoned by thousands of millions. No wonder that the travelling power of the swiftest molecule is but small, when its course is completely changed thousands of millions of times in a second.

The three kinds of diffusion also take place in liquids, but the relation between the rates at which they take place is not so simple as in the case of gases. The dynamical theory of liquids is not so well understood as that of gases, but the principal difference between a gas and a liquid seems to be that in a gas each molecule spends the greater part of its time in describing its free path, and is for a very small portion of its time engaged in encounters with other molecules, whereas in a liquid the molecule has hardly any free path, and is always in a state of close encounter with other molecules.

Hence in a liquid the diffusion of motion from one molecule to another takes place much more rapidly than the diffusion of the molecules themselves, for the same reason that it is more expeditious in a dense crowd to pass on a letter from hand to hand than to give it to a special messenger to work his way through the crowd. I have here a jar, the lower part of which contains a solution of copper sulphate, while the upper part contains pure water. It has been standing here since Friday, and you see how little progress the blue liquid has made in diffusing itself through the water above. The rate of diffusion of a solution of sugar has been carefully observed by Voit. Comparing his results with those of Loschmidt on gases, we find that about as much diffusion takes place in a second in gases as requires a day in liquids.

传导的实验之前，这种传递的速率是由黏滞系数的实验结果计算出来的。维也纳的斯特藩教授最近通过一个极其精巧的方法，成功地确定了空气的传导系数，他发现测量值和理论预测值惊人地符合。

所有这三种扩散——物质、动量和能量的扩散，都是由分子的运动完成的。分子的运动速率越大，在它与其他分子碰撞而使其运动方向发生改变之前所走的行程就越长，扩散的速率就越快。既然分子速率已知，通过扩散速率的实验，我们就可以确定一个分子在两次碰撞之间所走的平均距离。波恩的克劳修斯教授是第一个给出分子受激运动精确构想的人，他把这个距离叫做分子的平均自由程。根据洛施密特教授的扩散实验，我计算了 4 种常见气体的分子平均自由程。附表中列出了一个分子在两次碰撞之间的平均距离，这个距离非常之小，即使用最好的显微镜也不能分辨。它大概是光波长的 1/10，这样说你们就知道它有多小了。以分子的运动速率之大，行走这么短的距离，所用的时间当然是非常之短。我曾计算过一个分子每秒钟内碰撞的次数，结果也列在表中，都是几十亿次的水平。一秒钟之内方向要被改变几十亿次，难怪分子的运动速率虽大，却走不了多远。

这三种扩散过程在液体中也会发生，但是它们之间的速率关系就不像在气体中那么简单。液体的动力学理论不像气体的理论那么完善。看上去它们之间的根本区别是：相对而言，气体分子大部分时间是自由飞行，和其他分子发生碰撞的时间不多；液体分子则正好相反，平均自由程很短，总是在和附近的分子发生密切接触。

这样一来，液体中的动量扩散就比液体分子本身的扩散要快很多，这就好比在密集的人群中递送一封信，通过众人之手相传要比找一个专门的送信人穿越人群快一样。我这里有一个罐子，下半部装的是硫酸铜溶液，上半部装的是纯水。这个罐子从星期五就一直放在这里，而现在你们几乎看不出蓝色的硫酸铜扩散到上面的纯水中。沃伊特仔细计算了一种蔗糖溶液的扩散速率。把他的结果和洛施密特得到的气体扩散速率进行对比，我们发现在气体中一秒钟就能完成的扩散过程，在液体中则需要一整天。

The rate of diffusion of momentum is also slower in liquids than in gases, but by no means in the same proportion. The same amount of motion takes about ten times as long to subside in water as in air, as you will see by what takes place when I stir these two jars, one containing water and the other air. There is still less difference between the rates at which a rise of temperature is propagated through a liquid and through a gas.

In solids the molecules are still in motion, but their motions are confined within very narrow limits. Hence the diffusion of matter does not take place in solid bodies, though that of motion and heat takes place very freely. Nevertheless, certain liquids can diffuse through colloid solids, such as jelly and gum, and hydrogen can make its way through iron and palladium.

We have no time to do more than mention that most wonderful molecular motion which is called electrolysis. Here is an electric current passing through acidulated water, and causing oxygen to appear at one electrode and hydrogen at the other. In the space between, the water is perfectly calm, and yet two opposite currents of oxygen and of hydrogen must be passing through it. The physical theory of this process has been studied by Clausius, who has given reasons for asserting that in ordinary water the molecules are not only moving, but every now and then striking each other with such violence that the oxygen and hydrogen of the molecules part company, and dance about through the crowd, seeking partners which have become dissociated in the same way. In ordinary water these exchanges produce, on the whole, no observable effect, but no sooner does the electromotive force begin to act than it exerts its guiding influence on the unattached molecules, and bends the course of each toward its proper electrode, till the moment when, meeting with an unappropriated molecule of the opposite kind, it enters again into a more or less permanent union with it till it is again dissociated by another shock. Electrolysis, therefore, is a kind of diffusion assisted by electromotive force.

Another branch of molecular science is that which relates to the exchange of molecules between a liquid and a gas. It includes the theory of evaporation and condensation, in which the gas in question is the vapour of the liquid, and also the theory of the absorption of a gas by a liquid of a different substance. The researches of Dr. Andrews on the relations between the liquid and the gaseous state have shown us that though the statements in our own elementary text-books may be so neatly expressed that they appear almost self-evident, their true interpretation may involve some principle so profound that, till the right man has laid hold of it, no one ever suspects that anything is left to be discovered.

These, then, are, some of the fields from which the data of molecular science are gathered. We may divide the ultimate results into three ranks, according to the completeness of our knowledge of them.

To the first rank belong the relative masses of the molecules of different gases, and their velocities in metres per second. These data are obtained from experiments on the pressure and density of gases, and are known to a high degree of precision.

液体中动量的扩散速率也比气体中动量的扩散速率慢，但是没有差那么多。等量的运动在水中的衰减时间大概是空气中的10倍。让我来搅动一下这两个罐子，一个装的是水，另一个只有空气，大家看看结果是什么？液体和气体在温度传递上的速率差别还要更小一些。

固体中的分子也在不停地运动，但是它们的运动被限制在很小的范围内。所以物质扩散不会发生在固体中，但动量和能量扩散可以非常自由地进行。不过某些液体可以渗过像果冻和树胶一类的胶质固体，而氢气能够透过铁和钯。

时间有限，我们只能简单提一下电解这个最奇妙的分子运动现象。这里有电流通过酸性的水时，两个电极分别产生氧气和氢气。电极之间的水是完全静止的，但是其中必然有氧和氢的相对流动。克劳修斯研究了这一过程的物理学原理，给出理由认为普通水中的分子不但是运动的，而且相互碰撞的力量还很大，造成水分子中氧和氢的分离，分离的氧和氢在水中四处游动，寻找因同样原因而游离的其他对象来结合。在普通水中，这个交换过程在整体上没有造成可观测的效应。但是一旦电动势开始起作用，就对独立分子的运动产生定向影响，驱使游离分子向相应的电极运动，直到碰上异性游离分子而形成相对稳定的结合体。这个结合体还有可能再次被冲击离解。这样来看，电解是一种电动势协助下的分子扩散过程。

分子科学的另一个分支是关于液体和气体之间的分子交换。它既包括同一种物质气液两态之间的蒸发和凝结理论，也包括液体吸收不同种物质气体分子的理论。安德鲁斯博士关于气液两态之间关系的研究表明，虽然我们基本教科书中的结论看上去如此简洁，似乎是天经地义，但其实际的意义可能包含着非常深奥的原理。在有合适的人将其搞清楚之前，人们认为一切已经尽善尽美了。

以上就是分子科学取得数据成果的一些领域。我们根据完整的相关知识，把这些最终的成果分成三个等级。

第一个等级中有不同气体分子的相对质量以及它们以米每秒为单位的速度。这些数据是根据气体压强和密度的实验结果得出的，精度很高。

In the second rank we must place the relative size of the molecules of different gases, the length of their mean paths, and the number of collisions in a second. These quantities are deduced from experiments on the three kinds of diffusion. Their received values must be regarded as rough approximations till the methods of experimenting are greatly improved.

There is another set of quantities which we must place in the third rank, because our knowledge of them is neither precise, as in the first rank, nor approximate, as in the second, but is only as yet of the nature of a probable conjecture. These are the absolute mass of a molecule, its absolute diameter, and the number of molecules in a cubic centimetre. We know the relative masses of different molecules with great accuracy, and we know their relative diameters approximately. From these we can deduce the relative densities of the molecules themselves. So far we are on firm ground.

The great resistance of liquids to compression makes it probable that their molecules must be at about the same distance from each other as that at which two molecules of the same substance in the gaseous form act on each other during an encounter. This conjecture has been put to the test by Lorenz Meyer, who has compared the densities of different liquids with the calculated relative densities of the molecules of their vapours, and has found a remarkable correspondence between them.

Now Loschmidt has deduced from the dynamical theory the following remarkable proportion:—As the volume of a gas is to the combined volume of all the molecules contained in it, so is the mean path of a molecule to one-eighth of the diameter of a molecule.

Assuming that the volume of the substance, when reduced to the liquid form, is not much greater than the combined volume of the molecules, we obtain from this proportion the diameter of a molecule. In this way Loschmidt, in 1865, made the first estimate of the diameter of a molecule. Independently of him and of each other, Mr. Stoney in 1868, and Sir W. Thomson in 1870, published results of a similar kind, those of Thomson being deduced not only in this way, but from considerations derived from the thickness of soap bubbles, and from the electric properties of metals.

According to the table, which I have calculated from Loschmidt's data, the size of the molecules of hydrogen is such that about two million of them in a row would occupy a millimetre, and a million million million million of them would weigh between four and five grammes.

In a cubic centimetre of any gas at standard pressure and temperature there are about nineteen million million million molecules. All these numbers of the third rank are, I need not tell you, to be regarded as at present conjectural. In order to warrant us in putting any confidence in numbers obtained in this way, we should have to compare together a greater number of independent data than we have as yet obtained, and to show that they lead to consistent results.

在第二个等级中，我们应该归入的是不同种气体分子的相对尺寸、平均自由程和每秒钟的碰撞次数。这些参数的量值是从三种扩散的实验结果推导出来的。在实验方法得到极大改进之前，这些公认的结果只能被看作是大概的近似。

还有一批参数只能放进第三个等级，这是因为我们在这个等级中的相关知识只是基于可能的推测，既不像第一等级中的那么精确，也不像第二等级中的那样近似精确。这里面包括分子的绝对质量、绝对直径和 1 立方厘米中的分子数目。我们有不同分子的相对质量的准确结果，也大概知道它们的相对直径。由此我们可以得到分子的相对密度。这些都是确实有据的结果。

液体强大的抗压缩性似乎表明：其分子之间的距离已经很近，和同一物质的两个气态分子在碰撞时发生相互作用的距离差不多。洛伦茨 · 迈耶想了一个办法验证这个猜想，他对各种液体的密度和相应液体蒸气的相对密度的计算值进行比较，发现它们明显相关。

现在洛施密特根据动力学原理推导出一个不寻常的比例关系：气体的体积和该气体中所有分子体积之和的比值，等于分子平均自由程与其直径之比的 1/8。

假定气体液化后的体积，比相应分子的体积之和大不了多少，我们就可以通过这个比例关系得到分子的直径。1865 年洛施密特由此第一次估计出了分子的直径。斯托尼先生在 1868 年，汤姆孙爵士于 1870 年也都各自独立发表了类似的结果。后者除了以上的考虑，还考虑了与肥皂泡厚度以及金属电特性有关的结果。

表中的结果是我根据洛施密特的数据计算得到的。按照这个表，200 万个氢分子排成一列，只有 1 毫米长。100 万的四次方那么多的氢分子合起来的质量只有 4~5 克。

在标准压强和温度下，1 立方厘米的任何气体都含有 1.9×10^{19} 那么多的分子。这些量值都属于前面所说的第三个等级，不用我说你也知道，它们只是当前的推论。除非由独立方法得到的比现在多得多的数据经过比较后都趋向一致的结果，否则我们不能轻易接受以上的结果。

Thus far we have been considering molecular science as an inquiry into natural phenomena. But though the professed aim of all scientific work is to unravel the secrets of nature, it has another effect, not less valuable, on the mind of the worker. It leaves him in possession of methods which nothing but scientific work could have led him to invent, and it places him in a position from which many regions of nature, besides that which he has been studying, appear under a new aspect.

The study of molecules has developed a method of its own, and it has also opened up new views of nature.

When Lucretius wishes us to form a mental representation of the motion of atoms, he tells us to look at a sunbeam shining through a darkened room (the same instrument of research by which Dr. Tyndall makes visible to us the dust we breathe,) and to observe the motes which chase each other in all directions through it. This motion of the visible motes, he tells us, is but a result of the far more complicated motion of the invisible atoms which knock the motes about. In his dream of nature, as Tennyson tells us, he

"saw the flaring atom-streams
And torrents of her myriad universe,
Ruining along the illimitable inane,
Fly on to clash together again, and make
Another and another frame of things
For ever."

And it is no wonder that he should have attempted to burst the bonds of Fate by making his atoms deviate from their courses at quite uncertain times and places, thus attributing to them a kind of irrational free will, which on his materialistic theory is the only explanation of that power of voluntary action of which we ourselves are conscious.

As long as we have to deal with only two molecules, and have all the data given us, we can calculate the result of their encounter, but when we have to deal with millions of molecules, each of which has millions of encounters in a second, the complexity of he problem seems to shut out all hope of a legitimate solution.

The modern atomists have therefore adopted a method which is I believe new in the department of mathematical physics, though it has long been in use in the Section of Statistics. When the working members of Section F get hold of a Report of the Census, or any other document containing the numerical data of Economic and Social Science, they begin by distributing the whole population into groups, according to age, income-tax, education, religious belief, or criminal convictions. The number of individuals is far too great to allow of their tracing the history of each separately, so that, in order to reduce their labour within human limits, they concentrate their attention on a small number of artificial groups. The varying number of individuals in each group, and not the varying state of each individual, is the primary datum from which they work.

到现在为止，我们一直认为分子科学是对自然现象的一种探索。虽然所有科学工作的目的都是为了揭示自然的秘密，但是它还有另外一个至少同样重要的作用，就是对科学工作者心灵的影响。只有科学工作才能让他们发明并掌握新的方法，这使他们站在了一个新的高度，看到除了自己的研究领域之外，其他许多自然领域也呈现出新的面貌。

分子领域的研究发展了一套自己的方法，也打开了一扇观察自然的窗户。

为了让我们在头脑中建立一幅原子运动的图像，卢克莱修让我们观察一束射入暗室的日光（廷德尔博士用同样的设备展示了我们呼吸的灰尘的运动情况），在这束日光中我们可以看到向各个方向追逐乱窜的尘埃。他告诉我们，这些可见尘埃的运动，是更加复杂的不可见的原子运动不断将它们撞来撞去的结果。就像丁尼生告诉我们的一样，在卢克莱修梦想的自然中，他

“看见闪耀的原子流
和她在无穷宇宙中的激流，
在无尽的虚空中衰耗，
又飞来撞在一起，
创造一个又一个事物的构架，
永不止息。”

难怪他试图打破必然的枷锁，让他的原子在很不确定的时刻和地点改变轨迹，因而赋予它们一种非理性的自由意志，这就是他在物质理论中对我们所知的随机行为的产生所做的唯一解释。

如果我们只需要研究两个分子，而且知道所有的数据，我们就可以计算它们相互碰撞的结果。但实际上有极多的分子，每个分子每秒钟都要经历极多的碰撞，问题的复杂性似乎超越了任何合理解答的可能性。

现代原子论者采取了一种我认为在数学物理学科中是全新的方法，尽管它在统计部门中已经应用很久了。当F部门的工作人员得到一份人口普查报告，或者是其他包含经济和社会科学统计数据的文件时，他们先把整个人口按照年龄、所得税、教育水平、宗教信仰或犯罪记录分组。人口的数量太大，很难一一跟踪每个人的实际情况。为了在人力资源有限的情况下把工作量控制在合理范围，他们把注意力集中在少数几个人为划分出来的组群上，每个组群个体人数的变化，而非每个个体的状态变化是他们进行研究工作的最初数据。

This, of course, is not the only method of studying human nature. We may observe the conduct of individual men and compare it with that conduct which their previous character and their present circumstances, according to the best existing theory, would lead us to expect. Those who practise this method endeavour to improve their knowledge of the elements of human nature, in much the same way as an astronomer corrects the elements of a planet by comparing its actual position with that deduced from the received elements. The study of human nature by parents and schoolmasters, by historians and statesmen, is therefore to be distinguished from that carried on by registrars and tabulators, and by those statesmen who put their faith in figures. The one may be called the historical, and the other the statistical method.

The equations of dynamics completely express the laws of the historical method as applied to matter, but the application of these equations implies a perfect knowledge of all the data. But the smallest portion of matter which we can subject to experiment consists of millions of molecules, not one of which ever becomes individually sensible to us. We cannot, therefore, ascertain the actual motion of any one of these molecules, so that we are obliged to abandon the strict historical method, and to adopt the statistical method of dealing with large groups of molecules.

The data of the statistical method as applied to molecular science are the sums of large numbers of molecular quantities. In studying the relations between quantities of this kind, we meet with a new kind of regularity, the regularity of averages, which we can depend upon quite sufficiently for all practical purposes, but which can make no claim to that character of absolute precision which belongs to the laws of abstract dynamics.

Thus molecular science teaches us that our experiments can never give us anything more than statistical information, and that no law deduced from them can pretend to absolute precision. But when we pass from the contemplation of our experiments to that of the molecules themselves, we leave the world of chance and change, and enter a region where everything is certain and immutable.

The molecules are conformed to a constant type with a precision which is not to be found in the sensible properties of the bodies which they constitute. In the first place the mass of each individual molecule, and all its other properties, are absolutely unalterable. In the second place the properties of all molecules of the same kind are absolutely identical.

Let us consider the properties of two kinds of molecules, those of oxygen and those of hydrogen.

We can procure specimens of oxygen from very different sources—from the air, from water, from rocks or every geological epoch. The history of these specimens has been very different, and if, during thousands of years, difference of circumstances could produce difference of properties, these specimens of oxygen would show it.

这显然不是研究人类性质的唯一方法。我们可以观察个体的行为，也可以按照当前最好的理论，根据以往的特征预测目前情况下的行为，并把观察结果和预测的行为相比较。人们采用这种方法以增进他们对人类性质各个方面的认识，就像天文学家通过对比行星的实际位置和依据已知参数预测的位置之间的差别来修正轨道参数一样。父母、校长、历史学家、政治家对人性的研究，和登记员、制表人还有注重数据的政治家的研究是不同的。一个可以叫历史方法，另一个则是统计方法。

动力学方程完全是历史方法定律在物质研究上的应用，但应用这些方程需要知道所有的数据，然而哪怕是实验中物质最小的部分也包含着极多我们看不见的分子。我们不可能确定任何一个分子的实际运动状况，所以必须放弃严格的历史方法，采用统计方法来处理大量分子的情况。

应用在分子科学中的统计方法牵涉的数据是大量分子的参量的总和。在研究这类参量之间的关系时，我们碰上了一种新的规律，也就是平均值的规律。对于实际应用，依赖这些规律就足够了，但是它们不能给出理论动力学定律所具有的绝对精确性。

分子科学告诉我们：实验永远只能提供统计信息，从实验中总结的定律都没有绝对的精确性。但是当我们的关注点从实验转向分子本身时，我们就离开了充满偶然性和变化的世界，进入到一切都是确定不变的领域。

分子都精确地具有恒定不变的属性，这种属性是分子所组成的物体的可测宏观参数所不具备的。首先，所有分子的质量和所有其他性质都是绝对不变的。其次，同种物质所有分子的所有性质都是绝对相同的。

让我们来考虑一下氧和氢这两种分子的性质。

我们可以从空气、水和各个地质时代的岩石等不同来源制备氧气样品。这些样品形成的历史完全不同。如果在几千年的时间里，不同的环境能造就不同的性质，这些氧气样品应该就能表现出这些不同的性质。

In like manner we may procure hydrogen from water, from coal, or, as Graham did, from meteoric iron. Take two litres of any specimen of hydrogen, it will combine with exactly one litre of any specimen of oxygen, and will form exactly two litres of the vapour of water.

Now if, during the whole previous history of either specimen, whether imprisoned in the rocks, flowing in the sea, or careering through unknown regions with the meteorites, any modification of the molecules had taken place, these relations would no longer be preserved.

But we have another and an entirely different method of comparing the properties of molecules. The molecule, though indestructible, is not a hard rigid body, but is capable of internal movements, and when these are excited it emits rays, the wavelength of which is a measure of the time of vibration of the molecule.

By means of the spectroscope the wavelengths of different kinds of light may be compared to within one ten-thousandth part. In this way it has been ascertained, not only that molecules taken from every specimen of hydrogen in our laboratories have the same set of periods of vibration, but that light, having the same set of periods of vibration, is emitted from the sun and from the fixed stars.

We are thus assured that molecules of the same nature as those of our hydrogen exist in those distant regions, or at least did exist when the light by which we see them was emitted.

From a comparison of the dimensions of the buildings of the Egyptians with those of the Greeks, it appears that they have a common measure. Hence, even if no ancient author had recorded the fact that the two nations employed the same cubit as a standard of length, we might prove it from the buildings themselves. We should also be justified in asserting that at some time or other a material standard of length must have been carried from one country to the other, or that both countries had obtained their standards from a common source.

But in the heavens we discover by their light, and by their light alone, stars so distant from each other that no material thing can ever have passed from one to another, and yet this light, which is to us the sole evidence of the existence of these distant worlds, tells us also that each of them is built up of molecules of the same kinds as those which we find on earth. A molecule of hydrogen, for example, whether in Sirius or in Arcturus, executes its vibrations in precisely the same time.

Each molecule, therefore, throughout the universe, bears impressed on it the stamp of a metric system as distinctly as does the metre of the Archives at Paris, or the double royal cubit of the Temple of Karnac.

同样地，我们可以从水、煤炭或者像格雷姆那样从陨铁中制备出氢气样品。任取两升氢气样品，它都可以和任取的一升氧气样品反应，正好生成两升水蒸气。

如果在任何一个样品的整个历史中，不论它是固锁在岩石中，还是漂流在海洋里，或者是跟着陨石穿越着未知的区域，只要分子发生了任何变化，上面的关系都不能再保持。

我们有另一种完全不同的办法来比较分子的性质。分子虽然不能被毁灭，但也不是刚性实体，它有内部运动，当内部运动被激发时，分子发出射线，这个射线的波长表征了分子振动的周期。

光谱仪可以比较不同光的波长差别，精度可达万分之一。用这个办法，我们可以确认，不但我们实验室中每一个氢气样品中的分子具有同样的振动周期组合，就连太阳和其他恒星发射的光也存在同样的振动周期组合。

这样我们就确信了在茫茫的宇宙中也存在和我们的氢分子一样的分子，或者说至少在那些我们用来观察分子的光线放射出时，它们是存在的。

比较古埃及和古希腊的建筑规模，可以感觉到它们似乎使用了同样的度量标准。因此，尽管历史上没有留下两个国家都使用同样的肘长作为长度标准的记载，我们也许可以通过建筑本身证明这一点。我们也有理由认为，肯定是某个实物长度标准在某个时候被人从一个国家带到了另一个国家，或者两个国家从同一个来源得到了这个标准。

对天空中的星体，我们只是通过它们发射的光线证实它们的存在，它们之间的距离如此遥远，物质从一个恒星到另一个恒星的传递是完全不可能的，但是它们发射的光线，除了作为证明这些遥远天体存在的唯一证据以外，还告诉我们，组成每一个天体的分子都和我们在地球上发现的分子相同。比如，不论是天狼星还是大角星，上面的氢分子都以同样的周期振动。

宇宙中的每一个分子，都铭刻着一个度量系统的痕迹，它的独特清晰，就如同巴黎档案局的米原尺，或者卡尔纳克神庙的皇家双肘尺一样。

No theory of evolution can be formed to account for the similarity of molecules, for evolution necessarily implies continuous change, and the molecule is incapable of growth or decay, of generation or destruction.

None of the processes of Nature, since the time when Nature began, have produced the slightest difference in the properties of any molecule. We are therefore unable to ascribe either the existence of the molecules or the identity of their properties to the operation of any of the causes which we call natural.

On the other hand, the exact quality of each molecule to all others of the same kind gives it, as Sir John Herschel has well said, the essential character of a manufactured article, and precludes the idea of its being eternal and self existent.

Thus we have been led, along a strictly scientific path, very near to the point at which Science must stop. Not that Science is debarred from studying the internal mechanism of a molecule which she cannot take to pieces, any more than from investigating an organism which she cannot put together. But in tracing back the history of matter Science is arrested when she assures herself, on the one hand, that the molecule has been made, and on the other that it has not been made by any of the processes we call natural.

Science is incompetent to reason upon the creation of matter itself out of nothing. We have reached the utmost limit of our thinking faculties when we have admitted that because matter cannot be eternal and self-existent it must have been created.

It is only when we contemplate, not matter in itself, but the form in which it actually exists, that our mind finds something on which it can lay hold.

That matter, as such, should have certain fundamental properties—that it should exist in space and be capable of motion, that its motion should be persistent, and so on, are truths which may, for anything we know, be of the kind which metaphysicians call necessary. We may use our knowledge of such truths for purposes of deduction but we have no data for speculating as to their origin.

But that there should be exactly so much matter and no more in every molecule of hydrogen is a fact of a very different order. We have here a particular distribution of matter—a *collocation*—to use the expression of Dr. Chalmers, of things which we have no difficulty in imagining to have been arranged otherwise.

The form and dimensions of the orbits of the planets, for instance, are not determined by any law of nature, but depend upon a particular collocation of matter. The same is the case with respect to the size of the earth, from which the standard of what is called the metrical system has been derived. But these astronomical and terrestrial magnitudes are far inferior in scientific importance to that most fundamental of all standards which forms

任何进化论都不能解释分子的相似性，因为进化隐含着不断地变化，而分子既不生长也不腐朽，不能被制造也不能被摧毁。

自从自然界形成以来，自然界的所有过程都没能使任何分子的性质有丝毫的改变。所以我们不能把分子的存在或其性质的同一性归结为任何自然原因的作用。

另一方面，正像约翰·赫歇尔爵士说的那样，同类分子精确相同的性质是其所构成的物质的根本特征，这就排除了分子的永恒自存在性。

这样一来，我们就被引领着，沿着一条严格的科学道路，走到了科学的尽头。这不是说科学手段不能分解分子就不能用来研究分子的内部机制，就像不能因为科学不能组合成生物就不用她来研究生物一样。但在追溯物质的历史时，科学被困住了，她一方面确认分子已被创造，另一方面又确认任何自然界的过程都不能创造分子。

科学不能理解物质如何被无中生有地创造出来。当我们承认由于物质不能永恒自存在，因此它一定已经被创造出来的时候，我们已经到了自己思维能力的极限。

只有当我们思考物质的存在形式，而不是物质本身的时候，我们的思维才算找到了一些可以解决问题的线索。

我们知道，形而上学者认为必要的真理是：物质必须具有一些基本的性质——它应该存在于空间中，能够运动，运动应该是持续的等等。我们可以利用已知的与这些真理有关的知识进行推导，但是并无数据可以用来猜测这些真理的本源。

但是每一个氢分子的质量都正好是这么多，这个事实就完全是另一个性质了。这种质量的分布，或者用查默斯博士的词语——这种**配置**，实在是太特别了，让我们都不习惯。

比如，行星轨道的形式和尺度，也不是由任何自然定律决定的，而是取决于特殊的质量的布置。作为米制计量体系标准的地球的尺寸也是这样。但是这些天文和地理的量值在科学上的重要性都远不能和构成分子系统基础的那些最基本的标准相比。我们知道自然界过程一直在起作用，就算最后不摧毁，它们至少也要改变地球

the base of the molecular system. Natural causes, as we know, are at work, which tend to modify, if they do not at length destroy, all the arrangements and dimensions of the earth and the whole solar system. But though in the course of ages catastrophes have occurred and may yet occur in the heavens, though ancient systems may be dissolved and new systems evolved out of their ruins, the molecules out of which these systems are built—the foundation stones of the material universe—remain unbroken and unworn.

They continue this day as they were created, perfect in number and measure and weight, and from the ineffaceable characters impressed on them we may learn that those aspirations after accuracy in measurement, truth in statement, and justice in action, which we reckon among our noblest attributes as men, are ours because they are essential constituents of the image of Him who in the beginning created, not only the heaven and the earth, but the materials of which heaven and earth consist.

Table of Moncular Date

		Hydrogen	Oxygen	Carbonic oxide	Carbonic acid
Rank I	Mass of molecule (hydrogen = 1)	1	16	14	22
	Velocity (of mean square), metres per second at 0℃	1859	465	497	396
Rank II	Mean path, tenth-metres.	965	560	482	379
	Collisions in a second, (millions)	17750	7646	9489	9720
Rank III	Diameter, tenth-metre	5.8	7.6	8.3	9.3
	Mass, twenty-fifth-grammes.	46	736	644	1012

Table of Diffusion: $\frac{(\text{centimetre})^2}{\text{second}}$ measure

	Calculated	Observed	
H& O	0.7086	0.7214	Diffusion of matter observed by Loschmidt.
H& CO	0.6519	0.6422	
H& CO_2	0.5575	0.5558	
O& CO	0.1807	0.1802	
O& CO_2	0.1427	0.1409	
CO & CO_2	0.1386	0.1406	
H	1.2990	1.49	Diffusion of momentum Graham and Meyer.
O	0.1884	0.213	
CO	0.1748	0.212	
CO_2	0.1087	0.117	
Air		0.256	Diffusion of temperature observed by Stefan.
Copper		1.077	
Iron		0.183	
Cane sugar in water		0.00000365	Voit
Diffusion in a day		0.3144	
Salt in water		0.00000116	Fick

(**8**, 437-441; 1873)

和整个太阳系的秩序和尺寸。然而，尽管在历史上宇宙曾经发生过灾变，以后也可能再发生，虽然旧的体系可能被消灭，新的体系在旧体系的废墟上演化发展，但作为这些系统组成基础的分子——物质宇宙的基石，却不会被破坏和磨损。

它们被创造的时候是什么样，现在还是什么样，数量、大小和质量都没有丝毫改变，从铭记在它们身上的不可磨灭的特征，我们也许能够明白，我们对测量的精准、言语的真切、行为的正义这些人类最高尚品质的追求，是因为它们也是造物主形象的根本组成部分。他当初不仅创造了天和地，也创造了组成天和地的所有物质。

分子数据表

		氢气	氧气	一氧化碳	二氧化碳
第一等级	分子质量（氢气=1）	1	16	14	22
	0℃下的速度（均方根），米 / 秒	1859	465	497	396
第二等级	平均自由程，10^{-10}米	965	560	482	379
	每秒碰撞次数（百万次）	17750	7646	9489	9720
第三等级	直径，10^{-10}米	5.8	7.6	8.3	9.3
	质量，10^{-25}克	46	736	644	1012

扩散数据表：单位是 $\frac{(\text{厘米})^2}{\text{秒}}$

	计算值	测量值	
氢 & 氧	0.7086	0.7214	洛施密特观察的物质扩散
氢 & 一氧化碳	0.6519	0.6422	
氢 & 二氧化碳	0.5575	0.5558	
氧 & 一氧化碳	0.1807	0.1802	
氧 & 二氧化碳	0.1427	0.1409	
一氧化碳 & 二氧化碳	0.1386	0.1406	
氢	1.2990	1.49	格雷姆和迈耶的动量扩散
氧	0.1884	0.213	
一氧化碳	0.1748	0.212	
二氧化碳	0.1087	0.117	
空气		0.256	斯特藩观察的温度扩散
铜		1.077	
铁		0.183	
蔗糖在水溶液中		0.00000365	沃伊特
一天后扩散的距离		0.3144	
食盐在水溶液中		0.00000116	菲克

（何钧 翻译；鲍重光 审稿）

On the Dynamical Evidence of the Molecular Constitution of Bodies*

J. C. Maxwell

Editor's Note

Here James Clerk Maxwell offers an update on efforts to understand the properties of gases and liquids from first principles—what today is called statistical mechanics. Maxwell had made a decisive contribution with his kinetic theory of gases. Here he reports on other work, such as that of Rudolph Clausius, who introduced a theoretical quantity known as the "virial", representing an average over all particle pairs of the product of the distance between the particles and the force of attraction or repulsion. Maxwell also mentions the work of two other figures, an unknown graduate student named Johannes Diderik van der Waals and an American physicist, Josiah Willard Gibbs. Both later became leaders in understanding the fundamental behavior of material systems.

I

WHEN any phenomenon can be described as an example of some general principle which is applicable to other phenomena, that phenomenon is said to be explained. Explanations, however, are of very various orders, according to the degree of generality of the principle which is made use of. Thus the person who first observed the effect of throwing water into a fire would feel a certain amount of mental satisfaction when he found that the results were always similar, and that they did not depend on any temporary and capricious antipathy between the water and the fire. This is an explanation of the lowest order, in which the class to which the phenomenon is referred consists of other phenomena which can only be distinguished from it by the place and time of their occurrence, and the principle involved is the very general one that place and time are not among the conditions which determine natural processes. On the other hand, when a physical phenomenon can be completely described as a change in the configuration and motion of a material system, the dynamical explanation of that phenomenon is said to be complete. We cannot conceive any further explanation to be either necessary, desirable, or possible, for as soon as we know what is meant by the words configuration, motion, mass, and force, we see that the ideas which they represent are so elementary that they cannot be explained by means of anything else.

The phenomena studied by chemists are, for the most part, such as have not received a complete dynamical explanation.

---

* A lecture delivered at the Chemical Society, Feb. 18, by Prof. Clerk Maxwell, F. R. S.

关于物体分子构成的动力学证据*

麦克斯韦

编者按

在这篇文章中，克拉克·麦克斯韦介绍了从基本原理（现在的名字是统计力学）出发研究气体和液体性质的最新进展。此前，麦克斯韦提出了他的气体动力学理论，这是他作出的决定性的贡献。在本文中他报道了鲁道夫·克劳修斯等其他研究人员的工作。克劳修斯引入了“位力”这个理论物理量，用来表示所有粒子对的粒子间距离与粒子间引力或斥力乘积的总平均。麦克斯韦在这篇文章中还提到了另外两个人的工作，其中一个是当时还没什么名气的研究生约翰内斯·迪德里克·范德瓦尔斯，另一个是美国物理学家约西亚·威拉德·吉布斯，这两人后来都成了研究物质体系基本行为特征的领袖人物。

I

任何一种现象，一旦可以被视为某种适用于其他现象的普遍原理的实例，我们就可以说这种现象得到了解释。然而，解释包括很多不同层面，要视所采用原理的普适程度而定。因此，第一个观察到将水泼入火中这种现象的人，在发现结果总是相似的而且这些结果并不取决于水和火之间那瞬息万变、反复无常的不相容性时，便会获得一定程度的精神满足。这是一种最低层次的解释，在这种解释中，以上现象所涉及的分类还包含其他现象，而这些现象只是发生的时间和地点不同，所涉及的原理却是非常普适的，因此时间和地点并不属于能够决定自然过程的条件。另一方面，当某一物理现象能够被完全描述为物质系统结构或运动的变化时，我们就可以说这种现象得到了完整的动力学解释。我们想不出还有什么更进一步的解释是必要的、有用的或者可能的，因为一旦我们了解了结构、运动、质量和力这些词的意义，便会发现它们所表示的意义如此基本，以至于无法再用其他概念来解释它们。

化学家们研究的现象，大部分还没有得到完整的动力学解释。

* 2 月 18 日，克拉克·麦克斯韦教授（皇家学会会员）在化学学会所作的讲座。

Many diagrams and models of compound molecules have been constructed. These are the records of the efforts of chemists to imagine configurations of material systems by the geometrical relations of which chemical phenomena may be illustrated or explained. No chemist, however, professes to see in these diagrams anything more than symbolic representations of the various degrees of closeness with which the different components of the molecule are bound together.

In astronomy, on the other hand, the configurations and motions of the heavenly bodies are on such a scale that we can ascertain them by direct observation. Newton proved that the observed motions indicate a continual tendency of all bodies to approach each other, and the doctrine of universal gravitation which he established not only explains the observed motions of our system, but enables us to calculate the motions of a system in which the astronomical elements may have any values whatever.

When we pass from astronomical to electrical science, we can still observe the configuration and motion of electrified bodies, and thence, following the strict Newtonian path, deduce the forces with which they act on each other; but these forces are found to depend on the distribution of what we call electricity. To form what Gauss called a "construirbar Vorstellung" of the invisible process of electric action is the great desideratum in this part of science.

In attempting the extension of dynamical methods to the explanation of chemical phenomena, we have to form an idea of the configuration and motion of a number of material systems, each of which is so small that it cannot be directly observed. We have, in fact, to determine, from the observed external actions of an unseen piece of machinery, its internal construction.

The method which has been for the most part employed in conducting such inquiries is that of forming an hypothesis, and calculating what would happen if the hypothesis were true. If these results agree with the actual phenomena, the hypothesis is said to be verified, so long, at least, as some one else does not invent another hypothesis which agrees still better with the phenomena.

The reason why so many of our physical theories have been built up by the method of hypothesis is that the speculators have not been provided with methods and terms sufficiently general to express the results of their induction in its early stages. They were thus compelled either to leave their ideas vague and therefore useless, or to present them in a form the details of which could be supplied only by the illegitimate use of the imagination.

In the meantime the mathematicians, guided by that instinct which teaches them to store up for others the irrepressible secretions of their own minds, had developed with the utmost generality the dynamical theory of a material system.

很多化合物分子的图示或模型已经建立。它们都是化学家根据可以说明或解释的化学现象的几何关系努力猜想物质系统的结构而得到的结果。不过，这些图示只不过是对构成分子的不同成分束缚在一起的各种不同的紧密程度的符号表示而已，没有一位化学家敢说他能从中看出更多的东西。

另一方面，在天文学中，天体的结构和运动所具有的尺度是我们可以通过直接观测来确定的。牛顿证明了观测到的运动显示出所有物体之间都有不断彼此靠拢的趋势，他建立的万有引力定律不仅解释了观测到的我们自身所在系统的运动，而且使我们能够计算出具有任意量值的天文学对象所组成的系统的运动情况。

从天文学转向电学，我们仍然可以观测带电体的结构和运动，并遵循严格的牛顿式方法来推导出它们的相互作用力。但是我们发现这种力依赖于所谓的电的分布。因此，建立一种关于电作用不可见过程的、被高斯称为“统一表象”的理论，就成为这门科学的迫切需要。

在试图将动力学方法推广以用于对化学现象的解释时，我们必须形成关于大量物质系统的结构和运动的观念，而其中每一个物质系统都小到无法直接观测的程度。事实上，我们只能通过观测那些看不见的结构部分的外部作用来确定其内部结构。

主要用来进行此类研究的方法是先建立一个假说，然后计算如果该假说成立会产生什么结果。若这些结果与实际现象相吻合，我们就说这个假说得到了确证，至少在没有其他人提出与客观现象更一致的另一种假说之前，是可以这样说的。

之所以有如此多的物理理论依靠假说的方式建立起来，是因为理论家早期没有获得具有足够普适性的方法和术语来表达他们归纳出的结果。受此限制，他们要么将观点表达得含糊不清以至于毫无用处，要么用另一种形式表达观点，而这种形式的细节只能靠非常规的想象力来弥补。

与此同时，数学家在其本能（这种本能使他们将自身才智孕育出的丰硕果实供给他人）的驱使下已经发展出了具有充分普适性的关于物质系统的动力学理论。

Of all hypotheses as to the constitution of bodies, that is surely the most warrantable which assumes no more than that they are material systems, and proposes to deduce from the observed phenomena just as much information about the conditions and connections of the material system as these phenomena can legitimately furnish.

When examples of this method of physical speculation have been properly set forth and explained, we shall hear fewer complaints of the looseness of the reasoning of men of science, and the method of inductive philosophy will no longer be derided as mere guess-work.

It is only a small part of the theory of the constitution of bodies which has as yet been reduced to the form of accurate deductions from known facts. To conduct the operations of science in a perfectly legitimate manner, by means of methodised experiment and strict demonstration, requires a strategic skill which we must not look for, even among those to whom science is most indebted for original observations and fertile suggestions. It does not detract from the merit of the pioneers of science that their advances, being made on unknown ground, are often cut off, for a time, from that system of communications with an established base of operations, which is the only security for any permanent extension of science.

In studying the constitution of bodies we are forced from the very beginning to deal with particles which we cannot observe. For whatever may be our ultimate conclusions as to molecules and atoms, we have experimental proof that bodies may be divided into parts so small that we cannot perceive them.

Hence, if we are careful to remember that the word particle means a small part of a body, and that it does not involve any hypothesis as to the ultimate divisibility of matter, we may consider a body as made up of particles, and we may also assert that in bodies or parts of bodies of measurable dimensions, the number of particles is very great indeed.

The next thing required is a dynamical method of studying a material system consisting of an immense number of particles, by forming an idea of their configuration and motion, and of the forces acting on the particles, and deducing from the dynamical theory those phenomena which, though depending on the configuration and motion of the invisible particles, are capable of being observed in visible portions of the system.

The dynamical principles necessary for this study were developed by the fathers of dynamics, from Galileo and Newton to Lagrange and Laplace; but the special adaptation of these principles to molecular studies has been to a great extent the work of Prof. Clausius of Bonn, who has recently laid us under still deeper obligations by giving us, in addition to the results of his elaborate calculations, a new dynamical idea, by the aid of which I hope we shall be able to establish several important conclusions without much symbolical calculation.

在所有关于物体构成的假说中，最可信的无疑是这样一种假说：它除了假定物体构成物质系统之外绝无更多假定，并仅就观测到的现象在合理范围内所提供的有关物质系统的条件和联系等信息进行推导。

在合理地阐明和解释这种物理假说方法的实例之后，我们受到的对科学界人士推理不严谨的抱怨将会变得很少，并且相应的归纳哲学方法也不会再被人嘲笑为凭空臆测了。

迄今为止，物体构成理论中只有一小部分已经通过已知的事实被归纳为精确的结论。要通过合理的实验和严格的论证引导科学活动沿着十分合理的道路前进，就需要一种战略技巧，这种技巧可遇不可求，即使对于那些凭借原始观测与丰富联想进行科学研究的人来说也是一样。下面的事实并不会贬低科学先驱的成就，即他们在未知领域的开拓工作，这些开拓工作曾经一度被隔离于有着已确定的作用基础的交流体系之外，而这个交流体系正是任何科学持续发展的唯一保障。

要研究物体构成，我们从一开始就不得不处理看不见的微粒。因为不管我们对于分子和原子得到的最终结论将是什么，现在我们已经得到了实验上的证据，可以证明物体可以被分割成小到无法察觉的部分。

由此，如果我们细心地注意到微粒这个词只表示物体的一小部分，而且毫不涉及关于物质终极可分性的假设，我们就可以认为物体是由微粒组成的，我们还可以断言，在物体或者具有可观测尺度的物体的某一部分中，微粒的数量是非常大的。

下面我们还需要一种可以研究由大量微粒组成的物质系统的动力学方法，这就要形成一种关于微粒的结构、运动以及作用在粒子上的力的概念，并从动力学理论中推导出在系统的可见部分中可观测到的现象，尽管这些现象是由不可见微粒的运动和结构决定的。

进行这种研究必不可少的动力学原理是由多位动力学奠基人发展起来的，从伽利略、牛顿到拉格朗日和拉普拉斯；而将这些原理应用于分子研究，则很大程度上是波恩的克劳修斯教授的贡献，除了复杂的计算结果之外，他最近提出的一种新的动力学观点，给我们的研究工作带来了很大的帮助。借助这种观点，我希望我们可以不用大量的符号演算也能得到一些重要的结论。

The equation of Clausius, to which I must now call your attention, is of the following form: —

$$pV = \frac{2}{3}T - \frac{2}{3}\Sigma\Sigma\left(\frac{1}{2}Rr\right)$$

Here p denotes the pressure of a fluid, and V the volume of the vessel which contains it. The product pV, in the case of gases at constant temperature, remains, as Boyle's Law tells us, nearly constant for different volumes and pressures. This member of the equation, therefore, is the product of two quantities, each of which can be directly measured.

The other member of the equation consists of two terms, the first depending on the motion of the particles, and the second on the forces with which they act on each other.

The quantity T is the kinetic energy of the system, or, in other words, that part of the energy which is due to the motion of the parts of the system.

The kinetic energy of a particle is half the product of its mass into the square of its velocity, and the kinetic energy of the system is the sum of the kinetic energy of its parts.

In the second term, r is the distance between any two particles, and R is the attraction between them. (If the force is a repulsion or a pressure, R is to be reckoned negative.)

The quantity $\frac{1}{2}Rr$, or half the product of the attraction into the distance across which the attraction is exerted, is defined by Clausius as the virial of the attraction. (In the case of pressure or repulsion, the virial is negative.)

The importance of this quantity was first pointed out by Clausius, who, by giving it a name, has greatly facilitated the application of his method to physical exposition.

The virial of the system is the sum of the virials belonging to every pair of particles which exist in the system. This is expressed by the double sum $\Sigma\Sigma(\frac{1}{2}Rr)$, which indicates that the value of $\frac{1}{2}Rr$ is to be found for every pair of particles, and the results added together.

Clausius has established this equation by a very simple mathematical process, with which I need not trouble you, as we are not studying mathematics tonight. We may see, however, that it indicates two causes which may affect the pressure of the fluid on the vessel which contains it: the motion of its particles, which tends to increase the pressure, and the attraction of its particles, which tends to diminish the pressure.

We may therefore attribute the pressure of a fluid either to the motion of its particles or to a repulsion between them.

现在我必须提醒大家注意，克劳修斯方程具有如下形式：

$$pV = \frac{2}{3}T - \frac{2}{3}\Sigma\Sigma\left(\frac{1}{2}Rr\right)$$

其中，p 表示某种流体的压力，V 表示盛放流体的容器的体积。由玻意耳定律可知，对于恒温条件下的气体，乘积 pV 在不同的体积和压力下基本保持不变。可见，方程的左边，是两个直接观测量的乘积。

方程的另一边由两项组成，第一项由微粒的运动决定，第二项由微粒间的相互作用力决定。

T 这个量指系统的动能，或者换一种说法，是由于系统各部分的运动而具有的那部分能量。

微粒的动能是其质量与速率平方的乘积的一半，而系统的动能则是其各部分动能的总和。

在第二项中，r 表示任意两个微粒之间的距离，R 表示它们之间的吸引力。（如果作用力为排斥力或者压力，则 R 取负值。）

$\frac{1}{2}Rr$ 这个量，即引力与彼此吸引的微粒之间距离的乘积的一半，克劳修斯将其定义为引力的位力。（对于压力和排斥力，位力取负值。）

克劳修斯首次指出了这一物理量的重要性。通过给这一物理量指定名字，克劳修斯大大推动了他的方法在物理解释中的应用。

系统的位力是系统中每一对微粒之间位力的总和，可以表示为双重求和 $\Sigma\Sigma(\frac{1}{2}Rr)$，意思是要得到每一对微粒之间的 $\frac{1}{2}Rr$ 值，并将结果累加起来。

克劳修斯通过简单的数学过程建立了这一方程，这里我不需要以此来烦扰各位，因为今晚我们不是在研究数学问题。不过我们还是可以看到，方程指出有两个因素可能会影响流体对容纳它的容器的压力：流体中微粒的运动倾向于增大压力，而微粒间的吸引力倾向于减小压力。

由此，我们可以将流体的压力归结为流体中微粒的运动或者微粒间的排斥力。

Let us test by means of this result of Clausius the theory that the pressure of a gas arises entirely from the repulsion which one particle exerts on another, these particles, in the case of gas in a fixed vessel, being really at rest.

In this case the virial must be negative, and since by Boyle's Law the product of pressure and volume is constant, the virial also must be constant, whatever the volume, in the same quantity of gas at constant temperature. It follows from this that Rr, the product of the repulsion of two particles into the distance between them, must be constant, or in other words that the repulsion must be inversely as the distance, a law which Newton has shown to be inadmissible in the case of molecular forces, as it would make the action of the distant parts of bodies greater than that of contiguous parts. In fact, we have only to observe that if Rr is constant, the virial of every pair of particles must be the same, so that the virial of the system must be proportional to the number of pairs of particles in the system—that is, to the square of the number of particles, or in other words to the square of the quantity of gas in the vessel. The pressure, according to this law, would not be the same in different vessels of gas at the same density, but would be greater in a large vessel than in a small one, and greater in the open air than in any ordinary vessel.

The pressure of a gas cannot therefore be explained by assuming repulsive forces between the particles.

It must therefore depend, in whole or in part, on the motion of the particles.

If we suppose the particles not to act on each other at all, there will be no virial, and the equation will be reduced to the form

$$Vp = \frac{2}{3}T$$

If M is the mass of the whole quantity of gas, and c is the mean square of the velocity of a particle, we may write the equation—

$$Vp = \frac{1}{3}Mc^2,$$

or in words, the product of the volume and the pressure is one-third of the mass multiplied by the mean square of the velocity. If we now assume, what we shall afterwards prove by an independent process, that the mean square of the velocity depends only on the temperature, this equation exactly represents Boyle's Law.

But we know that most ordinary gases deviate from Boyle's Law, especially at low temperatures and great densities. Let us see whether the hypothesis of forces between the particles, which we rejected when brought forward as the sole cause of gaseous pressure, may not be consistent with experiment when considered as the cause of this deviation from Boyle's Law.

When a gas is in an extremely rarefied condition, the number of particles within a given distance of any one particle will be proportional to the density of the gas. Hence the virial

让我们用克劳修斯的结果来验证一下下面的理论：气体压力完全源于一个微粒施加在另一个微粒上的排斥力，而就恒容容器中的气体而言，这些微粒是真正静止的。

在上述情况中，位力必定为负值，又由玻意耳定律可知压力与体积的乘积为常数，则等量恒温的气体，不论其体积是多少，位力也一定是常数。由此可知，两个微粒之间的排斥力与距离的乘积 Rr 必为常数，换言之，排斥力必定与距离成反比。然而，牛顿已经指出这一定律不适用于分子间作用力的情况，因为它将使物体中相隔较远的部分间的相互作用比相邻的部分间的相互作用更强。实际上，我们只需注意到，如果 Rr 是常数，那么每一对微粒的位力都一定是相同的，于是系统的位力一定会正比于系统中微粒的对数——即正比于微粒数的平方，或者说正比于容器中气体数量的平方。根据这一定律，相同密度的气体在不同容器中的压力也是不同的，在较大容器中的压力会比在较小容器中的压力更大，而在开放的大气中的压力比在任何普通容器中的压力都大。

因此，气体的压力无法通过假定微粒间存在排斥力进行解释。

这样一来，气体的压力就一定是完全地或部分地依赖于微粒的运动。

如果我们假定微粒间没有任何相互作用，那么就没有位力存在，方程可简化为如下形式：

$$Vp = \frac{2}{3}T$$

如果用 M 表示气体的总质量，c^2 表示微粒速率的均方，我们就可以将方程写作——

$$Vp = \frac{1}{3}Mc^2,$$

或者用文字表述为，体积与压力的乘积是质量与速率均方乘积的 1/3。现在我们先假设，速率的均方仅取决于温度，那么这个方程恰好就是玻意耳定律，后面我们将通过独立的过程来证明。

但是我们知道大部分常见气体都会偏离玻意耳定律，尤其是在低温和高密度的状态下。现在我们再来看看刚才在作为产生气体压力的唯一原因被提出时，已经被我们否定的微粒间存在作用力的假说，在作为偏离玻意耳定律的原因时是否会与实验不一致。

当气体处于极稀薄的状态时，在任一微粒周围给定距离之内的微粒数量将正比于气体密度。因此，一个微粒作用于其他微粒而产生的位力将随密度变化，单位体

arising from the action of one particle on the rest will vary as the density, and the whole virial in unit of volume will vary as the square of the density.

Calling the density ρ, and dividing the equation by V, we get—

$$p = \frac{1}{3}\rho c^2 - \frac{2}{3}A\rho^2,$$

where A is a quantity which is nearly constant for small densities.

Now, the experiments of Regnault show that in most gases, as the density increases the pressure falls below the value calculated by Boyle's Law. Hence the viral must be positive; that is to say, the mutual action of the particles must be in the main attractive, and the effect of this action in diminishing the pressure must be at first very nearly as the square of the density.

On the other hand, when the pressure is made still greater the substance at length reaches a state in which an enormous increase of pressure produces but a very small increase of density. This indicates that the virial is now negative, or, in other words, the action between the particles is now, in the main, repulsive. We may therefore conclude that the action between two particles at any sensible distance is quite insensible. As the particles approach each other the action first shows itself as an attraction, which reaches a maximum, then diminishes, and at length becomes a repulsion so great that no attainable force can reduce the distance of the particles to zero.

The relation between pressure and density arising from such an action between the particles is of this kind.

As the density increases from zero, the pressure at first depends almost entirely on the motion of the particles, and therefore varies almost exactly as the pressure, according to Boyle's Law. As the density continues to increase, the effect of the mutual attraction of the particles becomes sensible, and this causes the rise of pressure to be less than that given by Boyle's Law. If the temperature is low, the effect of attraction may become so large in proportion to the effect of motion that the pressure, instead of always rising as the density increases, may reach a maximum, and then begin to diminish.

At length, however, as the average distance of the particles is still further diminished, the effect of repulsion will prevail over that of attraction, and the pressure will increase so as not only to be greater than that given by Boyle's Law, but so that an exceedingly small increase of density will produce an enormous increase of pressure.

Hence the relation between pressure and volume may be represented by the curve *A B C D E F G*, where the horizontal ordinate represents the volume, and the vertical ordinate represents the pressure.

积内的总位力将随密度的平方变化。

设密度为ρ，将方程两边同除以 V，我们得到——

$$p = \frac{1}{3}\rho c^2 - \frac{2}{3}A\rho^2,$$

其中 A 这个量在密度很小时近似为常数。

勒尼奥的实验表明，大多数气体在密度增大时，其压力会下降到低于玻意耳定律的计算值。因此位力必定是正的；也就是说，微粒间的相互作用必定以引力为主，而且这种作用减小压力的效果在开始的时候一定是近似与密度的平方有关。

另一方面，当压力继续增大，物质最终将达到另一种状态，即压力急剧增大但密度只有很小的增加。这表明此时位力为负值，或者换句话说，此时微粒间的相互作用主要为斥力。我们可以由此确定，在任何可感知的尺度中，两个微粒之间的相互作用都是十分微弱的。随着微粒彼此靠近，它们之间的相互作用首先表现为引力，引力达到最大值后开始减小，最终转变为极大的斥力，以至于没有任何可获得的力能够将微粒间的距离减小到零。

源于微粒间这种相互作用的压力与密度之间的关系就是这样的。

随着密度从零开始逐渐增大，一开始压力几乎完全取决于微粒的运动，因此几乎与玻意耳定律计算得到的压力精确地吻合。随着密度继续增加，微粒间相互吸引力的影响逐渐体现出来，这使得压力的增大小于根据玻意耳定律预计的值。如果温度很低，吸引力的影响会大到可以与运动的影响相抗衡的程度，那么压力就不再总是随着密度的增加而增大，而是在达到一个最大值后开始减小。

最终，随着微粒的平均距离继续减小，斥力的影响将胜过引力的影响，而压力将会增大到超出玻意耳定律所预计的值，而且此时很小的密度增加也会导致压力的急剧增大。

由此，压力与体积之间的关系可以用曲线 $A\ B\ C\ D\ E\ F\ G$ 来表示，其中横坐标代表体积，纵坐标代表压力。

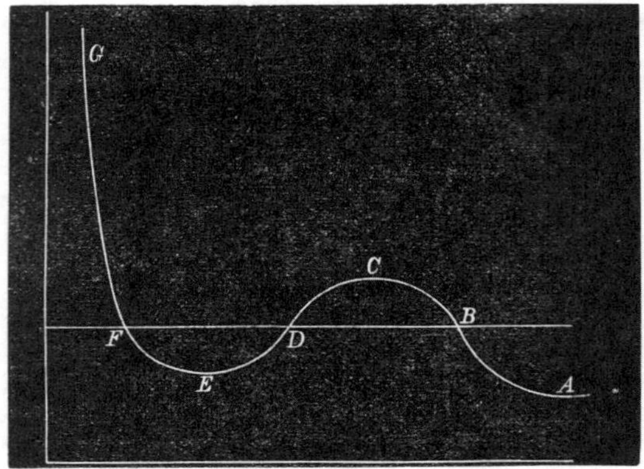

Fig. 1

As the volume diminishes, the pressure increases up to the point *C*, then diminishes to the point *E*, and finally increases without limit as the volume diminishes.

We have hitherto supposed the experiment to be conducted in such a way that the density is the same in every part of the medium. This, however, is impossible in practice, as the only condition we can impose on the medium from without is that the whole of the medium shall be contained within a certain vessel. Hence, if it is possible for the medium to arrange itself so that part has one density and part another, we cannot prevent it from doing so.

Now the points *B* and *F* represent two states of the medium in which the pressure is the same but the density very different. The whole of the medium may pass from the state *B* to the state *F*, not through the intermediate states *C D E*, but by small successive portions passing directly from the state *B* to the state *F*. In this way the successive states of the medium as a whole will be represented by points on the straight line *B F*, the point *B* representing it when entirely in the rarefied state, and *F* representing it when entirely condensed. This is what takes place when a gas or vapour is liquefied.

Under ordinary circumstances, therefore, the relation between pressure and volume at constant temperature is represented by the broken line *A B F G*. If, however, the medium when liquefied is carefully kept from contact with vapour, it may be preserved in the liquid condition and brought into states represented by the portion of the curve between *F* and *E*. It is also possible that methods may be devised whereby the vapour may be prevented from condensing, and brought into states represented by points in *B C*.

The portion of the hypothetical curve from *C* to *E* represents states which are essentially unstable, and which cannot therefore be realised.

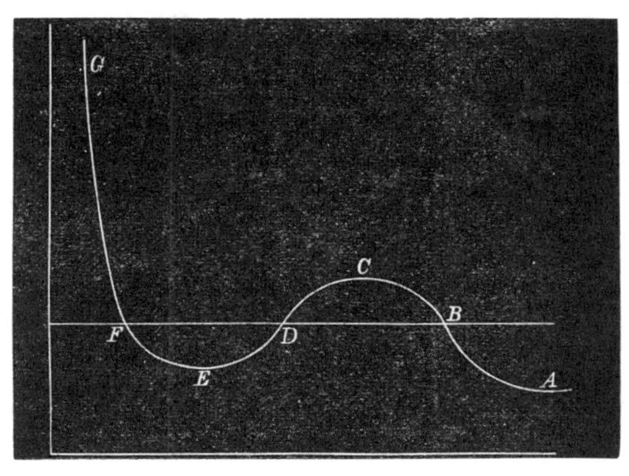

图 1

随着体积缩小，压力先增大到 C 点，然后减小到 E 点，最终将随体积缩小而无限增大。

到目前为止，我们始终假设在实验过程中介质的每一部分都具有相同的密度。不过，这在实际中是不可能的，我们能够从外部施加于介质的唯一限制，仅仅是确保整个介质都置于确定的容器之中。因此，如果介质有可能自身调整为各处密度不同，我们也无法阻止。

B 点和 F 点代表介质的两种状态，其压力相同但密度相差甚远。全部介质可从状态 B 到达状态 F，不经过中间状态 $C\,D\,E$，而是直接由状态 B 一点一点地转化为状态 F。这样，整个介质的连续变化状态对应于图中直线 $B\,F$ 上的点，B 点代表介质完全处于稀薄状态，F 则代表介质完全处于凝聚状态。这就是气体或者蒸气在液化过程中发生的现象。

因此，一般情况下，恒定温度时压力与体积的关系对应于图中的折线 $A\;B\;F\;G$。不过，如果在液化过程中小心地避免介质与蒸气接触，就可能可以使介质保持在液态，并达到图中 E 点和 F 点之间的曲线部分所代表的状态。也有可能可以设计出某种阻止蒸气凝聚的方法从而达到曲线 $B\,C$ 上的点所代表的状态。

假设的曲线中从 C 点到 E 点的部分代表本质上不稳定的状态，因此是无法实现的。

Now let us suppose the medium to pass from B to F along the hypothetical curve $B\ C\ D\ E\ F$ in a state always homogeneous, and to return along the straight line $F\ B$ in the form of a mixture of liquid and vapour. Since the temperature has been constant throughout, no heat can have been transformed into work. Now the heat transformed into work is represented by the excess of the area $F\ D\ E$ over $B\ C\ D$. Hence the condition which determines the maximum pressure of the vapour at given temperature is that the line $B\ F$ cuts off equal areas from the curve above and below.

The higher the temperature, the greater the part of the pressure which depends on motion, as compared with that which depends on forces between the particles. Hence, as the temperature rises, the dip in the curve becomes less marked, and at a certain temperature the curve, instead of dipping, merely becomes horizontal at a certain point, and then slopes upward as before. This point is called the critical point. It has been determined for carbonic acid by the masterly researches of Andrews. It corresponds to a definite temperature, pressure and density.

At higher temperatures the curve slopes upwards throughout, and there is nothing corresponding to liquefaction in passing from the rarest to the densest state.

The molecular theory of the continuity of the liquid and gaseous states forms the subject of an exceedingly ingenious thesis by Mr. Johannes Diderik van der Waals*, a graduate of Leyden. There are certain points in which I think he has fallen into mathematical errors, and his final result is certainly not a complete expression for the interaction of real molecules, but his attack on this difficult question is so able and so brave, that it cannot fail to give a notable impulse to molecular science. It has certainly directed the attention of more than one inquirer to the study of the Low-Dutch language in which it is written.

The purely thermodynamical relations of the different states of matter do not belong to our subject, as they are independent of particular theories about molecules. I must not, however, omit to mention a most important American contribution to this part of thermodynamics by Prof. Willard Gibbs†, of Yale College, U.S., who has given us a remarkably simple and thoroughly satisfactory method of representing the relations of the different states of matter by means of a model. By means of this model, problems which had long resisted the efforts of myself and others may be solved at once.

(**11**, 357-359; 1875)

* Over de continuiteit van den gas en vloeistof toestand. Leiden: A.W. Sijthoff, 1873.

† "A method of geometrical representation of the thermodynamic properties of substances by means of surfaces." *Transactions of the Connecticut Academy of Arts and Sciences*, vol. II, Part 2.

现在，让我们设想介质始终以均匀的状态沿着假设的曲线 $B\ C\ D\ E\ F$ 从 B 变化到 F，并以蒸气和液体混合的形式沿着直线 $F\,B$ 返回。因为温度一直是恒定的，所以其间不会有热能转化为功，而转化为功的热能对应于区域 $F\,D\,E$ 与 $B\,C\,D$ 的面积差。因此，在给定温度下使蒸气压力达到最大的条件，就是直线 $B\,F$ 与上方和下方曲线围成的面积相等。

温度越高，由运动决定的那部分压力与由微粒间作用力决定的部分相比就越大。因此，随着温度升高，曲线的凹陷变得不明显，当温度达到某一个值时，曲线不再凹陷而在某一点达到水平线的位置之后曲线就像前面那样上升了。这个点称为临界点。安德鲁斯已经用巧妙的研究方法确定了碳酸的临界点。临界点对应于确定的温度、压力和密度。

在更高的温度下，曲线始终是向上倾斜的，在从最稀薄到最稠密的状态变化过程中不发生液化。

关于液态与气态连续性的分子理论构成了一篇极具才华的论文的主题，论文的作者是莱顿大学的毕业生约翰内斯·迪德里克·范德瓦尔斯先生 *。尽管我觉得他在某些地方犯了一些数学错误，而且他的最终结果也确实不是对真实分子间相互作用的完整表达，但是他对这一难题所作的努力表现了他非凡的才华和勇气，以至于他的工作不可能不对分子科学产生重要的推动作用。确实有不止一位研究者受此驱使而去学习这篇文章撰写时所用的荷兰语。

物质不同状态间纯粹的热力学联系不属于我们的主题，因为它与这些关于分子的具体理论无关。不过，我决不能忘记提及一项极为重要的，由美国耶鲁大学的威拉德·吉布斯教授作出的热力学方面的贡献†，他利用一个模型给我们提供了一种极其简单又十分令人满意的表示不同物态间关联的方法。借助这个模型，那些长期以来阻碍我本人和其他人努力的难题可能立刻就会迎刃而解。

---

*《论液态和气态的连续性》，莱登赛特霍夫出版社，1873 年。

†《借助曲面表示物质热力学性质的几何描述法》，《康涅狄格艺术与科学学会学报》，第 2 卷，第 2 部分。

II

Let us now return to the case of a highly rarefied gas in which the pressure is due entirely to the motion of its particles. It is easy to calculate the mean square of the velocity of the particles from the equation of Clausius, since the volume, the pressure, and the mass are all measurable quantities. Supposing the velocity of every particle the same, the velocity of a molecule of oxygen would be 461 metres per second, of nitrogen 492, and of hydrogen 1,844, at the temperature 0℃.

The explanation of the pressure of a gas on the vessel which contains it by the impact of its particles on the surface of the vessel has been suggested at various times by various writers. The fact, however, that gases are not observed to disseminate themselves through the atmosphere with velocities at all approaching those just mentioned, remained unexplained, till Clausius, by a thorough study of the motions of an immense number of particles, developed the methods and ideas of modern molecular science.

To him we are indebted for the conception of the mean length of the path of a molecule of a gas between its successive encounters with other molecules. As soon as it was seen how each molecule, after describing an exceedingly short path, encounters another, and then describes a new path in a quite different direction, it became evident that the rate of diffusion of gases depends not merely on the velocity of the molecules, but on the distance they travel between each encounter.

I shall have more to say about the special contributions of Clausius to molecular science. The main fact, however, is, that he opened up a new field of mathematical physics by showing bow to deal mathematically with moving systems of innumerable molecules.

Clausius, in his earlier investigations at least, did not attempt to determine whether the velocities of all the molecules of the same gas are equal, or whether, if unequal, there is any law according to which they are distributed. He therefore, as a first hypothesis, seems to have assumed that the velocities are equal. But it is easy to see that if encounters take place among a great number of molecules, their velocities, even if originally equal, will become unequal, for, except under conditions which can be only rarely satisfied, two molecules having equal velocities before their encounter will acquire unequal velocities after the encounter. By distributing the molecules into groups according to their velocities, we may substitute for the impossible task of following every individual molecule through all its encounters, that of registering the increase or decrease of the number of molecules in the different groups.

By following this method, which is the only one available either experimentally or mathematically, we pass from the methods of strict dynamics to those of statistics and probability.

II

现在让我们回过头来看看极稀薄状态的气体，在这种状态下压力完全来源于气体微粒的运动。由于体积、压力和质量都是可观测量，因此很容易利用克劳修斯方程计算出微粒速率的均方。假设每种微粒具有相同的速率，则在 0℃时，氧分子的速率为每秒 461 米，氮分子为每秒 492 米，而氢分子为每秒 1,844 米。

已经有多位学者多次提出过用气体微粒对容器表面的撞击来解释气体对容器的压力。不过，我们一直不能解释为什么从未观测到气体以接近于上面所提到的速率散布到大气中的现象，直到克劳修斯通过对大量微粒的运动的全面研究发展了现代分子科学的方法和观念。

我们得益于他提出的一个概念，即一个气体分子在与其他分子相继发生两次碰撞之间所经过路程的平均长度。一旦看到每个气体分子是如何在经历了一段极短的路程后与另一个分子相撞，随后气体分子又沿着完全不同的方向开始新的旅程，那么很明显的就是：气体扩散的速率不仅依赖于分子的速度，还与分子在相继发生的碰撞之间途经的距离有关。

关于克劳修斯对分子科学的特殊贡献，我还有很多要说。最主要的是，他通过展示如何用数学方式处理含无数分子的运动系统从而开创了数学物理的新领域。

至少在早期的研究中，克劳修斯并没有试图去确定同一气体中的所有分子是否都具有相同的速率，或者如果它们的速率不等，其分布是否应遵循某种规律。他的第一个假说似乎假定所有分子的速率都是相等的。不过显而易见的是，如果碰撞发生在大量分子之间，那么即使它们的速率开始时是相等的，之后也会变得不相等，因为除了在某些几乎无法满足的条件下，两个碰撞前具有相同速率的分子在碰撞后总会获得不同的速率。通过将分子按其速率分组，我们就只需记录不同组内分子数量的增加或减少，而不必完成跟踪每一个分子的所有碰撞这项不可能完成的任务。

利用这种无论从实验角度还是数学角度来说都是唯一可行的方法，我们从严格的动力学方法转到了统计和概率的方法上。

When an encounter takes place between two molecules, they are transferred from one pair of groups to another, but by the time that a great many encounters have taken place, the number which enter each group is, on an average, neither more nor less than the number which leave it during the same time. When the system has reached this state, the numbers in each group must be distributed according to some definite law.

As soon as I became acquainted with the investigations of Clausius, I endeavoured to ascertain this law.

The result which I published in 1860 has since been subjected to a more strict investigation by Dr. Ludwig Boltzmann, who has also applied his method to the study of the motion of compound molecules. The mathematical investigation, though, like all parts of the science of probabilities and statistics, it is somewhat difficult, does not appear faulty. On the physical side, however, it leads to consequences, some of which, being manifestly true, seem to indicate that the hypotheses are well chosen, while others seem to be so irreconcilable with known experimental results, that we are compelled to admit that something essential to the complete statement of the physical theory of molecular encounters must have hitherto escaped us.

I must now attempt to give you some account of the present state of these investigations, without, however, entering into their mathematical demonstration.

I must begin by stating the general law of the distribution of velocity among molecules of the same kind.

If we take a fixed point in this diagram and draw from this point a line representing in direction and magnitude the velocity of a molecule, and make a dot at the end of the line, the position of the dot will indicate the state of motion of the molecule.

If we do the same for all the other molecules, the diagram will be dotted all over, the dots being more numerous in certain places than in others.

The law of distribution of the dots may be shown to be the same as that which prevails among errors of observation or of adjustment.

The dots in the diagram before you may be taken to represent the velocities of molecules, the different observations of the position of the same star, or the bullet-holes round the bull's eye of a target, all of which are distributed in the same manner.

The velocities of the molecules have values ranging from zero to infinity, so that in speaking of the average velocity of the molecules we must define what we mean.

当两个分子发生碰撞时，它们就会从原本所在组的一对转变成另一组的一对，而且在同时发生大量碰撞时，平均来看，同一时间段内进入某一个组的分子数量，不会多于或少于离开该组的分子数量。当系统达到这种状态时，分子在各组的分布一定符合某种确定的规律。

我在了解了克劳修斯的研究工作之后，便立即努力探求这一规律。

我在 1860 年发表的结果后来由路德维希·玻尔兹曼博士进行了更为严格的研究，他还将他的方法应用于对化合物分子运动的研究。数学研究，例如所有关于概率统计的科学部分，虽然有些困难，却不无裨益。从物理角度来说，数学方法的某些推论是非常正确的，似乎可以说明所选假说是正确的，而另一些却与已知的实验结果相违背，这使我们不得不承认，到目前为止，我们遗漏了关于分子碰撞物理理论完整表述中的某些核心部分。

现在我必须努力为各位介绍这方面研究的现状，不过，我们不会涉及其中的数学证明。

我必须从同种分子中速度分布的一般规律开始说起。

如果从图中选择一个固定的点，从这点出发画一条线来表示速度的方向和大小，并在线段末端画出端点，那么端点的位置就表示分子的运动状态。

如果我们对所有分子作同样处理，那么整张图上将画满了点，某些位置的点会比其他位置的点多。

可以看出，这些点的分布规律与观测或调节中经常出现的误差的分布规律是相同的。

你面前这张图中的点可以用来表示分子的速度，或者对同一颗恒星位置的不同观测结果，或者位于目标靶心周围的弹洞，上述这些具有相同的分布方式。

分子速率的取值范围是从零到无限大，因此说到分子的平均速率我们必须先定义我们指的是什么。

Fig. 2. Diagram of Velocities

The most useful quantity for purposes of comparison and calculation is called the "velocity of mean square". It is that velocity whose square is the average of the squares of the velocities of all the molecules.

This is the velocity given above as calculated from the properties of different gases. A molecule moving with the velocity of mean square has a kinetic energy equal to the average kinetic energy of all the molecules in the medium, and if a single mass equal to that of the whole quantity of gas were moving with this velocity, it would have the same kinetic energy as the gas actually has, only it would be in a visible form and directly available for doing work.

If in the same vessel there are different kinds of molecules, some of greater mass than others, it appears from this investigation that their velocities will be so distributed that the average kinetic energy of a molecule will be the same, whether its mass be great or small.

Here we have perhaps the most important application which has yet been made of dynamical methods to chemical science. For, suppose that we have two gases in the same vessel. The ultimate distribution of agitation among the molecules is such that the average kinetic energy of an individual molecule is the same in either gas. This ultimate state is also, as we know, a state of equal temperature. Hence the condition that two gases shall have the same temperature is that the average kinetic energy of a single molecule shall be the same in the two gases.

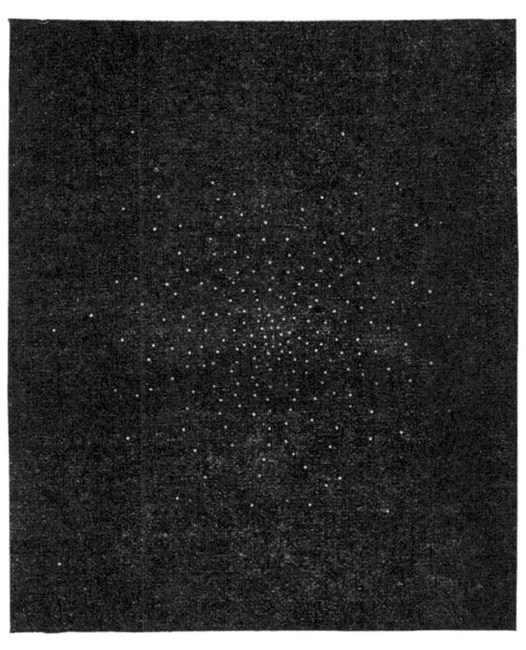

图 2. 速度图像

为了便于比较和计算，最有用的量叫作“均方速率”。这个量的平方是所有分子速率平方的平均值。

这就是上面给出的利用不同气体性质计算得到的速度。以均方速率运动的分子具有的动能等于介质内全体分子的平均动能，并且若以此速率运动的单个物体的质量等于气体的总质量，那么该物体与气体具有相同的动能，只是前者具有可见的形式并可直接用于做功。

如果同一容器中包含不同种类的分子，其中一些具有比较大的质量，那么根据这项研究，这些分子的速度分布方式应满足每种分子具有相同的平均动能，不管其质量是大还是小。

这里我们用到的可能是动力学方法最重要的应用，即将它应用于化学科学。比如，假定我们将两种气体置于同一容器中。分子运动的最终分布形式应满足两种气体中每一种的单个分子的平均动能相等。正如我们所知，这种最终态也是等温态。由此，两种气体具有相同温度的条件，就是两种气体中单个分子的平均动能相同。

Now, we have already shown that the pressure of a gas is two-thirds of the kinetic energy in unit of volume. Hence, if the pressure as well as the temperature be the same in the two gases, the kinetic energy per unit of volume is the same, as well as the kinetic energy molecule. There must, therefore, be the same number of molecules in unit of volume in the two gases.

This result coincides with the law of equivalent volumes established by Gay Lussac. This law, however, has hitherto tested on purely chemical evidence, the relative masses of the molecules of different substances having been deduced from the proportions in which the substances enter into chemical combination. It is now demonstrated on dynamical principles. The molecule is defined as that small portion of the substance which moves as one lump during the motion of agitation. This is a purely dynamical definition, independent of any experiments on combination.

The density of a gaseous medium, at standard temperature and pressure, is proportional to the mass of one of its molecules as thus defined.

We have thus a safe method of estimating the relative masses of molecules of different substances when in the gaseous state. This method is more to be depended on than those founded on electrolysis or on specific heat, because our knowledge of the conditions of the motion of agitation is more complete than our knowledge of electrolysis, or of the internal motions of the constituents of a molecule.

I must now say something about these internal motions, because the greatest difficulty which the kinetic theory of gases has yet encountered belongs to this part of the subject.

We have hitherto considered only the motion of the centre of mass of the molecule. We have now to consider the motion of the constituents of the molecule relative to the centre of mass.

If we suppose that the constituents of a molecule are atoms, and that each atom is what is called a material point, then each atom may move in three different and independent ways, corresponding to the three dimensions of space, so that the number of variables required to determine the position and configuration of all the atoms of the molecule is three times the number of atoms.

It is not essential, however, to the mathematical investigation to assume that the molecule is made up of atoms. All that is assumed is that the position and configuration of the molecule can be completely expressed by a certain number of variables.

Let us call this number n.

现在，我们已经指出，气体的压力是单位体积动能的2/3。因此，如果两种气体具有相同的压力和温度，那么它们每单位体积的动能以及每个分子的动能也是相同的。由此，两种气体单位体积内必定含有相同数量的分子。

这一结果与盖·吕萨克提出的等容定律是一致的。不过，到目前为止该定律只基于化学证据，不同物质的相对分子质量已经从物质参加化合反应时所占据的比例中推导出来了。现在已经用动力学原理证明了该定律。分子被定义为扰动运动中作为一个整体运动的物体的一小部分。这是一个纯粹的动力学定义，与任何化合实验无关。

在标准的温度和压力下，气态介质的密度正比于该气体按以上定义的一个分子的质量。

这样，我们就有一种可以估计气态时不同物质的相对分子质量的可靠方法。这种方法比那些建立在电解或者比热基础上的方法更可信赖，因为我们已经掌握的关于扰动运动条件方面的知识比关于电解或者分子结构内部运动的知识更完备。

现在我必须谈谈这些分子的内部运动，因为目前气体动力学理论遇到的最大困难就在这部分。

迄今为止我们考虑的只是分子质心的运动。现在我们不得不考虑分子组分相对于质心的运动。

如果我们假定分子是由原子构成的，并且每个原子都是所谓的质点，那么每个原子都可以在3个不同且独立的方向上运动，这对应于空间的3个维度，因此要确定分子中所有原子结构和位置所需的变量数就是原子个数的3倍。

不过，对数学研究来说，假定分子是由原子构成的并不是一个基本假设。全部的前提假设只是，分子的位置和结构可以用一定数量的变量完整地表达。

让我们将这个数设为n。

Of these variables, three are required to determine the position of the centre of mass of the molecule, and the remaining $n-3$ to determine its configuration relative to its centre of mass.

To each of the n variables corresponds a different kind of motion.

The motion of translation of the centre of mass has three components.

The motions of the parts relative to the centre of mass have $n-3$ components.

The kinetic energy of the molecule may be regarded as made up of two parts—that of the mass of the molecule supposed to be concentrated at its centre of mass, and that of the motions of the parts relative to the centre of mass. The first part is called the energy of translation, the second that of rotation and vibration. The sum of these is the whole energy of motion of the molecule.

The pressure of the gas depends, as we have seen, on the energy of translation alone. The specific heat depends on the rate at which the whole energy, kinetic and potential, increases as the temperature rises.

Clausius had long ago pointed out that the ratio of increment of the whole energy to that of the energy of translation may be determined if we know by experiment the ratio of the specific heat at constant pressure to that at constant volume.

He did not, however, attempt to determine *à priori* the ratio of the two parts of the energy, though he suggested, as an extremely probable hypothesis, that the average values of the two parts of the energy in a given substance always adjust themselves to the same ratio. He left the numerical value of this ratio to be determined by experiment.

In 1860 I investigated the ratio of the two parts of the energy on the hypothesis that the molecules are elastic bodies of invariable form. I found, to my great surprise, that whatever be the shape of the molecules, provided they are not perfectly smooth and spherical, the ratio of the two parts of the energy must be always the same, the two parts being in fact equal.

This result is confirmed by the researches of Boltzmann, who has worked out the general case of a molecule having n variables.

He finds that while the average energy of translation is the same for molecules of all kinds at the same temperature, the whole energy of motion is to the energy of translation as n to 3.

For a rigid body $n=6$, which makes the whole energy of motion twice the energy of energy of translation.

在这些变量中，有 3 个是用来确定分子质心位置的，而其余 n–3 个则用来确定相对于质心的分子结构。

n 个变量的每一个都对应于一种不同的运动。

质心的平移运动有 3 个分量。

各部分相对于质心的运动则包含 n–3 个分量。

分子的动能可以看作是由两部分组成——其中一部分是假定整个分子的质量集中于其质心所产生的，另一部分是各部分相对质心运动所产生的。第一部分称为平动动能，第二部分称为转动和振动动能。上述量的总和就是分子运动的总能量。

正如我们所知，气体的压力只由平动动能决定。比热则依赖于温度升高时动能与势能的总增加量与温度增量的比值。

克劳修斯早就指出，如果我们能够通过实验获得恒压比热与恒容比热的比值，就能确定总能量增量与平动动能增量的比值。

但是，他并没有试图先验性地确定两部分能量的比值，尽管他提出了一个极有可能成立的假说：对于一种给定的物质，两部分能量的平均值总是具有相同的比值。他将这一比值的具体数值留待实验确定。

1860 年，我基于分子是不变形的弹性体的假说研究了两部分能量的比值。我惊讶地发现，不论分子是何种形状，只要它们不是绝对光滑和完美球状，那么两部分能量的比值就一定总是相同的，实际上这两部分能量是相等的。

这一结果被玻尔兹曼的研究工作所证实，他给出了具有 n 个变量的分子在一般情况下的结果。

他发现，相同温度下任何种类的分子都具有相同的平均平动动能，运动的总能量与平动动能之比为 n : 3。

对于刚体来说 n=6，这使得运动的总能量为平动动能的 2 倍。

But if the molecule is capable of changing its form under the action of impressed forces, it must be capable of storing up potential energy, and if the forces are such as to ensure the stability of the molecule, the average potential energy will increase when the average energy of internal motion increases.

Hence, as the temperature rises, the increments of the energy of translation, the energy of internal motion, and the potential energy are as 3, $(n-3)$, and e respectively, where e is a positive quantity of unknown value depending on the law of the force which binds together the constituents of the molecule.

When the volume of the substance is maintained constant, the effect of the application of heat is to increase the whole energy. We thus find for the specific heat of a gas at constant volume—

$$\frac{1}{2J}\frac{p_0 V_0}{273^\circ}(n+e)$$

where p_0 and V_0 are the pressure and volume of unit of mass at zero centigrade, or 273° absolute temperature, and J is the dynamical equivalent of heat. The specific heat at constant pressure is

$$\frac{1}{2J}\frac{p_0 V_0}{273^\circ}(n+2+e)$$

In gases whose molecules have the same degree of complexity the value of n is the same, and that of e *may* be the same.

If this is the case, the specific heat is inversely as the specific gravity, according to the law of Dulong and Petit, which is, to a certain degree of approximation, verified by experiment.

But if we take the actual values of the specific heat as found by Regnault and compare them with this formula, we find that $n + e$ for air and several other gases cannot be more than 4.9. For carbonic acid and steam it is greater. We obtain the same result if we compare the ratio of the calculated specific heats

$$\frac{2+n+e}{n+e}$$

with the ratio as determined by experiment for various gases, namely, 1.408.

And here we are brought face to face with the greatest difficulty which the molecular theory has yet encountered, namely, the interpretation of the equation $n + e = 4.9$.

If we suppose that the molecules are atoms—mere material points, incapable of rotatory energy or internal motion—then n is 3 and e is zero, and the ratio of the specific heats is 1.66, which is too great for any real gas.

但是，如果分子能够在外力作用下改变形状，它就一定可以储存势能，而如果外力是能保证分子的稳定性的那种，当内部运动的平均能量增加时，平均势能也会增加。

因此，随着温度升高，平动动能、内部运动能量和势能的增量分别为 3，$(n-3)$ 和 e，其中 e 是取值为正的未知量，其数值取决于将分子各组成部分束缚在一起的力的定律。

当物体的体积保持不变时，加热的效果是增加总能量。我们由此给出恒容气体的比热——

$$\frac{\mathrm{I}}{2\mathrm{J}}\frac{p_0V_0}{273^\circ}(n+e)$$

其中 p_0 和 V_0 是零摄氏度或绝对温度 273° 时单位质量的压力和体积，而 J 则是热的动力学当量。恒压条件下的比热为

$$\frac{\mathrm{I}}{2\mathrm{J}}\frac{p_0V_0}{273^\circ}(n+2+e)$$

对于那些有相同复杂程度的分子，n 值是相同的，e 值**可能**是相同的。

如果情况就是如此，那么根据杜隆–珀蒂定律，比热与比重成反比，这是一条在一定精度范围内已经被实验证明了的定律。

但是，如果我们采用勒尼奥所发现的实际比热值并将其与此公式进行比较就会发现，对空气和其他几种气体，$n+e$ 的值不会超过 4.9。而对碳酸和水蒸气来说这个值则大一些。如果将计算所得的比热之比值

$$\frac{2+n+e}{n+e}$$

与实验确定的比值相比，我们得到相同的结果，即 1.408。

此时，我们将面临分子理论中最大的困难，也就是，如何解释等式 $n+e=4.9$。

如果我们假设分子是不可分割的质点——没有转动能量或者内部运动——那么 n 为 3 而 e 为 0，比热的比值是 1.66，而这个值对于任何实际气体来说都太大了。

But we learn from the spectroscope that a molecule can execute vibrations of constant period. It cannot therefore be a mere material point, but a system capable of changing its form. Such a system cannot have less than six variables. This would make the greatest value of the ratio of the specific heats 1.33, which is too small for hydrogen, oxygen, nitrogen, carbonic oxide, nitrous oxide, and hydrochloric acid.

But the spectroscope tells us that some molecules can execute a great many different kinds of vibrations. They must therefore be systems of a very considerable degree of complexity, having far more than six variables. Now, every additional variable introduces an additional amount of capacity for internal motion without affecting the external pressure. Every additional variable, therefore, increases the specific heat, whether reckoned at constant pressure or at constant volume.

So does any capacity which the molecule may have for storing up energy in the potential form. But the calculated specific heat is already too great when we suppose the molecule to consist of two atoms only. Hence every additional degree of complexity which we attribute to the molecule can only increase the difficulty of reconciling the observed with the calculated value of the specific heat.

I have now put before you what I consider to be the greatest difficulty yet encountered by the molecular theory. Boltzmann has suggested that we are to look for the explanation in the mutual action between the molecules and the etherial medium which surrounds them. I am afraid, however, that if we call in the help of this medium, we shall only increase the calculated specific heat, which is already too great.

The theorem of Boltzmann may be applied not only to determine the distribution of velocity among the molecules, but to determine the distribution of the molecules themselves in a region in which they are acted on by external forces. It tells us that the density of distribution of the molecules at a point where the potential energy of a molecule is ψ, is proportional to $e^{-\frac{\psi}{\kappa\theta}}$ where θ is the absolute temperature, and κ is a constant for all gases. It follows from this, that if several gases in the same vessel are subject to an external force like that of gravity, the distribution of each gas is the same as if no other gas were present. This result agrees with the law assumed by Dalton, according to which the atmosphere may be regarded as consisting of two independent atmospheres, one of oxygen, and the other of nitrogen; the density of the oxygen diminishing faster than that of the nitrogen, as we ascend.

This would be the case if the atmosphere were never disturbed, but the effect of winds is to mix up the atmosphere and to render its composition more uniform than it would be if left at rest.

Another consequence of Boltzmann's theorem is, that the temperature tends to become equal throughout a vertical column of gas at rest.

但是，通过光谱仪我们了解到，分子可以进行周期固定的振动。因此分子不可能只是一个质点，而应是一个结构可以改变的体系。这一体系的变量不可能少于 6 个。这使得比热比值的最大值是 1.33，而这个结果对于氢气、氧气、氮气、碳氧化物、氮氧化物和氢氯酸来说都太小了。

但是光谱仪告诉我们某些分子可以进行多种不同类型的振动。因此它们必然是具有相当复杂程度的体系，具有的变量数远大于 6。那么，每增加一个变量，都会在不影响外部压力的前提下引入一些附加的内部运动的能力。因此，不论是在恒压还是恒容条件下计算，每增加一个变量都会使比热增加。

分子以势能形式储存能量的能力也是如此。但是，当我们假定分子仅由两个原子组成，计算得到的比热值就已经太大了。于是，分子的复杂程度每增加一点，都只会加大使比热的观测值与计算值相一致的难度。

现在我已将我认为的分子理论遇到的最大困难呈现在各位面前。玻尔兹曼曾经建议，应该从分子与围绕在其周围的以太介质的相互作用中寻求解释。不过，要是我们借助于这种介质的话，恐怕只会使已经过大的比热计算值变得更大。

玻尔兹曼定理不仅可以用于确定分子的速度分布，还可以用于确定在外力作用下分子自身在一个区域中的分布。它告诉我们，在分子势能为 ψ 的一点，分子的分布密度正比于 $e^{-\frac{\psi}{\kappa\theta}}$，其中 θ 为绝对温度，而 κ 对于所有气体都是常数。由此可知，如果同一容器中的几种气体受到外力（例如重力）作用，每一种气体的分布与没有其他气体存在时是一样的。这一结果与道尔顿提出的定律是一致的，根据这个定律，大气可以被视为是由两种独立的气体组成，一种为氧气，另一种为氮气；随着海拔升高，氧气的浓度比氮气的浓度减小得更快些。

以上是大气丝毫不受扰动的情况，而风的影响会将大气混匀，使其组成比保持静止时更均匀。

玻尔兹曼定理的另一推论是，静止状态的气体在垂直方向上温度趋于一致。

In the case of the atmosphere, the effect of wind is to cause the temperature to vary as that of a mass of air would do if it were carried vertically upwards, expanding and cooling as it ascends.

But besides these results, which I had already obtained by a less elegant method and published in 1866, Boltzmann's theorem seems to open up a path into a region more purely chemical. For if the gas consists of a number of similar systems, each of which may assume different states having different amounts of energy, the theorem tells us that the number in each state is proportional to $e^{-\frac{\psi}{\kappa\theta}}$ where ψ is the energy, θ the absolute temperature, and κ a constant.

It is easy to see that this result ought to be applied to the theory of the states of combination which occur in a mixture of different substances. But as it is only during the present week that I have made any attempt to do so, I shall not trouble you with my crude calculations.

I have confined my remarks to a very small part of the field of molecular investigation. I have said nothing about the molecular theory of the diffusion of matter, motion, and energy, for though the results, especially in the diffusion of matter and the transpiration of fluids are of great interest to many chemists, and though from them we deduce important molecular data, they belong to a part of our study the data of which, depending on the conditions of the encounter of two molecules, are necessarily very hypothetical. I have thought it better to exhibit the evidence that the parts of fluids are in motion, and to describe the manner in which that motion is distributed among molecules of different masses.

To show that all the molecules of the same substance are equal in mass, we may refer to the methods of dialysis introduced by Graham, by which two gases of different densities may be separated by percolation through a porous plug.

If in a single gas there were molecules of different masses, the same process of dialysis, repeated a sufficient number of times, would furnish us with two portions of the gas, in one of which the average mass of the molecules would be greater than in the other. The density and the combining weight of these two portions would be different. Now, it may be said that no one has carried out this experiment in a sufficiently elaborate manner for every chemical substance. But the processes of nature are continually carrying out experiments of the same kind; and if there were molecules of the same substance nearly alike, but differing slightly in mass, the greater molecules would be selected in preference to form one compound, and the smaller to form another. But hydrogen is of the same density, whether we obtain it from water or from a hydrocarbon, so that neither oxygen nor carbon can find in hydrogen molecules greater or smaller than the average.

对于大气来说，风的影响导致温度变化，例如，若一定质量的气体垂直向上运动，那么它在上升过程中会膨胀并会冷却。

然而，除了给出这些我已经于 1866 年就以不那么精巧的方法获得并发表的结果之外，玻尔兹曼的定理似乎还开辟了一条通向更纯粹的化学领域的道路。因为，如果气体是由大量相似的系统构成，每一个系统都可以假定为具有不同能量的不同状态，定理告诉我们每一状态中系统的数量正比于 $e^{-\frac{\psi}{\kappa\theta}}$，其中 ψ 表示能量，θ 表示绝对温度，而 κ 是一个常数。

容易看出，这一结果应当被用于发生在不同物质混合体中的组合的状态理论。不过，由于这项工作是我这个星期才刚刚开始的，我还不想以我粗糙的计算结果来搅扰诸位。

我一直将论题限制在分子研究领域的一个很小的范围内。我并没有提及关于物质扩散、运动以及能量的分子理论，因为，尽管这些结果，尤其是关于物质扩散和液体蒸发的结果，对很多化学家来说极有吸引力，并且从这些结果中我们可以推导出重要的分子数据，但这些数据依赖于两个分子碰撞的条件，因而属于我们研究工作中必然具有很多假设性的部分。我认为，还是展示一下能表明流体各部分运动的证据，并且描述一下此运动在不同质量的分子中的分布情况更好些。

为表明同种物质的所有分子都具有相同的质量，我们就要谈到格雷姆引入的透析法，用这种方法可以通过一个多孔塞进行过滤，将两种不同密度的气体分离。

如果在一种气体中存在不同质量的分子，将同样的透析过程足够多次地重复之后，我们就可以得到两部分气体，其中一部分的平均分子质量比另一部分的大。这两部分气体有不同的密度和化合量。现在，也许可以说，没有谁曾以足够精细的方式对每一种化学物质进行过这一实验。然而自然过程在持续不断地进行着这种类型的实验。如果同种物质具有十分相似但质量略有差别的分子，那么其中质量较大的分子将会被优先选择来形成一种化合物，而质量较小的分子则形成另一种。但是，无论是从水中还是从碳氢化合物中得到的氢都具有相同的密度，因此，氧和碳都不能在氢分子中找到大于或小于平均质量的个体。

The estimates which have been made of the actual size of molecules are founded on a comparison of the volumes of bodies in the liquid or solid state, with their volumes in the gaseous state. In the study of molecular volumes we meet with many difficulties, but at the same time there are a sufficient number of consistent results to make the study a hopeful one.

The theory of the possible vibrations of a molecule has not yet been studied as it ought, with the help of a continual comparison between the dynamical theory and the evidence of the spectroscope. An intelligent student, armed with the calculus and the spectroscope, can hardly fail to discover some important fact about the internal constitution of a molecule.

The observed transparency of gases may seem hardly consistent with the results of molecular investigations.

A model of the molecules of a gas consisting of marbles scattered at distances bearing the proper proportion to their diameters, would allow very little light to penetrate through a hundred feet.

But if we remember the small size of the molecules compared with the length of a wave of light, we may apply certain theoretical investigations of Lord Rayleigh's about the mutual action between waves and small spheres, which show that the transparency of the atmosphere, if affected only by the presence of molecules, would be far greater than we have any reason to believe it to be.

A much more difficult investigation, which has hardly yet been attempted, relates to the electric properties of gases. No one has yet explained why dense gases are such good insulators, and why, when rarefied or heated, they permit the discharge of electricity, whereas a perfect vacuum is the best of all insulators.

It is true that the diffusion of molecules goes on faster in a rarefied gas, because the mean path of a molecule is inversely as the density. But the electrical difference between dense and rare gas appears to be too great to be accounted for in this way.

But while I think it right to point out the hitherto unconquered difficulties of this molecular theory, I must not forget to remind you of the numerous facts which it satisfactorily explains. We have already mentioned the gaseous laws, as they are called, which express the relations between volume, pressure, and temperature, and Gay Lussac's very important law of equivalent volumes. The explanation of these may be regarded as complete. The law of molecular specific heats is less accurately verified by experiment, and its full explanation depends on a more perfect knowledge of the internal structure of a molecule than we as yet possess.

目前已有的对分子实际大小的估计都建立在将物体固态或液态时的体积与气态时的体积相比较的基础之上。在对分子体积的研究中我们遇到了很多困难，但同时也得到了很多可靠的结果，这使研究充满了希望。

对于分子可能的振动，相关理论的研究尚未开始，它本应在不断将动力学理论与光谱证据相比较的帮助下展开。用光谱工具和计算法武装起来的才智之士，一定会发现一些与分子内部结构有关的重要事实。

观测到的气体的透明度看来似乎很难与分子研究的结果一致。

一种认为气体是由间隔距离与其自身直径符合适当比例的分散小球组成的气体分子模型，只能允许极少量的光穿透 100 英尺的距离。

不过，要是我们还记得与光的波长相比分子的尺寸很微小，我们就会采用瑞利勋爵关于波与微小球体间相互作用的某些理论研究。他的研究表明，气体的透明度如果只受分子存在的影响，就会比我们有任何理由能够去相信的还要大得多。

一项几乎还没有人尝试过的更加困难的研究，是与气体的电学性质有关。还没有人能够解释为什么稠密的气体是非常良好的绝缘体，为什么气体在稀释或加热时可以放电，而完全的真空却是最好的绝缘体。

的确，稀薄气体中的分子扩散会进行得更快一些，因为分子运动的平均路程反比于密度。但是稠密气体与稀薄气体的电学差异似乎非常大以至于无法从这个角度来解释。

虽然我认为指出这种分子理论目前尚无法克服的困难是应该的，但我绝对不会忘记提醒诸位它已经令人满意地解释了大量问题。我们已经提到过气体定律，正如我们所说，它表达了气体的体积、压力与温度之间的关系，以及盖·吕萨克的极其重要的等容定律。可以认为关于这些定律的解释是完整的。分子比热定律已被实验相对粗略地验证了，而对它的完整解释则有赖于拥有比今天更完备的关于分子内部结构的知识。

But the most important result of these inquiries is a more distinct conception of thermal phenomena. In the first place, the temperature of the medium is measured by the average kinetic energy of translation of a single molecule of the medium. In two media placed in thermal communication, the temperature as thus measured tends to become equal.

In the next place, we learn how to distinguish that kind of motion which we call heat from other kinds of motion. The peculiarity of the motion called heat is that it is perfectly irregular; that is to say, that the direction and magnitude of the velocity of a molecule at a given time cannot be expressed as depending on the present position of the molecule and the time.

In the visible motion of a body, on the other hand, the velocity of the centre of mass of all the molecules in any visible portion of the body is the observed velocity of that portion, though the molecules may have also an irregular agitation on account of the body being hot.

In the transmission of sound, too, the different portions of the body have a motion which is generally too minute and too rapidly alternating to be directly observed. But in the motion which constitutes the physical phenomenon of sound, the velocity of each portion of the medium at any time can be expressed as depending on the position and the time elapsed; so that the motion of a medium during the passage of a sound-wave is regular, and must be distinguished from that which we call heat.

If, however, the sound-wave, instead of traveling onwards in an orderly manner and leaving the medium behind it at rest, meets with resistances which fritter away its motion into irregular agitations, this irregular molecular motion becomes no longer capable of being propagated swiftly in one direction as sound, but lingers in the medium in the form of heat till it is communicated to colder parts of the medium by the slow process of conduction.

The motion which we call light, though still more minute and rapidly alternating than that of sound, is, like that of sound, perfectly regular, and therefore is not heat. What was formerly called Radiant Heat is a phenomenon physically identical with light.

When the radiation arrives at a certain portion of the medium, it enters it and passes through it, emerging at the other side. As long as the medium is engaged in transmitting the radiation it is in a certain state of motion, but as soon as the radiation has passed through it, the medium returns to its former state, the motion being entirely transferred to a new portion of the medium.

Now, the motion which we call heat can never of itself pass from one body to another unless the first body is, during the whole process, hotter than the second. The motion of radiation, therefore, which passes entirely out of one portion of the medium and enters another, cannot be properly called heat.

不过，由这些探索得到的最重要的结果是关于热现象更加明确的概念。首先，介质的温度由介质中单个分子的平均平动动能来度量。在两种介质有热交换时，以此确定的温度有变得一致的倾向。

其次，我们知道了如何将所谓的热运动与其他类型的运动区分开。我们所谓的热运动的特性是完全无规则；也就是说，在任一给定时刻，一个分子的速度的方向和大小不可能用时间和现在分子所处的位置表示。

另一方面，在物体的可视运动中，物体的任一可视部分所含全部分子的质心速度，就是所观测到的该部分的速度。不过，分子也可能由于被加热而具有不规则的扰动。

同样地，在声音的传播过程中，物体不同部分的运动通常由于过于细微和过于快速变化而难以被直接观测。不过，在产生声音这一物理现象的运动过程中，介质中每一部分在任意时刻的速度都可以用位置和所经过的时间表示出来；因此在声波传播过程中介质的运动是规则的，必然不同于我们所说的热运动。

然而，如果声波不是以整齐有序的方式向前传播并使其所经之处的介质归于静止，而是遇到阻力从而将运动耗损在无规则的激发中，这种无规则的分子运动就无法再以声音的形式在一个方向上快速地传播，而只能以热的形式留在介质中，直到通过缓慢的传导过程流向介质较冷的部分。

被我们称为光的那种运动，尽管比声音运动更加细微和快速变化，但也和声音一样是非常规则的运动，因此不是热运动。以前被称为热辐射的物理现象在物理本质上与光是一样的。

辐射在到达介质的某一部分后，就会进入并穿过该部分，再从另一端出现。该部分介质在传导辐射时处于一种特定的运动状态，而一旦辐射穿过该部分介质，该部分介质便会回到先前的状态，而运动完全转移到了介质中新的部分。

现在，我们称为热的这种运动，不会自发地从一个物体进入另一个物体，除非在整个传导过程中，第一个物体总是比第二个物体的温度高一些。因此，从介质的一部分完全流出再进入另一部分的辐射运动，就不适宜被称作热了。

We may apply the molecular theory of gases to test those hypotheses about the luminiferous ether which assume it to consist of atoms or molecules.

Those who have ventured to describe the constitution of the luminiferous ether have sometimes assumed it to consist of atoms or molecules.

The application of the molecular theory to such hypotheses leads to rather startling results.

In the first place, a molecular ether would be neither more nor less than a gas. We may, if we please, assume that its molecules are each of them equal to the thousandth or the millionth part of a molecule of hydrogen, and that they can traverse freely the interspaces of all ordinary molecules. But, as we have seen, an equilibrium will establish itself between the agitation of the ordinary molecules and those of the ether. In other words, the ether and the bodies in it will tend to equality of temperature, and the ether will be subject to the ordinary gaseous laws as to pressure and temperature.

Among other properties of a gas, it will have that established by Dulong and Petit, so that the capacity for heat of unit of volume of the ether must be equal to that of unit of volume of any ordinary gas at the same pressure. Its presence, therefore, could not fail to be detected in our experiments on specific heat, and we may therefore assert that the constitution of the ether is not molecular.

(**11**, 374-377; 1875)

我们可以用气体分子理论去检验那些假定光以太是由原子或分子组成的假说。

那些敢于描绘光以太组成的人们也曾假定它是由原子或分子组成的。

将分子理论应用于这类假说会导致相当令人惊讶的结果。

首先，以太分子只会是气体本身。如果我们愿意的话，我们可以假定每个以太分子都是一个氢分子的千分之一或百万分之一，并且它们可以自由穿越任何常见分子的间隙。但是，正如我们已经看到的，普通分子的扰动与以太分子的扰动之间会自发地建立平衡。换句话说，以太与其中的物体会趋于温度相同的状态，那么以太就会服从诸如压力和温度等常见的气体定律。

以太除了具有气体的性质以外，根据杜隆和珀蒂建立的定律，相同压力下，单位体积以太的热容与单位体积任何普通气体的热容相等。因此，在比热实验中不可能检测不到以太的存在，从而我们可以断言以太不是由分子构成的。

（王耀杨 翻译；李芝芬 审稿）

The Law of Storms

J. J. Murphy

Editor's Note

Waterspout phenomena have been known since the time of ancient Greece, but their explanation was still being debated. Here Joseph John Murphy offers a new explanation for cyclones. He notes that air warms and expands whenever water vapour within it condenses into droplets, as the condensation releases latent heat. In a column of air, such a process may be self-amplifying: the expanding air lowers the pressure on air beneath it, causing that air to expand, cool and trigger further condensation. There is some basis in Murphy's idea, especially as waterspouts tend to occur where cold air exists over warmer bodies of water. But the dynamics causing waterspout formation, usually in connection with cumulus clouds high above, are considerably more complex.

I have to thank you for publishing, in *Nature* of Dec. 2, 1875, my letter in reply to M. Faye's theory of cyclones, and I have now to submit some remarks on his theory of waterspouts.

I understand him to maintain that the dark part of the waterspout, which we see, contains a core of transparent air, which is descending at the centre of a vortex, and that the dark visible external part is a cloud formed by an ascending counter-current.

All this is unproved, and I think baseless. No dynamical reason can be assigned why there should be a downward current at the centre of the vortex. If the waterspout is formed in a vortex, which I think probable, though I am not certain of it, the vortical motion will produce not a downward but an upward current at its centre, in consequence of the diminution of barometric pressure, due to the air being thrown to the circumference by the centrifugal force. We see such upward currents formed in the little dust-whirlwinds that form themselves over streets and roads in windy weather.

Further, if M. Faye's theory were true, and if the waterspout were transparent at the centre, it could not be so well defined and solid as it usually is, nor could it be formed so rapidly.

The true theory of waterspouts is expounded in Espy's "Philosophy of Storms", a work which, notwithstanding its great error of denying the rotation of cyclones, made an era in meteorology, and, so far as I am aware, is not yet superseded.

风暴定律

墨菲

编者按

在古希腊时代人们就知道海龙卷这种现象，不过关于该现象的解释一直存在争议。在这篇文章中，约瑟夫·约翰·墨菲对龙卷风提出了一种新的解释。他认为，当空气中的水汽凝结成小水滴时，空气就会变热并膨胀，因为凝结过程释放了潜热。在一个空气柱中，这种过程会自身放大：空气的膨胀使其下方的空气压强降低，导致下方的空气发生膨胀、冷却并进一步引起水汽的凝结。墨菲的观点有一定的事实根据，特别是海龙卷经常发生在那些较温暖的洋面上存在冷空气的地方。不过，海龙卷的发生通常与高空积云有关，其动力学成因是相当复杂的。

感谢你们在 1875 年 12 月 2 日的《自然》上发表了我对费伊的龙卷风理论的回信，现在我对他的海龙卷理论再作一些评论。

我的理解是他坚持以下两点：第一，我们看到的海龙卷的黑色部分包含一个在涡旋中心位置正在下沉的透明空气核；第二，可以看到的黑色部分的外围是由一股上升的空气流形成的云。

这些观点还都没有得到证实，而且，我觉得也没有事实根据。他的理论没能说明在涡旋中心会有一股向下运动的气流的动力学原因。如果海龙卷是涡旋式的（我认为这是很可能的，尽管还不能确定），那么涡旋运动就会在涡旋中心产生一个向上而不是向下运动的气流，因为在离心力作用下空气向外围运动从而导致中心气压降低。在有风的天气里，我们可以在街道上小规模的沙尘旋风中看到这种上升的气流。

此外，如果费伊的理论是正确的，并且海龙卷的中心是透明的，那么海龙卷就不会像它通常表现出的那样坚实而清晰，而且也不会那么迅速地形成。

埃斯皮的《风暴原理》一书阐述了关于海龙卷的准确理论。尽管该理论存在否认龙卷风旋转这一重大错误，但它却开创了气象学的新纪元，并且就我所知，到现在为止它依然是无可替代的。

When vapour is condensed into water, forming cloud, the latent heat of the vapour is liberated and expands the air. A simple calculation shows that, after deducting the destroyed volume of the condensed vapour, the increased volume of the air due to this expansion is between four and five times as great as the volume of the vapour before condensation. If, then, the air is nearly saturated with moisture, and the temperature in a state of convective equilibrium for dry air (that is to say, when the difference between the temperatures of any two strata is that due to the difference of their pressures), and condensation begins in any column of air, the effect of liberating this heat will be to make the air of that column warmer and lighter than the air at corresponding heights in the surrounding columns. What follows is from Espy's work, page 44:—

> "It begins, by its diminished specific gravity, to rise, and then, if all circumstances are favourable, the cloud will increase as it ascends, and finally become of so great perpendicular depth, that by its less specific gravity the air below it, in consequence of diminished pressure, will so expand and cool by expansion, as to condense the vapour in it; and this process may go on so rapidly that the visible cone may appear to descend to the surface of the sea or earth from the place where it first appears, in about one or two seconds. The terms here employed must not be understood to mean that the cloud actually descends; it appears to the spectator to descend, but this is an optical deception, arising from new portions of invisible vapour constantly becoming condensed, while all the time the individual particles are in rapid motion upwards."

To this I will add as very probable, if not quite certain, that the rarefaction thus caused at the waterspout will produce an inflow of air from all sides, and this will produce a vortex at the centre; this again, by its centrifugal force, will increase the rarefaction, and thus will intensify the effect. But the commencement of the waterspout is in the way described by Espy in the above extract.

(**13**, 187; 1875)

Joseph John Murphy: Old Forge, Dunmurry, Co. Antrim, Dec. 12, 1875.

当水汽凝结为水滴而形成云时，水汽中的潜热就被释放出来并使空气膨胀。一个简单的计算说明：扣除水汽凝结造成的体积减小，潜热释放使空气膨胀而造成的空气体积增加大约是水汽凝结前空气体积的 4~5 倍。如果此时空气湿度接近饱和，而且温度正好能使干空气处于对流平衡状态（也就是说，此时任意两个气层之间的温度差都取决于它们之间的气压差），并且在某气柱中水汽发生了凝结，那么水汽凝结释放的潜热就会使该气柱中的空气比周围同高度的空气更暖更轻。接下来便会像埃斯皮的著作中第 44 页所说的：

“这部分空气由于比重减小而开始上升，如果此时环境条件都很适宜，那么随着空气的上升，云团会逐渐增大，最终其垂直厚度会变得非常大。这就会使得云团下比重较小的空气因为气压降低而膨胀变冷，从而使其中的水汽发生凝结。这个过程可能会进行得非常迅速，以至于可见的锥形云团在大概一两秒内就从其最初出现的地方下降到海面或者地面处。我们不能把这句话理解成云团真的在下降；在观察者看来云团似乎在下降，但这只是种视觉错觉，实际上是因为空气中的气体分子在快速上升的同时，不断有新的不可见水汽持续地发生凝结。”

对此，我想提出一个虽不是十分肯定但可能性很大的补充：海龙卷中的空气稀薄会使周围的空气从各个方向流入，使龙卷中心形成涡旋。在离心力作用下，龙卷中心的空气更加稀薄，从而加强了这种作用。不过，海龙卷最初的形成还是以上述摘录中埃斯皮描述的方式进行的。

（刘明 翻译；王鹏云 审稿）

On the Telephone, an Instrument for Transmitting Musical Notes by Means of Electricity

J. Munro

Editor's Note

Elisha Gray, *Nature* here reports, recently presented a paper describing a means for sending musical notes over long distances by electricity—in short, a telephone. Gray also showed how telegraphic messages could be transmitted this way. The device works by interrupting the electrical current at one end in a desired pattern, and then detecting a similar interruption at the other end, and linking this signal to a reed or box for producing sound. Gray demonstrated that as many as eight messages may be sent simultaneously. He invented his technique at nearly the same time as—some believe before—Alexander Graham Bell, who is widely credited as the inventor of the telephone.

MR. Elisha Gray recently read a paper before an American Society explaining his apparatus for transmitting musical notes by electricity. He showed experimentally how, by means of a current of electricity in a single wire, a number of notes could be reproduced simultaneously at a great distance, and how by this means also a number of telegraphic messages could be transmitted at once along a wire and separately received at the other end. One of Mr. Gray's apparatuses was exhibited in London at the last *soirée* of the Society of Telegraph Engineers by the president, Mr. Latimer Clark. The principle of the apparatus is as follows:—

A vibrating reed is caused to interrupt the electric current entering the wire a certain number of times per second and the current so interrupted at the sending end sets a similar reed vibrating at the distant end.

The sending reed is ingeniously maintained in constant vibration by a pair of intermittent electro-magnets which are magnetised and demagnetised by the vibrating reed itself.

Thus in Fig. 1 (which represents the transmitting part of the telephone and its connections for a single note), the current from the magnet battery flowing in the direction of the small arrow passes through the pair of electro-magnets A to the terminal *r* of the reed R, and thence by the spring contact *b* and the wire *bz* to the battery again, completing its circuit without passing through the other pair of electro-magnets B, which are not therefore magnetised. The reed R is consequently pulled over by the electro-magnets A. But on this taking place the spring contact *b* is broken and the circuit is no longer completed through *bz* but through the electro-magnets B, which are consequently magnetised, and

电话：一种利用电流传送音符的仪器

芒罗

编者按

《自然》的这篇文章报道的是伊莱沙·格雷最新提交的一篇论文，论文中描述了一种利用电流长距离传送音符的方法，简言之就是电话。另外，格雷还展示了如何通过这种方法传送电报信息。该装置的工作原理是：在一端以某种设计好的模式中断电流，然后在另一端探测到类似的中断电流信号，并将该信号连接到一个簧片或者盒子中来产生声音。格雷展示的这种装置可以同时传送多达 8 条信息。他差不多是在亚历山大·格雷厄姆·贝尔发明电话的同时发明了这种技术，甚至有些人认为格雷的发明比贝尔还早，不过，人们更普遍认为贝尔是电话的发明者。

最近，伊莱沙·格雷先生向美国某学会宣读了一篇论文，介绍了他发明的利用电流传送音符的设备，他用实验验证了如何利用单一导线中的电流使大量音符在很远的距离处能够同时重现，以及如何利用同样的方式使大量电报信息迅速通过一根导线传送而在接收端分别接收到信息。电报工程师协会会长拉蒂默·克拉克先生在该协会最近的一次晚宴（在伦敦）上展示了格雷先生的设备中的一种。这台设备的工作原理如下。

用一个振动簧片按每秒钟若干次中断进入电线的电流，那么这种在信号发送端的断续电流就会使得位于远程接收端的簧片产生类似的振动。

发送端的簧片通过一对间歇式电磁铁来精确地保持不断振动，而振动簧片本身又可以使电磁铁被磁化和退磁。

因此在图 1（示意了电话的发送部分及其产生单一音符所对应的连接）中，电流由磁铁电池出发，按照小箭头的方向经过一对电磁铁 A 流向簧片 R 的末端 *r*，而后，通过弹簧触点 *b* 沿 *bz* 又回到电池，没有经过另一对电磁铁 B 就完成了回路，因此 B 没有被磁化。结果簧片 R 被拉向磁化了的电磁铁 A。但是这将导致弹簧触点 *b* 断开，回路不再通过 *bz* 闭合，而是通过电磁铁 B 闭合，随之它被磁化并通过感应把簧片引向 B，因此簧片又被弹回到中间的位置，这样 *b* 点又被连接上，而电磁铁

tend by their induction on the reed to neutralise that of B. The reed therefore springs back to its intermediary position, but in so doing the contact at *b* is again made and the electro-magnets B again short-circuited and the reed pulled over (or rather *assisted* over, for it has its own resilience or spring) towards A; so this goes on keeping the reed in vibration between the electro-magnets and alternately making and breaking the spring contact *b* and also that of *a*, the number of contacts per second being dependant on the vibrating period of the reed.

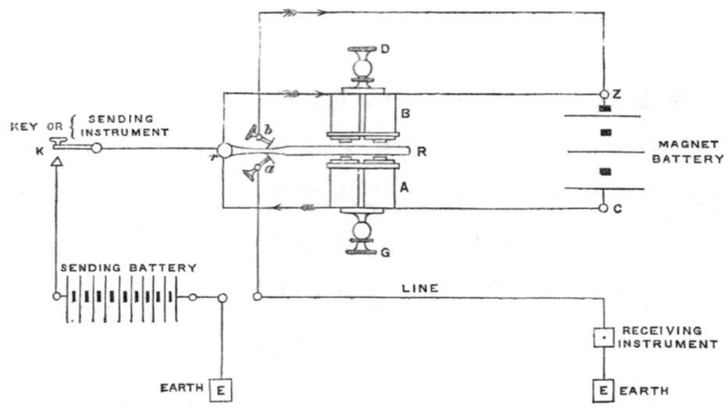

Fig. 1

While this is going on the reed of course emits its musical note. Two Leclanché or bichromate cells are sufficient to work the transmitter and give a good note. The spring contact *b* is to be adjusted by the screw there seen until the note emitted by the reed is both loud and pure. The magnets A and B are adjustable to or from the reed by the milled heads G and D.

The spring contact *a* just mentioned belongs properly to the line circuit. It is the intermittent contact which interrupts the current sent into the line. As will be seen from the diagram the circuit of the sending battery is made through the key K, the reed, and the spring contact *a*. On holding down the key K the current flows into the line, being interrupted, however, by the contact *a* as many times per second as the reed vibrates, and this intermittent current flowing to earth at the distant station, s made to elicit a corresponding note from the receiving apparatus there.

The receiving instruments are of two kinds, electro-magnetic and physiological.

In the first there is a plain double electro-magnet with a steel tongue having one end rigidly fixed to one pole, the other end being free to vibrate under the other pole. This stands over a wooden pipe closed at one end. Thus in Fig. 2 *t*T is the steel tongue fixed at *t* and free at T, while P is the sounding-pipe. The received current, coming from the line and passing through the electro-magnet M to earth, sets the tongue vibrating, and the pipe gives forth the same note as the reed at the sending station. Ten Daniell cells working through 1,000 ohms, give a good strong note, especially when the receiver is held in the hand close to the head. The screw *a*, Fig. 1, must be adjusted to give the best effect.

B 又被短路，弹簧重被拉向 A（或者更确切地说是因为簧片自身的弹力或弹性促使簧片偏向）；如此往复使得簧片 R 在一对电磁铁之间不停地振动，使弹簧触点 *a* 和 *b* 交替断开和接通，且每秒钟的接通次数取决于簧片 R 的振动周期。

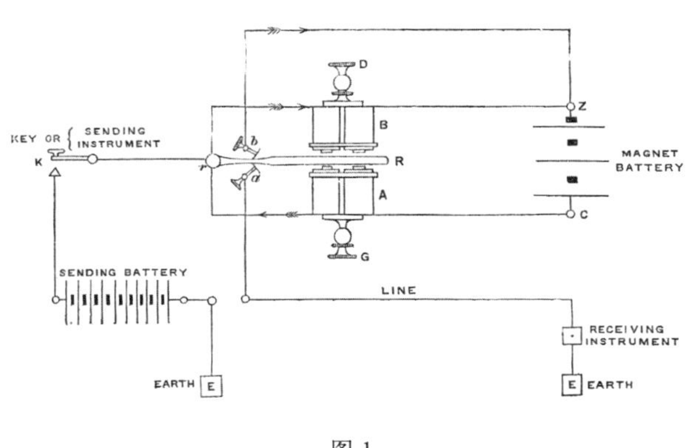

图 1

在上述的振动过程中簧片发出音符，两个勒克朗谢电池或者重铬酸盐电池足以启动信号发送器并能发出令人愉快的音符。通过螺丝可以调整弹簧触点 *b* 直到簧片发出的音符足够响亮和纯正。电磁铁 A 和 B 相对于簧片的位置可以用滚压了纹边的调节头 G 和 D 进行调整。

刚提到的弹簧触点 *a* 属于电话线路中的触点，它与簧片的间歇性接触不断地中断电话线路中的电流信号。正如图 1 中所示的，发送电池组的电路包括开关 K，簧片 R 和弹簧触点 *a*。接通开关按钮 K，电流流入电话线路，但随着簧片的振动，弹簧触点 *a* 会每秒多次断开而中断电流，这些间歇式的电流会流向远处的接地端，s 用于从接收设备引导出相应的信号。

信号接收设备分为两种，一种是电磁型的，另一种是生理型的。

第一类设备中有一个简单的双电磁铁，电磁铁上的钢舌一端被刚性地固定在一个电极上，另一端位于另一个电极下方，可以自由振动。这些竖立在一个一端封闭的木制管子上。如图 2 所示，*t*T 为钢舌，其中 *t* 为固定端，T 为自由端，P 为发音管。接收到的电流从电话线路流经电磁铁 M 最后到接地端，该电流使钢舌振动，于是发音管发出音符，这与发送端簧片发出的音符完全相同。10 个丹聂耳电池可以负载 1,000 欧姆的电阻，使设备发出强劲而优美的信号，尤其是当接收器被拿在手里靠近头部的时候。为了得到最好的效果，必须校正图 1 中的螺丝 *a*。

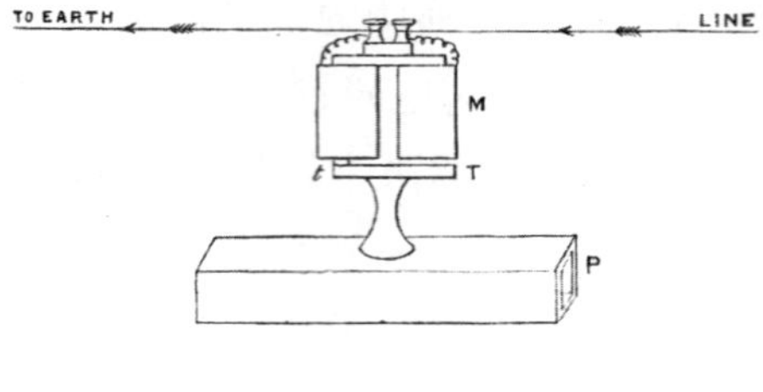

Fig. 2

The other receiving instrument is the most interesting of the two. It consists of a small induction coil used in conjunction with a peculiar sounding-box, as shown in Fig. 3.

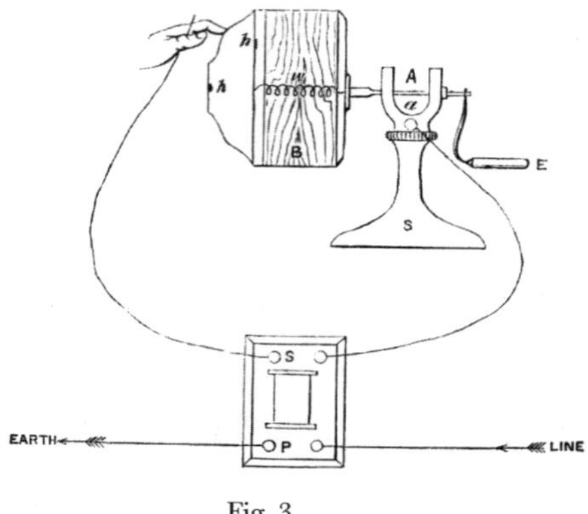

Fig. 3

Here the line-current is passed to earth through the primary circuit P of the small induction coil, and the induced current is led to the sounding-box. This consists of a flat hollow cylindrical wooden box B, covered by a convoluted face of sheet zinc with two air holes *hh*, perforated in it, this box is attached to a metal axle A, turning in forked iron bearings, insulated from but supported by an iron stand S. By this means the sounding-box can be revolved by the ebony handle E. The zinc face is connected across the empty interior of the box by a wire W to the metal bearings on the other side. One end of the secondary circuit of the induction coil is to be connected to the metal bearing by the terminal *a*, and the other to a short bare wire held in the left hand. On then striking a finger of the hand holding the wire smartly across the zinc face, the proper note is sounded by the box; or, what is more convenient, on turning the box by the insulated handle and keeping the point of the finger rubbing on its face, the note is heard. The rough under side of the finger pressed pretty hard on the bulging part of the face is best. The instant the current is put on by the sending key K, Fig. 1, the dry rasp of the skin on the zinc-surface becomes changed into a musical note.

These "sounders" can be made to receive indifferently a variety of notes. I have under my care at present a telephone with four transmitters tuned to give the four notes of the

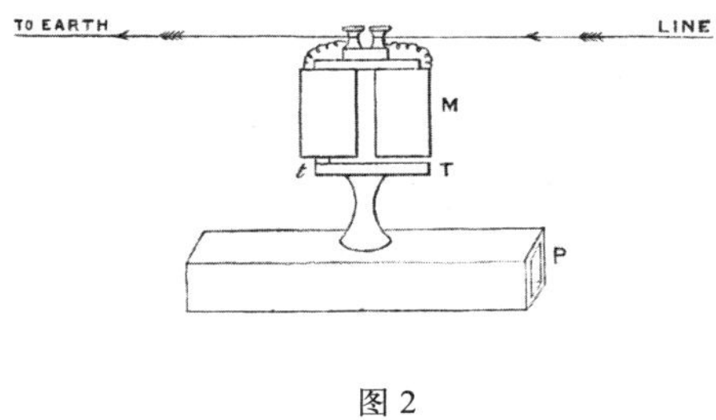

图 2

另一类接收设备是两类中最有意思的，它包括一个很小的感应线圈，这个感应线圈用来连接一个特殊的发音盒，如图 3 所示。

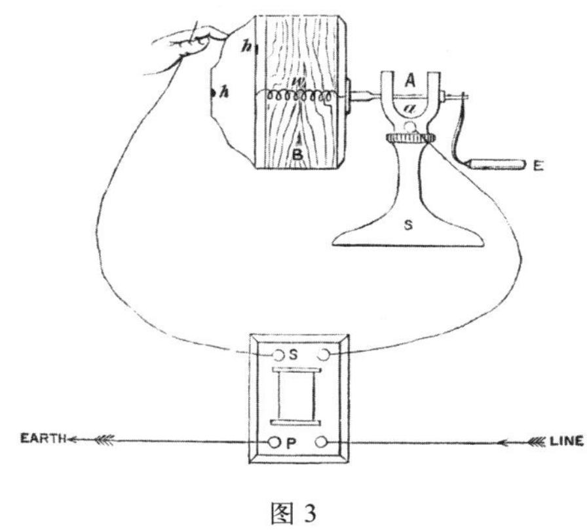

图 3

在这里，线路中的电流是通过带有小感应线圈的初始电路 P 后接地，感应电流通向发音盒。发音盒包括一个扁平的空心木质圆柱体 B，圆柱体表面缠绕着金属锌薄片，其中打有两个通气孔 *hh*，这个发音盒与金属轴 A 相连，A 可在铁制叉状支座上旋转，由铁架 S 支撑但与 S 之间是绝缘的。用这种方式可以通过黑檀木手柄 E 来转动发音盒，锌片表面通过金属丝 W 穿过中空的发音盒内部与另一端的金属支座连接。感应线圈次级电路的一端在接线点 *a* 与金属支架相连，另一端连着一根较短的裸线，用左手握住。接下来用握线的手的一个手指轻快地划过锌片表面，发音盒就会发出特定的音符；或者，更简便的方法是通过绝缘的手柄转动发音盒，使手指尖不停地摩擦锌片表面，这样也可以听到音符。被手指用力压住的锌片表面越粗糙，产生的声音越好。一旦通过发送端的开关 K（见图 1）接通电流，皮肤在锌片表面发出的单调声响就会转化成音符。

这些“发声器”可以制作成各种音符的接收器，目前我已经有一部具有 4 个发送器、能发出 4 种普通和弦音符的电话，以及两台能够准确解析任意一个音符或者

common chord, and two receivers, which interpret equally well any one of these notes or all together. But sounders are also made in the same way which will emit only one special note, and so are sensible only to the corresponding current. It is by their means that the telephone can be applied to multiplex telegraphy. As many as eight transmitters may be set to interrupt the line current according to the vibrations of eight different tuning-forks, and the resultant current can be made by means of eight special receivers to reproduce the same number of corresponding notes at the distant station. The current is controlled by eight keys at the sending end and sifted by eight sounders at the receiving end, each sounder being sensitive only to those portions of the current affected by its corresponding transmitter. The superimposed effect of the eight keys and transmitters on the line current can all be separately interpreted at the receiving end. Thus eight messages might be transmitted simultaneously along one wire in the same direction. It would seem hitherto, however, that this method of telegraphy by the telephone is inferior to the ordinary methods in point of speed of signalling, and in the length of circuit which can be worked by a given battery power.

(**14**, 30-32; 1876)

同时解析所有音符的接收器。但是发声器同样也可以做成只能发出一种特定音符，因此只对相应的电流敏感。这就意味着电话可以应用在多路电报上。最多可以有 8 个发送器，依据 8 个不同音叉的振动情况来中断各自电路中的电流，这个合成的电流可以通过远处接收端的 8 个特殊接收器来重现数量相同的对应音符。电流由发送端的 8 个开关控制，并通过接收端的 8 个发声器来过滤筛选，每一个发声器仅对与它相对应的发送器产生的电流敏感。发送端的 8 个开关及发送器对于线路电流所加的影响在接收端可以被分别解析出来。因此 8 种信息可以沿着一条电线向同一个方向同步传送。但是就目前来看，这种用电话发送电报的方式在信号传输速度方面和电池可以负载的线路长度方面不如普通的方法。

（胡雪兰 翻译；赵见高 审稿）

Maxwell's Plan for Measuring the Ether

Editor's Note

One of the most remarkable of Maxwell's scientific exploits was a scheme for telling the velocity of the Earth and the Solar System as a whole relative to the luminiferous ether, supposed at the time to be necessary for the propagation of electromagnetic waves. Maxwell's proposal to D. P. Todd, director of the Nautical Almanac office in Washington, D.C., was that accurate measurements of the rotation of Jupiter's satellites around their planet would allow this relative velocity through the ether to be derived. If Maxwell had been able to execute this plan (for which in Todd's opinion the data were not yet sufficiently accurate), he would have discovered that the ether is irrelevant to the propagation of electromagnetic waves and indeed does not exist—the foundation for Einstein's Theory of Special Relativity.

"On a Possible Mode of Detecting a Motion of the Solar System through the Luminiferous Ether". By the late Prof. J. Clerk Maxwell. In a letter to Mr. D. P. Todd, Director of the *Nautical Almanac* Office, Washington, U.S. Communicated by Prof. Stoke, Sec. R.S.

Mr. Todd has been so good as to communicate to me a copy of the subjoined letter, and has kindly permitted me to make any use of it.

As the notice referred to by Maxwell in the *Encyclopaedia Britannica* is very brief, being confined to a single sentence, and as the subject is one of great interest, I have thought it best to communicate the letter to the Royal Society.

From the researches of Mr. Huggins on the radial component of the relative velocity of our sun and certain stars, the coefficient of the inequality which we might expect as not unlikely, would be only something comparable with half a second of time. This, no doubt, would be a very delicate matter to determine. Still, for anything we know *à priori* to the contrary, the motion might be very much greater than what would correspond to this; and the idea has a value of its own, irrespective of the possibility of actually making the determination.

In his letter to me Mr. Todd remarks, "I regard the communication as one of extraordinary importance, although (as you will notice if you have access to the reply which I made) it is likely to be a long time before we shall have tables of the satellites of Jupiter sufficiently accurate to put the matter to a practical test."

麦克斯韦测量以太的计划

编者按

麦克斯韦最突出的科学成就之一是，他制定了地球和太阳系作为一个整体相对于以太的速度的测量计划。当时，以太被认为是电磁波传播的必要条件。麦克斯韦向华盛顿特区航海历书处的办公室主任托德建议，通过精确测量木星卫星围绕木星的公转可以得到太阳系整体相对于以太的速度。如果麦克斯韦当时真的实施了这个计划（托德认为该计划所需的数据还不够精确），那么他会发现，以太和电磁波的传播没有任何关系，而且实际上以太根本就不存在——这便是爱因斯坦的狭义相对论的基础。

《论一种探测太阳系在以太中运动的可能方式》，作者是已故的克拉克·麦克斯韦教授。这篇文章出现在麦克斯韦写给美国华盛顿**航海历书处**的办公室主任托德先生的一封信中，由皇家学会的秘书斯托克教授宣读。

托德先生十分慷慨地给了我一份附信的拷贝，并且大方地允许我使用它。

因为麦克斯韦提到的《大英百科全书》中的参考资料十分简略，仅仅只有一句话。而这个题目又十分吸引人，所以我觉得最好还是在皇家学会宣读这封信。

根据哈金斯先生对太阳和某些星体的相对速度径向分量的研究，系数上的差别（我们暂且认为结果可靠）大约只有半秒。毫无疑问，这对实验测量来说是一个很精细的问题。而且这和我们**先验的**观点相反，我们先验的观点中的运动比这里涉及的大很多。如果不考虑实际测量的可操作性，这个想法还是有它自身的价值的。

在托德先生给我的信中，他评论到，“虽然（如您将在我的答复中看到的那样）将这个计划付诸实施需要足够精确的木星卫星的数据表，而且这恐怕需要很长时间才能获得，但是我仍然认为这封信极其重要。”

I have not thought it expedient to delay the publication of the letter on the chance that something bearing on the subject might be found among Maxwell's papers.

(Copy)

Cavendish Laboratory,

Cambridge,

19th March, 1879

Sir,

I have received with much pleasure the tables of the satellites of Jupiter which you have been so kind as to send me, and I am encouraged by your interest in the Jovial system to ask you if you have made any special study of the apparent retardation of the eclipses as affected by the geocentric position of Jupiter.

I am told that observations of this kind have been somewhat put out of fashion by other methods of determining quantities related to the velocity of light, but they afford the *only* method, so far as I know, of getting any estimate of the direction and magnitude of the velocity of the sun with respect to the luminiferous medium. Even if we were sure of the theory of aberration, we can only get differences of position of stars, and in the terrestrial methods of determining the velocity of light, the light comes back along the same path again, so that the velocity of the earth with respect to the ether would alter the time of the double passage by a quantity depending on the square of the ratio of the earth's velocity to that of light, and this is quite too small to be observed.

But if J E is the distance of Jupiter from the earth, and l the geocentric longitude, and if l' is the longitude and λ the latitude of the direction in which the sun is moving through ether with velocity v, and if V is the velocity of light and t the time of transit from J to E,

$$\mathrm{JE} = [\,\mathrm{V} - v\cos\lambda\cos(l - l')\,]\,t$$

By a comparison of the values of t when Jupiter is in different signs of the zodiac, it would be possible to determine l' and $v\cos\lambda$.

I do not see how to determine λ, unless we had a planet with an orbit very much inclined to the ecliptic. It may be noticed that whereas the determination of V, the velocity of light, by this method depends on the differences of J E, that is, on the diameter of the earth's orbit, the determination of $v\cos\lambda$ depends on J E itself, a much larger quantity.

But no method can be made available without good tables of the motion of the satellites, and as I am not an astronomer, I do not know whether, in comparing the observations with the tables of Damoiseau, any attempt has been made to consider the term in $v\cos\lambda$.

将这封信推迟到可以在麦克斯韦的论文中找到相关内容的时候再发表，我认为并不适宜。

（原信的拷贝）

卡文迪什实验室，
剑桥，
1879 年 3 月 19 日

先生：

我非常高兴能收到您寄给我的木星卫星的数据表。受您对木星系统的兴趣的启发，我想问您有没有对由木星相对于地心的位置而造成的木星卫星蚀的明显延迟作过专门的研究。

我得知，在其他测量与光速有关的物理量的方法面前，这种观测已经有些过时了。但是，据我所知，这种观测给我们提供了估算太阳相对于以太介质的速度的方向和大小的**唯一**方法。即便我们坚信像差理论，我们得到的也只是恒星位置的差别。并且按照在地球上测量光速的方法，光按原路返回，地球相对于以太的速度会改变这一往返的时间。但是，这个时间上的改变量依赖于地球速度与光速之比的平方，因此，这个量太小，以至于无法观测。

但是，如果 J E 表示木星与地球之间的距离，l 表示地心经度，l' 和 λ 分别表示太阳在以太中运动方向的经度和纬度，v 表示太阳在以太中运动的速度，V 表示光速，t 表示光从木星到地球的传播时间，

$$\mathrm{JE} = [\,\mathrm{V} - v\cos\lambda\cos(l - l')\,]\,t$$

通过比较木星在黄道不同位置时的 t，就有可能确定 l' 和 $v\cos\lambda$。

除非有一个行星的轨道向黄道极大倾斜，不然我就无法知道怎样确定 λ。我们应该注意到，尽管用这种方法测量的光速 V 也与 JE 的变化有关，也就是说，依赖于地球轨道的直径，但是 $v\cos\lambda$ 却依赖于一个更大的量——JE 本身。

然而，没有精确的卫星运动数据就不会有任何行之有效的方法。我本人并不是天文学家，我不知道这些观测与达穆瓦索的数据表相比，是否曾经尝试过将 $v\cos\lambda$ 中的各项考虑在内。

I have, therefore, taken the liberty of writing to you, as the matter is beyond the reach of any one who has not made a special study of the satellites.

In the article E [ether] in the ninth edition of the "Encyclopaedia Britannica", I have collected all the facts I know about the relative motion of the ether and the bodies which move in it, and have shown that nothing can be inferred about this relative motion from any phenomena hitherto observed, except the eclipses, &c., of the satellites of a planet, the more distant the better.

If you know of any work done in this direction, either by yourself or others, I should esteem it a favour to be told of it.

Believe me,

Yours faithfully,

(Signed) J. Clerk Maxwell

(**21**, 314-315; 1880)

因此，我十分冒昧地给您写这封信。这个问题对没有专门研究过卫星的人来说，实在是勉为其难。

在《大英百科全书》第 9 版关于以太的文章中，我收集了所有我知道的关于以太以及在其中运动的物体的相对运动的资料，从这些资料中我发现，由目前观测到的实验现象，除了行星离我们越远，越有利于卫星蚀等现象以外，从其他任何已观测到的相关现象中都不能推断出有关这种相对运动的结论。

如果您知道有任何人，不论是您还是别人，做过这方面的工作，都请您告诉我，我将不胜感激。

相信我，

您忠实的，

（签名）克拉克 · 麦克斯韦

（王静 翻译；鲍重光 审稿）

Clerk Maxwell's Scientific Work

P. G. Tait

Editor's Note

James Clerk Maxwell died in November 1879. In this essay four months later, Scottish physicist Peter Guthrie Tait, who had known Maxwell from childhood, paid tribute to his accomplishments. Maxwell was producing influential, original work before the age of twenty. In 1864 he published a landmark paper giving the first complete statement of his theory of electricity and magnetism. It explained electromagnetic phenomena without recourse to action at a distance, and provided a unified view of what light is. Guthrie notes that the facility of Maxwell's thinking did not always translate into effective lectures. While the treatises he wrote were models of clarity, his extemporaneous lectures gave free rein to his imagination in a way that taxed his audiences.

AT the instance of Sir W. Thomson, Mr. Lockyer, and others I proceed to give an account of Clerk Maxwell's work, necessarily brief, but I hope sufficient to let even the non-mathematical reader see how very great were his contributions to modern science. I have the less hesitation in undertaking this work that I have been intimately acquainted with him since we were schoolboys together.

If the title of mathematician be restricted (as it too commonly is) to those who possess peculiarly ready mastery over symbols, whether they try to understand the significance of each step or no, Clerk Maxwell was not, and certainly never attempted to be, in the foremost rank of mathematicians. He was slow in "writing out", and avoided as far as he could the intricacies of analysis. He preferred always to have before him a geometrical or physical representation of the problem in which he was engaged, and to take all his steps with the aid of this: afterwards, when necessary, translating them into symbols. In the comparative paucity of symbols in many of his great papers, and in the way in which, when wanted, they seem to grow full-blown from pages of ordinary text, his writings resemble much those of Sir William Thomson, which in early life he had with great wisdom chosen as a model.

There can be no doubt that in this habit, of constructing a mental representation of every problem, lay one of the chief secrets of his wonderful success as an investigator. To this were added an extraordinary power of penetration, and an altogether unusual amount of patient determination. The clearness of his mental vision was quite on a par with that of Faraday; and in this (the true) sense of the word he was a mathematician of the highest order.

But the rapidity of his thinking, which he could not control, was such as to destroy, except for the very highest class of students, the value of his lectures. His books and his written

克拉克·麦克斯韦的科学工作

泰特

编者按

1879 年 11 月，詹姆斯·克拉克·麦克斯韦逝世。在 4 个月后的这篇短文中，苏格兰物理学家彼得·格思里·泰特（他从小就认识麦克斯韦）热情称颂了麦克斯韦的成就。麦克斯韦在 20 岁之前就开始发表有影响力的原创研究论文。1864 年，他发表了一篇首次完整阐述其电磁理论的论文。在这篇里程碑式的论文中，麦克斯韦抛开一定距离外的实际效应去解释电磁现象，并就光的本质提出了一个统一性的观点。格思里认为，麦克斯韦深刻的思考并没有通过他那些颇有影响的演讲全部体现出来。不过，与他那堪称逻辑清晰之典范的专著不同，他的即席演讲在某种程度上则更自由地展现了他那对听众来说过于跳跃的想象力。

应汤姆孙爵士、洛克耶先生以及其他一些人士的要求，我将对麦克斯韦的科学工作进行介绍。介绍必然是简略的，但我希望足以让即使没有数学背景的读者也能了解他对现代科学的伟大贡献。我与麦克斯韦在学生时代就已经熟识，因此，我毫不犹豫地接受了这个任务。

如果说数学家的头衔只属于那些不论是否试图理解每一步的意义，都对符号了如指掌的人（事实通常就是这样），那么克拉克·麦克斯韦就不是、也从来不试图成为这样的一流数学家。他总是不慌不忙地“完稿”，并且尽可能避免复杂的分析。他更喜欢以几何或物理的形式表示自己所研究的问题，并且总是借助于以下方式来完成下一步的研究：在必要时，将这些表示转化成符号的形式。在他众多的伟大著作中所出现的较少的符号，在需要的时候似乎又能拓展成为直接触及主题的成熟的篇章，从这些方面来看，他的著作与威廉·汤姆孙爵士的十分相似，他在早年就很有远见卓识地把爵士当作自己的榜样。

毫无疑问，为每个问题建立思维上的表示，这个习惯是麦克斯韦作为一名研究者能够获得巨大成功的主要秘诀之一。除此之外，他还拥有卓越的洞察力和非凡的意志力。他的思路清晰，堪比法拉第。从数学家一词的这种（真正的）意义来说，麦克斯韦就是一位最高层次的数学家。

然而，麦克斯韦的思维之敏捷，连他自己都无法控制，这使得他的讲座只能被极高层次的学生所接受，而其他人则很难从中受益。他的著作和演讲稿（通常是对

addresses (always gone over twice in MS.) are models of clear and precise exposition; but his *extempore* lectures exhibited in a manner most aggravating to the listener the extraordinary fertility of his imagination.

His original work was commenced at a very early age. His first printed paper, "*On the Description of Oval Curves, and those having a Plurality of Foci*", was communicated for him by Prof. Forbes to the Royal Society of Edinburgh, and inserted in the "*Proceedings*" for 1846, before he reached his fifteenth year. He had then been taught only a book or two of Euclid, and the merest elements of Algebra. Closely connected with this are three unprinted papers, of which I have copies (taken in the same year), on "*Descartes' Ovals*", "*The Meloid and Apioid*", and "*Trifocal Curves*". All of these, which are drawn up in strict geometrical form and divided into consecutive propositions, are devoted to the properties of plane curves whose equations are of the form

$$mr + nr' + pr'' + \cdots = constant$$

r, r', r'', &c., being the distances of a point on the curve from given fixed points, and m, n, p, &c., mere numbers. Maxwell gives a perfectly general method of tracing all such curves by means of a flexible and inextensible cord. When there are but two terms, if m and n have the same sign we have the ordinary Descartes' Ovals, if their signs be different we have what Maxwell called the Meloid and the Apioid. In each case a simple geometrical method is given for drawing a tangent at any point, and some of the other properties of the curves are elegantly treated.

Clerk Maxwell spent the years 1847–1850 at the University of Edinburgh, without keeping the regular course for a degree. He was allowed to work during this period, without assistance or supervision, in the Laboratories of Natural Philosophy and of Chemistry: and he thus experimentally taught himself much which other men have to learn with great difficulty from lectures or books. His reading was very extensive. The records of the University Library show that he carried home for study, during these years, such books as Fourier's *Théorie de la Chaleur*, Monge's *Géometrie Descriptive*, Newton's *Optics*, Willis' *Principles of Mechanism*, Cauchy's *Calcul Différentiel*, Taylor's *Scientific Memoirs*, and others of a very high order. These were *read through*, not merely consulted. Unfortunately no list is kept of the books consulted in the Library. One result of this period of steady work consists in two elaborate papers, printed in the *Transactions of the Royal Society of Edinburgh*. The first (dated 1849) "*On the Theory of Rolling Curves*", is a purely mathematical treatise, supplied with an immense collection of very elegant particular examples. The second (1850) is "*On the Equilibrium of Elastic Solids*". Considering the age of the writer at the time, this is one of the most remarkable of his investigations. Maxwell reproduces in it, by means of a special set of assumptions, the equations already given by Stokes. He applies them to a number of very interesting cases, such as the torsion of a cylinder, the formation of the large mirror of a reflecting telescope by means of a partial vacuum at the back of a glass plate, and the theory of Örsted's apparatus for the compression of water. But he

手稿的再一次重温）是表述清晰精确的典范；但是他的**即席**演讲却因为极富想象力的风格而使听众难以理解。

麦克斯韦最初的工作在他年轻时就已经着手展开。他发表的第一篇论文《论椭圆曲线及多焦点椭圆曲线》由福布斯教授代他在爱丁堡皇家学会的会议上宣读，并收录在《爱丁堡皇家学会会刊》中。当时是1846年，麦克斯韦尚不满15岁。那时的他只学过一两本欧几里德的书和最基本的代数基础。紧接着，他又写了另外3篇没有发表的文章，我有这几篇文章的拷贝，它们分别是：《笛卡尔椭圆》、《芜菁科昆虫形曲线和芹亚科植物形曲线》、《三焦点曲线》。这些建立在严格的几何形式上并且被分成论题连贯的文章，可以用来研究具有以下形式的方程所描述的平面曲线的性质：

$$mr + nr' + pr'' + \cdots = \text{常量}$$

r, r', r'' 等是从一个给定的固定点到曲线上某一点的距离，m, n, p 等仅仅是一些数字。麦克斯韦用容易弯曲但不能伸展的绳子，给出了一种理想的绘制这样曲线的一般方法。当方程中只有 m 和 n 两项时，如果两者同号，就得到普通的笛卡尔椭圆，如果两者异号，就得到麦克斯韦所谓的芜菁科昆虫形曲线和芹亚科植物形曲线。在每种情况下，麦克斯韦都给出了在任意一点作切线的简单几何方法，并且很好地处理了曲线其他方面的性质。

1847~1850年，克拉克·麦克斯韦就读于爱丁堡大学，在此他并不需要去学那些学位要求的常规课程，而是被允许在既无人帮助也无人指导的条件下在自然哲学与化学实验室工作。因此，他在实验中自学了很多东西，而这些是其他人很难从书本中或课堂上学到的。他的阅读非常广泛。大学图书馆的记录显示，他在这些年中借回家研读的书有傅立叶的《热力学理论》、蒙日的《几何学说明》、牛顿的《光学》、威利斯的《力学原理》、柯西的《微分计算》、泰勒的《科学回忆录》等高水平著作。这些书他全部**通读**，而不是仅仅翻阅一下。很可惜图书馆没有保存他在馆内阅览的书单。麦克斯韦这段时间持续学习的成果之一是发表在《爱丁堡皇家学会会报》上的两篇详细论文。第一篇《论曲线滚动理论》(1849年),这是一篇纯数学的论文,文章给出了大量简洁而恰当的例子。第二篇是《论弹性固体的平衡》(1850年)。考虑到作者当时的年龄，这可以被认为是他最不寻常的研究之一。麦克斯韦运用一系列特殊的假设，重新构造了已经由斯托克斯给出的方程。他将这些方程应用到许多非常有趣的情况中，比如圆柱体的扭曲、通过在玻璃板后形成局部真空的方法实现反射式望远镜的巨大镜面的构造、奥斯特的水压缩装置的理论。此外，他还将其方程应用于张力（向一个垂直穿过透明板的圆柱体施加力偶后在透明平板中产生的）

also applies his equations to the calculation of the strains produced in a transparent plate by applying couples to cylinders which pass through it at right angles, and the study (by polarised light) of the doubly-refracting structure thus produced. He expresses himself as unable to explain the permanence of this structure when once produced in isinglass, gutta percha, and other bodies. He recurred to the subject twenty years later, and in 1873 communicated to the Royal Society his very beautiful discovery of the *temporary* double refraction produced by shearing in viscous liquids.

During his undergraduateship in Cambridge he developed the germs of his future great work on "Electricity and Magnetism" (1873) in the form of a paper "On Faraday's Lines of Force", which was ultimately printed in 1856 in the "Trans. of the Cam. Phil. Soc." He showed me the MS. of the greater part of it in 1853. It is a paper of great interest in itself, but extremely important as indicating the first steps to such a splendid result. His idea of a fluid, incompressible and without mass, but subject to a species of friction in space, was confessedly adopted from the analogy pointed out by Thomson in 1843 between the steady flow of heat and the phenomena of statical electricity.

Other five papers on the same subject were communicated by him to the *Philosophical Magazine* in 1861–1862, under the title *Physical Lines of Force*. Then in 1864 appeared his great paper "*On a Dynamical Theory of the Electromagnetic Field*". This was inserted in the *Philosophical Transactions*, and may be looked upon as the first complete statement of the theory developed in the treatise on *Electricity and Magnetism*.

In recent years he came to the conclusion that such analogies as the conduction of heat, or the motion of the mass-less but incompressible fluid, depending as they do on Laplace's equation, were best symbolised by the quaternion notation with Hamilton's ∇ operator; and in consequence, in his work on electricity, he gives the expressions for all the more important physical quantities in their quaternion form, though without employing the calculus itself in their establishment. I have discussed in another place (*Nature*, vol. VII, p. 478) the various important discoveries in this remarkable work, which of itself is sufficient to secure for its author a foremost place among natural philosophers. I may here state that the main object of the work is to do away with "action at a distance," so far at least as electrical and magnetic forces are concerned, and to explain these by means of stresses and motions of the medium which is required to account for the phenomena of light. Maxwell has shown that, on this hypothesis, the velocity of light is the ratio of the electro-magnetic and electro-static units. Since this ratio, and the actual velocity of light, can be determined by absolutely independent experiments, the theory can be put at once to an exceedingly severe preliminary test. Neither quantity is yet fairly known within about 2 or 3 percent, and the most probable values of each certainly agree more closely than do the separate determinations of either. There can now be little doubt that Maxwell's theory of electrical phenomena rests upon foundations as secure as those of the undulatory theory of light. But the life-long work of its creator has left it still in its infancy, and it will probably require for its proper development the services of whole generations of mathematicians.

的计算，以及对同样产生的双折射结构的研究（用偏振光）。他无法解释为什么在云母、杜仲胶以及其他物体中一旦产生了这种结构就会持续存在。20 年以后，他又重新回到这个题目，并于 1873 年向皇家学会宣读了他非常美妙的发现——由黏性液体中的切变造成的**暂时**双折射。

在剑桥读本科时，麦克斯韦完成了最终于 1856 年发表在《剑桥哲学学会学报》上的论文《论法拉第力线》，这篇文章是他后来的伟大著作《电磁学》（1873 年）的雏形。1853 年，他曾给我看过这篇论文的大部分手稿。这篇论文本身就十分有趣，但更重要的是它显示了走向未来辉煌成就的第一步。麦克斯韦关于流体的观点是，流体不可压缩，没有质量，但是要克服空间中的某种摩擦力，这无疑采用了汤姆孙于 1843 年从稳定热流和静电现象之间类推出的结果。

关于这个主题还有另外 5 篇文章，以《论物理力线》为题发表在 1861~1862 年间的《哲学杂志》上。1864 年，他的伟大著作《电磁场的动力学理论》问世了。这篇文章被收录在《自然科学会报》上，可以将其看作对《电磁学》专著中所阐述理论的第一次完整论述。

近年，他得出了以下结论：对于这些遵循拉普拉斯方程的类似物理量，比如热传导或没有质量却不可压缩的流体运动来说，含有哈密顿算子 ∇ 的四维表示法是最佳的表示方式。接着，他在电学研究中给出了所有更重要的物理量的四维形式，然而他并没有在建立表示方法的同时进行计算。我在别处（《自然》，第 7 卷，第 478 页）讨论过这项伟大工作中的多个重要发现，这本身就足以使其作者跻身最伟大的科学家行列。我可以在这里说，这项工作的目标就是，至少在涉及电磁力时，弄清“超距作用”，并且要通过光现象所需的介质应力和运动来解释它们。麦克斯韦指出，在这种假设下，光速为电磁单元与静电单元之比。因为这个比值和光速都可以由完全独立的实验确定，所以上面的理论可以用极为严格的初级实验来检验。但是，目前这两个物理量都还没能在 2% ~ 3% 的误差范围内被清楚地认识，每个量的最可几数值当然比分散的数值更加集中。现在毫无疑问，麦克斯韦电现象理论建立的基础和光的波动学说的基础一样可信。但是，这个理论在其创建者的毕生努力下也还是处在初级阶段，它的合理发展可能还需要整整一代数学家的努力。

This was not the only work of importance to which he devoted the greater part of his time while an undergraduate at Cambridge. For he had barely obtained his degree before he read to the Cambridge Philosophical Society a remarkable paper *On the Transformation of Surfaces by Bending*, which appears in their *Transactions* with the date March 1854. The subject is one which had been elaborately treated by Gauss and other great mathematicians, but their methods left much to be desired from the point of view of simplicity. This Clerk Maxwell certainly supplied; and to such an extent that it is difficult to conceive that any subsequent investigator will be able to simplify the new mode of presentation as much as Maxwell simplified the old one. Many of his results, also, were real additions to the theory; especially his treatment of the *Lines of Bending*. But the whole matter is one which, except in its almost obvious elements, it is vain to attempt to popularise.

The next in point of date of Maxwell's greatest works is his "Essay on the Stability of the Motion of Saturn's Rings", which obtained the Adam's Prize in the University of Cambridge in 1857. This admirable investigation was published as a pamphlet in 1859. Laplace had shown in the *Mécanique Céleste* that a uniform solid ring cannot revolve permanently about a planet; for, even if its density were so adjusted as to prevent its splitting, a slight disturbance would inevitably cause it to fall in. Maxwell begins by finding what amount of *want* of uniformity would make a solid ring stable. He finds that this could be effected by a satellite rigidly attached to the ring, and of about $4\frac{1}{2}$ times its mass:—but that such an arrangement, while not agreeing with observation, would require extreme artificiality of adjustment of a kind not elsewhere observed. Not only so, but the materials, in order to prevent its behaving almost like a liquid under the great forces to which it is exposed, must have an amount of rigidity far exceeding that of any known substance.

He therefore dismisses the hypothesis of solid rings, and (commencing with that of a ring of equal and equidistant satellites) shows that a continuous liquid ring cannot be stable, but may become so when broken up into satellites. He traces in a masterly way the effects of the free and forced waves which must traverse the ring, under various assumptions as to its constitution; and he shows that the only system of rings which can dynamically exist must be composed of a very great number of separate masses, revolving round the planet with velocities depending on their distances from it. But even in this case the system of Saturn cannot be permanent, because of the mutual actions of the various rings. These mutual actions must lead to the gradual spreading out of the whole system, both inwards and outwards:—but if, as is probable, the outer ring is much denser than the inner ones, a very small increase of its external diameter would balance a large change in the inner rings. This is consistent with the progressive changes which have been observed since the discovery of the rings. An ingenious and simple mechanism is described, by which the motions of a ring composed of equal satellites can be easily demonstrated.

Another subject which he treated with great success, as well from the experimental as from the theoretical point of view, was the Perception of Colour, the Primary Colour Sensations, and the Nature of Colour Blindness. His earliest paper on these subjects bears

以上并不是麦克斯韦利用在剑桥读本科期间的大部分时间来完成的唯一重要的工作。因为直到他在剑桥哲学学会的会议上宣读了一篇著名的论文《论弯曲引起的表面变换》，并于 1854 年 3 月发表在《剑桥哲学学会学报》上，他才获得了学位。高斯和其他一些伟大的数学家都曾经详细研究过这个课题，但是从简洁的角度来看，他们的方法还有很多需要改进的地方。克拉克·麦克斯韦无疑完成了后续的简化工作，并且将其简化到了这样的程度：很难想象任何一个后继研究者在简化现有的新模型时，能达到像麦克斯韦简化旧模型时那样的程度。他的很多结果，是对原有理论的丰富，尤其是他对**弯曲线**所作的处理。但问题是，除了其中几乎显而易见的部分，这些结果都没有实现普及。

麦克斯韦的下一个伟大著作是《关于土星环运动稳定性的评论》，这篇文章在 1857 年获得了剑桥大学的亚当斯奖。这项令人称赞的研究在 1859 年被印成了小册子。拉普拉斯在他的《天体力学》中指出，均匀的固体环不可能持久地围绕行星转动，因为，即使环的密度调整到可以使其避免分裂的程度，一个微小的扰动也会不可避免地导致其塌陷。麦克斯韦从寻找使一个固体环稳定**所需的**均匀度开始。他发现，可以通过将一个卫星与这个环作刚性连接来实现环的稳定，其中卫星的质量等于环质量的 $4\frac{1}{2}$ 倍。但是由于这种方法与观测不符，于是就需要极精巧的调节方式，而这种调节方式也没有在其他地方看到过。不但如此，为了避免环出现类似暴露在强力下的液体那样的表现，环的材料必须具有足够的硬度，而这种硬度远远超过了任何已知材料的硬度。

因此，麦克斯韦放弃了固体环的假设，开始设想一个由等距的相同行星组成的环。他指出，连续的液体环无法保持稳定，但是当它破裂成多个卫星的时候就可能达到稳定状态。他极为巧妙地论述了在各种假设的环结构中必须穿过环的或自由或受迫的波的影响，并且指出，唯一一种能够动态稳定存在的环形结构必须由大量分离的物质组成，这些物质围绕行星转动，其速度取决于它们到行星的距离。然而，即使在这种情况下，由于不同环之间的相互作用，土星系统也不可能持久稳定。因为这种相互作用必然会导致整个系统向内外两个方向扩散。但是，如果外环密度比内环密度大，外环直径很小的增大就能平衡内环直径很大的改变。这与环被发现以来所观察到的不断变化是一致的。麦克斯韦描述了一个独特而又简洁的机制，用它可以很容易地说明由相同卫星组成的环的运动。

麦克斯韦的另一个研究课题是颜色的感知、基础色觉以及色盲的本质。这项工作无论是从实验角度还是从理论角度来看都极为成功。他最早关于这些问题的文章诞生于 1855 年，第 7 篇则发表于 1872 年。“由于他在颜色组成方面的研究和其他光

date 1855, and the seventh has the date 1872. He received the Rumford Medal from the Royal Society in 1860, "For his Researches on the Composition of Colours and other optical papers". Though a triplicity about colour had long been known or suspected, which Young had (most probably correctly) attributed to the existence of three sensations, and Brewster had erroneously* supposed to be objective, Maxwell was the first to make colour-sensation the subject of actual measurement. He proved experimentally that any colour C (given in intensity of illumination as well as in character) may be expressed in terms of three arbitrarily chosen standard colours, X, Y, Z, by the formula

$$C = aX + bY + cZ$$

Here a, b, c are numerical coefficients, which may be positive or negative; the sign = means "matches", + means "superposed", and – directs the term to be taken to the other side of the equation.

These researches of Maxwell's are now so well known, in consequence especially of the amount of attention which has been called to the subject by Helmholtz' great work on Physiological Optics, that we need not farther discuss them here.

The last of his greatest investigations is the splendid Series on the Kinetic Theory of Gases, with the closely connected question of the sizes, and laws of mutual action, of the separate particles of bodies. The Kinetic Theory seems to have originated with D. Bernoulli; but his successors gradually reverted to statical theories of molecular attraction and repulsion, such as those of Boscovich. Herapath (in 1847) seems to have been the first to recall attention to the Kinetic Theory of gaseous pressure. Joule in 1848 calculated the average velocity of the particles of hydrogen and other gases. Krönig in 1856 (*Pogg. Ann.*) took up the question, but he does not seem to have advanced it farther than Joule had gone; except by the startling result that the weight of a mass of gas is only half that of its particles when at rest.

Shortly afterwards (in 1859) Clausius took a great step in advance, explaining, by means of the kinetic theory, the relations between the volume, temperature and pressure of a gas, its cooling by expansion, and the slowness of diffusion and conduction of heat in gases. He also investigated the relation between the length of the mean free path of a particle, the number of particles in a given space, and their least distance when in collision. The special merit of Clausius' work lies in his introduction of the processes of the theory of probabilities into the treatment of this question.

Then came Clerk Maxwell. His first papers are entitled "Illustrations of the Dynamical Theory of Gases", and appeared in the *Phil. Mag.* in 1860. By very simple processes he treats the collisions of a number of perfectly elastic spheres, first when all are of the same mass, secondly when there is a mixture of groups of different masses. He thus verifies

---

* All we can positively say to be erroneous is some of the principal arguments by which Brewster's view was maintained, for the subjective character of the triplicity has not been absolutely *demonstrated*.

学方面的论文”，麦克斯韦在1860年获得了皇家学会颁发的拉姆福德奖章。虽然人们知道或者猜测出颜色的三原色已经有很长一段时间了，例如杨曾经（很可能正确地）把三原色归因于存在三种主观色觉，而布鲁斯特错误地*猜测三原色是客观的，但是麦克斯韦却是第一个将实际测量引入色觉这个课题的人。他在实验中证明了，任何颜色C（以照度和特性的形式给出）都可以用三种任选的标准颜色X，Y，Z按照下面的公式表示出来：

$$\mathrm{C} = a\mathrm{X} + b\mathrm{Y} + c\mathrm{Z}$$

其中，a，b，c是数值系数，可以取正也可以取负；等号表示“匹配”，加号表示“叠加”，减号表示把这一项移到方程的另一边。

现在麦克斯韦的这些研究已经是众所周知的了，特别是后来亥姆霍兹关于生理光学的伟大工作使得这个研究课题吸引了很多注意力，因此，我们就不需要在这里进一步详细讨论它们了。

麦克斯韦最后一项伟大的研究是他那套卓越的关于气体动力学的丛书。这套丛书的内容与物体离散粒子的大小和相互作用定律等问题密切相关。动力学理论似乎最早起源于伯努利，但是他的后继者逐渐回归到关于粒子吸引和排斥的统计理论上，就如博斯科维克所做的那样。赫拉帕斯（1847年）似乎是第一个重新注意到气体压力动力学理论的人。焦耳在1848年计算了氢气和其他气体微粒的平均速度。科隆尼格在1856年（《波根多夫年鉴》）考虑到了这个问题，但是除了得到气体的重量只有静止气体微粒的一半这一惊人结果之外，他似乎并没有进一步发展焦耳的结果。

在不久之后（1859年），克劳修斯取得了巨大的进展。他用动力学理论成功地解释了气体体积、温度、压强之间的关系，膨胀造成的冷却，以及气体中缓慢的热扩散和热传导。他还研究了气体微粒的平均自由程长度、给定空间中的粒子数以及粒子碰撞过程中的最小距离这三者之间的关系。克劳修斯工作的最大价值在于，他在处理这个问题时，引入了概率论的方法。

然后就是克拉克·麦克斯韦。他第一篇论文的题目是《气体动力学理论图示》，于1860年发表在《哲学杂志》上。他通过一个非常简单的过程来处理许多完全弹性小球的碰撞。他首先研究了所有小球的质量都相等的情况，然后研究了含有不同质

* 我们只能说布鲁斯特的观点所使用的主要论据是错误的，而三原色的主观性还没有完全被证实。

Gay-Lussac's law, that the number of particles per unit volume is the same in all gases at the same pressure and temperature. He explains gaseous friction by the transference to and fro of particles between contiguous strata of gas sliding over one another, and shows that the coefficient of viscosity is independent of the density of the gas. From Stokes' calculation of that coefficient he gave the first deduced approximate value of the mean length of the free path; which could not, for want of data, be obtained from the relation given by Clausius. He obtained a closely accordant value of the same quantity by comparing his results for the kinetic theory of diffusion with those of one of Graham's experiments. He also gives an estimate of the conducting power of air for heat; and he shows that the assumption of non-spherical particles, which during collision change part of their energy of translation into energy of rotation, is inconsistent with the known ratio of the two specific heats of air.

A few years later he made a series of valuable experimental determinations of the viscosity of air and other gases at different temperatures. These are described in *Phil. Trans.* 1866; and they led to his publishing (in the next volume) a modified theory, in which the gaseous particles are no longer regarded as perfectly elastic, but as repelling one another according to the law of the inverse fifth power of the distance. This paper contains some very powerful analysis, which enabled him to simplify the mathematical theory for many of its most important applications. Three specially important results are given in conclusion, and they are shown to be independent of the particular mode in which gaseous particles are supposed to act on one another. These are:—

1. In a mixture of particles of two kinds differing in amounts of mass, the average energy of translation of a particle must be the same for either kind. This is Gay Lussac's Law already referred to.
2. In a vertical column of mixed-gases, the density of each gas at any point is ultimately the same as if no other gas were present. This law was laid down by Dalton.
3. Throughout a vertical column of gas gravity has no effect in making one part hotter or colder than another; whence (by the dynamical theory of heat) the same must by true for all substances.

Maxwell has published in later years several additional papers on the Kinetic Theory, generally of a more abstruse character than the majority of those just described. His two latest papers (in the *Phil. Trans.* and *Camb. Phil. Trans.* of last year) are on this subject:— one is an extension and simplification of some of Boltzmann's valuable additions to the Kinetic Theory. The other is devoted to the explanation of the motion of the radiometer by means of this theory. Several years ago (*Nature*, vol. XII, p. 217), Prof. Dewar and the writer pointed out, and demonstrated experimentally, that the action of Mr. Crookes' very beautiful instrument was to be explained by taking account of the increased length of the mean free path in rarefied gases, while the then received opinions ascribed it either to evaporation or to a quasi-corpuscular theory of radiation. Stokes extended the explanation to the behaviour of disks with concave and convex surfaces, but the subject was not at all fully investigated from the theoretical point of view till Maxwell took it up. During the last ten years of his life he had no rival to claim concurrence with him in the whole wide domain

量小球的情况。他由此证明了盖·吕萨克定律，即在同温同压下，所有气体单位体积内的粒子数相等。他把气体的摩擦力解释为相邻气层之间微粒的相对运动，并且指出黏滞系数和气体的密度无关。他从斯托克斯对黏滞系数的计算出发，第一次推出了气体平均自由程的近似值，由于缺少数据，这个数值是无法从克劳修斯给出的关系中得到的。他把自己用扩散动力学理论计算出的结果和格雷厄姆一个实验的结果进行比较，发现二者能够很好地吻合。麦克斯韦还估算出了空气的热导率，并且指出，非球形粒子在碰撞中能将部分平动动能转化为转动动能的假设与已知的空气的两种比热之比不符。

几年后，他做了一系列很有价值的实验，来确定不同温度下空气和其他气体的黏性。这些工作的结果发表在1866年的《自然科学会报》上。在该杂志接下来的一卷上，他又发表了修正后的理论，此理论中不再把气体粒子视为完全弹性小球，而是认为粒子之间存在着与相互距离的五次方成反比的排斥作用。这篇论文包含一些非常有力的分析，这使得他可以为了理论的重要应用而进行数学理论上的简化。结论中给出了3个特别重要的结果，并且这些结果和气体粒子之间的相互作用模型无关。它们是：

1. 在由两种质量不同的粒子组成的混合物中，两种粒子的平均平动动能一定相等。这是盖·吕萨克定律已经提到过的。
2. 在装有混合气体的立柱容器中，每一种气体在任一点的密度最终都将相同，就好像没有其他气体存在一样。这个定律是道尔顿建立的。
3. 在整个装有气体的立柱容器中，重力并没有使某一部分的温度高于或低于另外一部分。因此（根据热动力学理论），这个规律应该适用于所有物质。

在之后的几年中，麦克斯韦发表了另外几篇关于动力学理论的文章，这些文章比上面提到的工作中的大多数都更加深奥。他最近的两篇论文发表在去年的《自然科学会报》和《剑桥哲学学报》上：一篇是对玻耳兹曼在动力学理论上的重要补充的推广和简化；另一篇文章中，他用这个理论解释了辐射计的运转。几年之后，杜瓦教授等人（《自然》，第12卷，第217页）指出并且通过实验证实了：考虑到稀薄气体中平均自由程的增加，无论是采用蒸气辐射理论还是准颗粒辐射理论都可以解释克鲁克斯先生非常精巧的仪器的作用。斯托克斯将其推广，用于解释具有凹凸表面的圆盘的行为，但是从理论的角度来看，这个课题在麦克斯韦着手之前并没有得到充分的研究。在麦克斯韦生命的最后十年里，他没有遇到能够在分子力学的广阔领域内和他平起平坐的对手，然而在更深奥的电学领域，倒是有两三个人能与他

of molecular forces, and but two or three in the still more recondite subject of electricity.

"Every one must have observed that when a slip of paper falls through the air, its motion, though undecided and wavering at first, sometimes becomes regular. Its general path is not in the vertical direction, but inclined to it at an angle which remains nearly constant, and its fluttering appearance will be found to be due to a rapid rotation round a horizontal axis. The direction of deviation from the vertical depends on the direction of rotation.... These effects are commonly attributed to some accidental peculiarity in the form of the paper...." So writes Maxwell in the *Cam. and Dub. Math. Jour.* (May, 1854), and proceeds to give an exceedingly simple and beautiful explanation of the phenomenon. The explanation is, of course, of a very general character, for the complete working out of such a problem appears to be, even yet, hopeless; but it is thoroughly characteristic of the man, that his mind could never bear to pass by any phenomenon without satisfying itself of at least its general nature and causes.

In the same volume of the *Math. Journal* there is an exceedingly elegant "problem" due to Maxwell, with his solution of it. In a note we are told that it was "suggested by the contemplation of the structure of the crystalline lens in fish". It is as follows:—

A transparent medium is such that the path of a ray of light within it is a given circle, the index of refraction being a function of the distance from a given point in the plane of the circle. Find the form of this function, and show that for light of the same refrangibility—

1. The path of *every ray within the medium* is a circle.
2. All the rays proceeding from any point in the medium will meet accurately in another point.
3. If rays diverge from a point without the medium and enter it through a spherical surface having that point for its centre, they will be made to converge accurately to a point within the medium.

Analytical treatment of this and connected questions, by a novel method, will be found in a paper by the present writer (*Trans. R.S.E.* 1865).

Optics was one of Clerk Maxwell's favourite subjects, but of his many papers on various branches of it, or subjects directly connected with it, we need mention only the following:—

"On the General Laws of Optical Instruments" (*Quart. Math. Jour.* 1858)
"On the Cyclide" (*Quart. Math. Journal*, 1868)
"On the best Arrangement for Producing a Pure Spectrum on a Screen" (*Proc. R.S.E.* 1868)
"On the Focal Lines of a Refracted Pencil" (*Math. Soc. Proc.* 1873)

A remarkable paper, for which he obtained the Keith Prize of the *Royal Society of Edinburgh*, is entitled "On Reciprocal Figures, Frames, and Diagrams of Forces." It is published in the *Transactions* of the Society for 1870. Portions of it had previously appeared in the *Phil. Mag.* (1864).

相提并论。

“每个人都会注意到，一张纸片在空气中飘落时，虽然一开始摇摆不定，但是它的运动会趋于规则。它通常的路径并不是沿垂直方向，而是与垂直方向成一个角度，这个角度基本是一个常数。我们会发现纸片一开始的飘动是围绕一条水平轴快速转动。偏离垂直轴的方向取决于转动的方向……。这些结果通常被归因于纸张形状的某些偶然特性……”麦克斯韦在 1854 年 5 月发表于《剑桥与都柏林数学杂志》上的文章中这样写道，他想对这个现象作出非常简洁而漂亮的解释。当然，这个解释十分笼统，因为即使现在看来，完全解决这个问题也是希望渺茫。但这正是麦克斯韦的性格，他的思想决不容忍自己与任何连一般性质及成因都得不到满意解释的现象擦肩而过。

在《剑桥与都柏林数学杂志》的同一卷中，麦克斯韦提出了一个极其精彩的“问题”，并且自己作出了解答。我们从一则记录中得知，这个工作是“在思考鱼的晶状体结构时受到的启发”。内容如下：

所谓介质是透明的就是指光线在此介质中的传播路径是一个特定的圆，介质某一点的折射率是这一点到圆平面中给定点距离的函数。我找到了这个函数的形式，并且发现对于具有相同折射性质的光线来说——

1. **每一条光线在介质中的**路径都是一个圆。
2. 从介质中任意一点发出的光线，都会在另外一点精确相遇。
3. 如果光线在介质外的某一点发散，并且经由一个以此发散点为球心的球面进入介质，那么这些光线将精确地会聚到介质中的某一点上。

我本人在一篇文章（《爱丁堡皇家学会会报》，1865 年）中，用一种新颖的方法对这个问题及相关问题进行了分析。

光学是克拉克·麦克斯韦最喜欢的课题之一，但是在他关于光学不同分支或者直接与光学相关的大量文章中，我们只需提及下面这些：

《论光学仪器的普遍规律》（《数学季刊》，1858 年）

《论四次圆纹曲面》（《数学季刊》，1868 年）

《论在屏幕上生成纯光谱的最佳方案》（《爱丁堡皇家学会会刊》，1868 年）

《论折射光束的焦线》（《数学学会会刊》，1873 年）

麦克斯韦还有一篇意义重大的文章，题目是《论力的对应线图、框架和图解》，为此他获得了爱丁堡皇家学会的基思奖。文章于 1870 年发表在《爱丁堡皇家学会会报》上，文章中的一部分之前已经在《哲学杂志》（1864 年）上出现过。

The triangle and the polygon of forces, as well as the funicular polygon, had long been known; and also some corresponding elementary theorems connected with hydrostatic pressure on the faces of a polyhedron; but it is to Rankine that we owe the full principle of diagrams, and reciprocal diagrams, of frames and of forces. Maxwell has greatly simplified and extended Rankine's ideas: on the one hand facilitating their application to practical problems of construction, and on the other hand extending the principle to the general subject of stress in bodies. The paper concludes with a valuable extension to three dimensions of Sir George Airy's "Function of Stress".

His contributions to the *Proceedings of the London Mathematical Society* were numerous and valuable. I select as a typical specimen his paper on the forms of the stream-lines when a circular cylinder is moved in a straight line, perpendicular to its axis, through an infinitely extended, frictionless, incompressible fluid (vol. III, p. 224). He gives the complete solution of the problem; and, with his usual graphical skill, so prominent in his great work on Electricity, gives diagrams of the stream-lines, and of the paths of individual particles of the fluid. The results are both interesting and instructive in the highest degree.

In addition to those we have mentioned we cannot recall many pieces of *experimental* work on Maxwell's part:—with two grand exceptions. The first was connected with the determination of the British Association Unit of Electric Resistance, and the closely associated measurement of the ratio of the electrokinetic to the electrostatic unit. In this he was associated with Professors Balfour Stewart and Jenkin. The Reports of that Committee are among the most valuable physical papers of the age; and are now obtainable in a book-form, separately published. The second was the experimental verification of Ohm's law to an exceedingly close approximation, which was made by him at the Cavendish Laboratory with the assistance of Prof. Chrystal.

In his undergraduate days he made an experiment which, though to a certain extent physiological, was closely connected with physics. Its object was to determine why a cat always lights on its feet, however it may be let fall. He satisfied himself, by pitching a cat gently on a mattress stretched on the floor, giving it different initial amounts of rotation, that it instinctively made use of the conservation of Moment of Momentum, by stretching out its body if it were rotating so fast as otherwise to fall head foremost, and by drawing itself together if it were rotating too slowly.

I have given in this journal (vol. XVI, p. 119) a detailed account of his remarkable elementary treatise on "Matter and Motion", a work full of most valuable materials, and worthy of most attentive perusal not merely by students but by the foremost of scientific men.

His "Theory of Heat", which has already gone through several editions, is professedly elementary, but in many places is probably, in spite of its admirable definiteness, more difficult to follow than any other of his writings. In intrinsic importance it is of the same high order as his "Electricity", but as a whole it is *not* an elementary book. One of the

力的三角法则、多边形法则以及索状多边形法则早已为人们所熟知，一些和多面体表面静压有关的基础理论也是如此。但是兰金认为，我们缺少一套关于力的线图、框架、图解以及力本身的完整法则。麦克斯韦极大地简化并推广了兰金的观点：一方面，他使这个理论在建筑学实际问题上的应用更加方便；另一方面，他将这个理论推广到了物体中的压力这一一般主体。在文章结尾，麦克斯韦将乔治·艾里爵士的“压力函数”推广到了三维的情形，这个推广很有意义。

他为《伦敦数学学会会刊》贡献了大量有价值的稿件。我选取他的一篇文章作为其中的典范。这篇文章讨论了当一个圆柱体沿着垂直于轴的直线穿过一个不可压缩、无摩擦且可无限扩展的液体时液体中流线的形式（第 3 卷，第 224 页）。他给出了这个问题的完整解答，并且利用他在电学巨著中常用的图解技巧给出了流线和液体中单个粒子路径的图形。

除了上面提到的那些工作，麦克斯韦没有多少**实验性的**工作能被我们铭记，但是有两个工作是例外。第一个是关于确定电阻英制单位以及与其密切相关的电动力学单位与静电学单位之比的测量。这个工作是麦克斯韦和鲍尔弗·斯图尔特教授、詹金教授合作完成的。相关委员会的报告是这个时期最有价值的物理学论文之一，现在这些报告已经集结成册并且单独出版了。第二项工作是在实验上以极高的精度验证了欧姆定律。这项工作是他在卡文迪什实验室由克里斯托尔教授协助完成的。

麦克斯韦在读本科期间做了一个实验，虽然从某种程度上说这是一个生理学实验，但是它也和物理学密切相关。实验的目的是解释为什么猫在落地的时候总是能保持脚先着地。实验中他把猫轻轻地抛到毯子上，抛掷时，让猫具有不同的初始转动。麦克斯韦得到了满意的结果：猫在空中的时候，本能地利用了角动量守恒。如果给它的初始转动过快，它就会把身体伸展开，避免头先着地；相反地，如果给它的初始转动过慢，它就会缩成一团，最后总是能避免头先着地。

我曾经在贵刊上（第 16 卷，第 119 页）详细介绍了麦克斯韦著名的关于《物质和运动》的基础论文。他的这项工作有许多重要的结果，因此值得每一个人——不仅仅是学生，还包括一流的科学家——仔细研读。

他的《热学理论》已经出了好几版。尽管他自称这部书很基础，尽管书的思路的确非常清晰，但是其中许多地方可能比作者的其他任何著作都更难理解。这部书本身的重要性堪比他的《电学》，但是总体上却**不是**一本基础读物。克拉克·麦克斯

few knowable things which Clerk Maxwell did not know, was the distinction which most men readily perceive between what is easy and what is hard. What *he called* hard, others would be inclined to call altogether unintelligible. In the little book we are discussing there is matter enough to fill two or three large volumes without undue dilution (perhaps we should rather say, *with the necessary dilution*) of its varied contents. There is nothing flabby, so to speak, about anything Maxwell ever wrote: there is splendid muscle throughout, and an adequate bony structure to support it. "Strong meat for grown men" was one of his favourite expressions of commendation; and no man ever more happily exposed the true nature of the so-called "popular science" of modern times than he did when he wrote of "the forcible language and striking illustrations by which those who are past hope of being even beginners [in science] are prevented from becoming conscious of intellectual exhaustion before the hour has elapsed."

To the long list of works attached to Maxwell's name in the Royal Society's Catalogue of Scientific Papers may now be added his numerous contributions to the latest edition of the "Encyclopaedia Britannica"—Atom, Attraction, Capillarity, &c. Also the laborious task of preparing for the press, with copious and very valuable original notes, the "Electrical Researches of the Hon. Henry Cavendish." This work has appeared only within a month or two, and contains many singular and most unexpected revelations as to the early progress of the science of electricity. We hope shortly to give an account of it.

The works which we have mentioned would of themselves indicate extraordinary activity on the part of their author, but they form only a fragment of what he has published; and when we add to this the further statement, that Maxwell was always ready to assist those who sought advice or instruction from him, and that he has read over the proof-sheets of many works by his more intimate friends (enriching them by notes, always valuable and often of the quaintest character), we may well wonder how he found time to do so much.

Many of our readers must remember with pleasure the occasional appearance in our columns of remarkably pointed and epigrammatic verses, usually dealing with scientific subjects, and signed $\frac{dp}{dt}$*. The lines on Cayley's portrait, where determinants, roots of −1, space of *n* dimensions, the 27 lines on a cubic surface, &c., fall quite naturally into rhythmical English verse; the admirable synopsis of Dr. Ball's Treatise on Screws; the telegraphic love-letter with its strangely well-fitting *volts* and *ohms*; and specially the "Lecture to a Lady on Thomson's Reflecting Galvanometer", cannot fail to be remembered. No living man has shown a greater power of condensing the whole marrow of a question into a few clear and compact sentences than Maxwell shows in these verses. Always having a definite object, they often veiled the keenest satire under an air of charming innocence and *naïve* admiration. Here are a couple of stanzas from unpublished pieces of a similar kind:—first, some ghastly thoughts by an excited evolutionist—

* This *nom de plume* was suggested to him by me from the occurrence of his initials in the well-known expression of the second Law of Thermodynamics (for whose establishment on thoroughly valid grounds he did so much) $\frac{dp}{dt}$ = J. C. M.

韦不知道的少数几个显而易见的事情之一就是，难与易的区别，而这是多数人都能分清的。**他所谓的**难事，其他人会认为是根本无法理解的。我们正在谈论的这本薄薄的书的内容不用过度展开（可能我们应该说，**经过必要的展开**），其内容就足以写满两三卷书。可以说，麦克斯韦写的东西从不松散拖沓：他的文章只有健美的肌肉和适量的用以支撑的骨架。“成人的强健肌肉”是麦克斯韦最喜欢的表达称赞的说法。“过来人总是希望自己保持那种［科学上的］初学者的状态，对他们来说，有力的语言和精彩的图示就是能在时间消逝之前避免灵感和智慧枯竭的灵药。”当麦克斯韦写出上面这段话时，恐怕没有人比他更乐于揭示现代所谓“流行科学”的真实本质了。

在皇家学会科学文献目录中，有麦克斯韦署名的工作已经可以列出一长串了。现在，应该还要加上他对《大英百科全书》中原子、吸引作用、毛细作用等方面所作的很多贡献。另外，他还为编写出版《亨利·卡文迪什电学研究》付出了辛勤的劳动，整理了丰富而珍贵的手稿。这项工作是在最近的一两个月才问世的，书中记录着电学早期发展历程带给我们的很多意想不到的非凡启迪。我们希望不久之后能介绍一下这项工作。

上面提到的工作已经显示了作者超常的科研能力，然而这些只是他已发表著作的冰山一角。如果我还补充说，麦克斯韦总是乐于帮助那些寻求建议或指引的人，而且他通读过很多好友的著作的校样（在上面所作的注释使其更加丰富，往往起到画龙点睛的作用），大家可能会觉得十分惊讶，他哪来那么多时间完成这么多事情。

很多读者一定还记得在我们的专栏中偶尔出现的那些非常尖锐的讽刺小诗，通常都是关于科学主题的，并且署名$\frac{dp}{dt}$*；在凯莱肖像画上的诗句中由行列式、–1的根、n维空间以及三次曲面上的27条线等很自然地组成的一首充满韵律的英文小诗；为鲍尔博士关于旋量的论文写的令人赞叹的简介；用出奇得体的**伏特**和**欧姆**组成的电报情书；特别是那篇《为女士所作的关于汤姆孙反射检流计的演讲》，所有这些都让人无法忘怀。当今世上没有人可以超越麦克斯韦在小诗中表现出的用几个清晰而简洁的句子就把问题的精髓概括出来的能力。这些小诗总是具有明确的目标，但是又把最尖锐的讽刺隐藏在迷人的纯真和纯朴的赞美之中。这里有两段没有发表的类似这种风格的片段：首先，是一个狂热的革命者的可怕想法——

* 这个笔名是我建议他取的，因为我在著名的热力学第二定律$\frac{dp}{dt}$ = J. C. M.中发现了他名字的首字母缩写（而他本人也为这个定律能够建立在坚实的基础之上做了许多工作）。

To follow my thoughts as they go on,
Electrodes I'd place in my brain;
Nay, I'd swallow a live entozöon,
New feelings of life to obtain—

next on the non-objectivity of Force—

Both Action and Reaction now are gone;
Just ere they vanished
Stress joined their hands in peace, and made them one,
Then they were banished.

It is to be hoped that these scattered gems may be collected and published, for they are of the very highest interest, as the work during leisure hours of one of the most piercing intellects of modern times. Every one of them contains evidence of close and accurate thought, and many are in the happiest form of epigram.

I cannot adequately express in words the extent of the loss which his early death has inflicted not merely on his personal friends, on the University of Cambridge, on the whole scientific world, but also, and most especially, on the cause of common sense, of true science, and of religion itself, in these days of much vain-babbling, pseudo-science, and materialism. But men of his stamp never live in vain; and in one sense at least they cannot die. The spirit of Clerk Maxwell still lives with us in his imperishable writings, and will speak to the next generation by the lips of those who have caught inspiration form his teachings and example.

(**21**, 317-321; 1880)

我跟随着自己的感觉，
我要把电极放进我的脑子里；
要不我就吞下活生生的寄生虫，
我的生命会有崭新的感受——

另一个，是关于力的非客观性——

作用力和反作用力都消失了；
就在他们消失之前，
压力让他们静静地携起手来，合二为一，
然后，他们被放逐天涯。

人们希望能够把这些散落的宝石结集出版，因为它们是现代最敏锐的智者中的一员闲暇时完成的作品，而又是如此有趣。每一首诗都证明了作者缜密的思维，并且很多都是以讽刺诗那种诙谐的手法写成的。

我实在无法用语言表达麦克斯韦的早逝是多么巨大的损失，受到损失的不仅仅是他的朋友、剑桥大学和整个科学界，特别是，在充满空谈、伪科学和物质主义的今天，人们对常识、真科学以及宗教本身的探究也会因此受到巨大的损失。然而，脚踏实地的人决不会生活在空谈中，至少从某种意义上讲，这样的人不会从世界上消失。麦克斯韦的精神会在他不朽的著作中与我们同在，并且这种精神会由那些受过他的教诲并以他为榜样的人，传承给下一代。

（王静 翻译；鲍重光 审稿）

Density of Nitrogen

Rayleigh

Editor's Note

Measurements of atomic weights of the elements—the weights, relative to hydrogen, of equal quantities—revealed that these were often close to whole numbers. William Prout suggested in 1815–1816 that hydrogen might thus be the building block of all atoms. In 1888 Lord Rayleigh at Cambridge determined the atomic weight of nitrogen, and found that the gas obtained from ammonia was slightly lighter, by one thousandth, than atmospheric nitrogen. Here Rayleigh appeals for an explanation of the discrepancy, which he suspects might be due to an impurity. Rayleigh later teamed up with William Ramsay of University College London, and in 1894 they announced a new, unreactive element in air, which they named argon after the Greek for "idle".

I am much puzzled by some recent results as to the density of *nitrogen*, and shall be obliged if any of your chemical readers can offer suggestions as to the cause. According to two methods of preparation I obtain quite distinct values. The relative difference, amounting to about 1/1,000 part, is small in itself; but it lies entirely outside the errors of experiment, and can only be attributed to a variation in the character of the gas.

In the first method the oxygen of atmospheric air is removed in the ordinary way by metallic copper, itself reduced by hydrogen from the oxide. The air, freed from CO_2 by potash, gives up its oxygen to copper heated in hard glass over a large Bunsen, and *then* passes over about a foot of red-hot copper in a furnace. This tube was used merely as an indicator, and the copper in it remained bright throughout. The gas then passed through a wash-bottle containing sulphuric acid, thence again through the furnace over *copper oxide*, and finally over sulphuric acid, potash, and phosphoric anhydride.

In the second method of preparation, suggested to me by Prof. Ramsay, everything remained unchanged, except that the *first* tube of hot copper was replaced by a wash-bottle containing liquid *ammonia*, through which the air was allowed to bubble. The ammonia method is very convenient, but the nitrogen obtained by means of it was 1/1,000 part *lighter* than the nitrogen of the first method. The question is, to what is the discrepancy due?

The first nitrogen would be too heavy, if it contained residual oxygen. But on this hypothesis something like 1 percent would be required. I could detect none whatever by means of alkaline pyrogallate. It may be remarked the density of this nitrogen agrees closely with that recently obtained by Leduc, using the same method of preparation.

氮气的密度

瑞利

编者按

元素原子量——相对于同等数量的氢的质量——的测定结果显示，各种元素原子量的数值通常都接近整数。1815~1816 年间，威廉·普劳特提出氢原子可能是组成其他原子的基本单元。1888 年，剑桥的瑞利勋爵测定了氮的原子量，结果发现通过氨得到的氮气比空气中的氮气轻千分之一。在这篇文章中，瑞利对这种差异提出了一种解释，他怀疑该差异可能是因为某种杂质造成的。后来，瑞利与伦敦大学学院的威廉姆·拉姆齐一起合作，并于 1894 年宣布发现空气中存在一种没有反应活性的新元素，他们根据希腊语中的“懒惰”一词将其命名为氩。

我对近来一些关于**氮气**密度的结果感到很困惑，如果你们读者中有熟悉化学的人可以提供相关原因的建议，我将不胜感激。用两种不同的制备方法，我得到了显然不同的数值。相对差别大约是 1/1,000，虽然这一差别本身并不大，但是这完全不属于实验误差，因此只能归因于气体性质的不同。

在第一种制备方法中，我们用普通的方法除去空气中的氧气，其中用到了通过氢气还原氧化铜制得的金属铜。首先是用钾碱除去空气中的 CO_2，然后使气体与硬质玻璃管中已经用本生灯加热的铜反应，以除去空气中的氧气，**接着**再使气体通过加热炉中的一块大约一英尺长的红热的铜。这个管道仅仅是用作指示剂，其中的铜应该始终保持亮红色。然后将气体通过一个盛有硫酸的洗气瓶，之后再使其通过加热炉中的**氧化铜**，最后依次通过硫酸、钾碱和磷酸酐。

第二种制备方法是拉姆齐教授向我建议的。与第一种方法相比，只有**第一个**加热铜的玻璃管被一个装有液**氨**的洗气瓶代替，这一洗气瓶允许空气在通过时产生气泡，除此之外再没有任何改变。这种氨法非常简便，但是通过这种方法得到的氮气比采用第一种方法制得的氮气**轻** 1/1,000。那么，产生这种差异的原因是什么呢？

用第一种方法制得的氮气中如果还有残余的氧气，那么其密度就会更大。但是按照这种假设，制得的氮气中就要有大约 1% 的氧气，但是我用碱性焦棓酸盐进行检测并没有发现任何氧气。值得注意的是，这种氮气的密度与最近勒迪克用同一种方法制得的氮气的密度非常接近。

On the other hand, can the ammonia-made nitrogen be too light from the presence of impurity? There are not many gases lighter than nitrogen, and the absence of hydrogen, ammonia, and water seems to be fully secured. On the whole it seemed the more probable supposition that the impurity was hydrogen, which in this degree of dilution escaped the action of the copper oxide. But a special experiment appears to exclude this explanation.

Into nitrogen prepared by the first method, but before its passage into the furnace tubes, one or two thousandths by volume of hydrogen were introduced. To effect this in a uniform manner the gas was made to bubble through a small hydrogen generator, which could be set in action under its own electromotive force by closing an external contact. The rate of hydrogen production was determined by a suitable galvanometer enclosed in the circuit. But the introduction of hydrogen had not the smallest effect upon the density, showing that the copper oxide was capable of performing the part desired of it.

Is it possible that the difference is independent of impurity, the nitrogen itself being to some extent in a different (dissociated) state?

I ought to have mentioned that during the fillings of the globe, the rate of passage of gas was very uniform, and about 2/3 litre per hour.

(**46**, 512-513; 1892)

Rayleigh: Terling Place, Witham, September 24.

另一方面，通过氨法制得的氮气会不会由于存在杂质而更轻呢？比氮气轻的气体并不是很多，而这种制备方法似乎可以完全保证得到的氮气中不含氢气、氨气和水蒸气。总的来看，最有可能的一种假设就是这种氮气中含有氢气，在此种稀释度下氢气逃过了与氧化铜的反应。但是，一个特殊的实验似乎又排除了这一解释。

在将第一种方法制得的氮气通入到炉管之前先向其中加入 1/1,000 ~ 2/1,000 体积的氢气。为了保证这一过程均匀稳定，我们使气体以冒泡的方式通过一个小型氢气发生器，可以通过切断一个外部接触来使这一氢气发生器在其自身的电动势下开始工作。通过一个连接在电路中的适当的检流计可以测定氢气产生的速率。但是，氢气的引入对氮气的密度没有丝毫的影响，这表明氧化铜能非常好地发挥除去氢气的作用。

那么氮气密度的这种差异会不会与其是否含有杂质无关，而是由于氮气本身在某种程度上就处于不同的状态（解离）呢？

另外，我还要说明一下，在气体注入球形容器的过程中，气体通过的速率非常平稳，大约是每小时 2/3 升。

（李世媛 翻译；李芝芬 审稿）

On a New Kind of Rays*

W. C. Röntgen

Editor's Note

This is an English translation of Wilhelm Conrad Röntgen's German report, in December 1895, of the discovery of X-rays. While experimenting with a cathode ray tube (also called a Crookes' or Lenard's tube, after earlier investigators), in which electrons or "kathode rays" are accelerated by electric fields, Röntgen found that the tube emits radiation that penetrates black paper and induces fluorescence in a screen on the other side. The rays also can penetrate matter and produce photographic images of "buried" objects such as bones. Röntgen deduces that these rays are not cathode rays, but seem instead to be akin to ultraviolet rays, yet with much greater penetrating power. He called them X-rays simply "for the sake of brevity".

(1) A discharge from a large induction coil is passed through a Hittorf's vacuum tube, or through a well-exhausted Crookes' or Lenard's tube. The tube is surrounded by a fairly close-fitting shield of black paper; it is then possible to see, in a completely darkened room, that paper covered on one side with barium platinocyanide lights up with brilliant fluorescence when brought into the neighbourhood of the tube, whether the painted side or the other be turned towards the tube. The fluorescence is still visible at two metres distance. It is easy to show that the origin of the fluorescence lies within the vacuum tube.

(2) It is seen, therefore, that some agent is capable of penetrating black cardboard which is quite opaque to ultra-violet light, sunlight, or arc-light. It is therefore of interest to investigate how far other bodies can be penetrated by the same agent. It is readily shown that all bodies possess this same transparency, but in very varying degrees. For example, paper is very transparent; the fluorescent screen will light up when placed behind a book of a thousand pages; printer's ink offers no marked resistance. Similarly the fluorescence shows behind two packs of cards; a single card does not visibly diminish the brilliancy of the light. So, again, a single thickness of tinfoil hardly casts a shadow on the screen; several have to be superposed to produce a marked effect. Thick blocks of wood are still transparent. Boards of pine two or three centimetres thick absorb only very little. A piece of sheet aluminium, 15 mm thick, still allowed the X-rays (as I will call the rays, for the sake of brevity) to pass, but greatly reduced the fluorescence. Glass plates of similar thickness behave similarly; lead glass is, however, much more opaque than glass free from lead. Ebonite several centimetres thick is transparent. If the hand be held before the fluorescent screen, the shadow shows the bones darkly, with only faint outlines of the surrounding tissues.

---

* By W. C. Röntgen. Translated by Arthur Stanton from the *Sitzungsberichte der Würzburger Physik-medic. Gesellschaft*, 1895.

论一种新型的射线*

伦琴

编者按

此文译自威廉·康拉德·伦琴 1895 年 12 月的一份关于发现了 X 射线的德文报告。当伦琴用阴极射线管（早期的科研人员也把它称作克鲁克斯管或莱纳德管，管中的电子或“阴极射线”被电场加速）进行实验时，他发现阴极射线管发射的射线能够穿透黑色的纸并在其另一侧的屏幕上显示出荧光。这种射线还可以穿透物质，人们可以利用它拍摄出像骨骼这样的“被遮挡的”物质的照片。伦琴推测这种射线不属于阴极射线，它似乎更接近紫外线，但具有更强的穿透力。“为了简便起见”，他把这类射线称为“X 射线”。

(1) 让大号感应线圈中产生的放电通过希托夫真空管，或者通过抽成真空的克鲁克斯管或莱纳德管。管子用黑纸包裹严实。在完全黑暗的房间里，将一面涂有铂氰酸钡的纸放在管子旁边，不论朝向管子的是涂有铂氰酸钡的一面还是没有涂的那面，纸都会被鲜艳的荧光照亮。在 2 米之外，这种荧光依然可见。很显然，荧光来源于真空管中。

(2) 由此可见，某些射线能够穿透这种紫外光、太阳光和电弧光都几乎不能透过的黑纸板。这就引起了人们研究这种射线到底能够多大程度地穿透其他物质的兴趣。很容易就能证明，所有物质对这种射线都是透明的，只是透明的程度大不相同。比如，纸是非常透明的，即使将荧光屏置于一本 1,000 页厚的书后面，我们仍然会在荧光屏上看到亮光，印刷油墨也不会造成明显的阻挡。类似地，单独一张卡片不会明显减弱光的强度，即使在两叠卡片后面我们也仍然能看到亮光。同样，单张锡箔纸的遮挡几乎不会使荧光屏上出现阴影，要产生明显的遮挡效果就必须重叠许多张锡箔纸。厚木块对于这种射线也是透明的。2~3 厘米厚的松木板的吸收效果非常微弱。15 毫米厚的铝板也能使 X 射线（为了简便起见，我将称这种射线为 X 射线）透过，但是能够大幅度地减弱荧光。玻璃板的作用与厚度相近的铝板类似，不过，含铅的玻璃对这种射线的阻挡效果比不含铅的玻璃更强。几厘米厚的硬质橡胶也是透明的。如果把手放在荧光屏前，屏幕上就会显示出骨骼的黑影，而周围组织则只有模糊的轮廓。

---

* 作者为伦琴。由阿瑟·斯坦顿译自 1895 年的《维尔茨堡物理–医学学会会刊》。

Water and several other fluids are very transparent. Hydrogen is not markedly more permeable than air. Plates of copper, silver, lead, gold, and platinum also allow the rays to pass, but only when the metal is thin. Platinum 0.2 mm thick allows some rays to pass; silver and copper are more transparent. Lead 1.5 mm thick is practically opaque. If a square rod of wood 20 mm in the side be painted on one face with white lead, it casts little shadow when it is so turned that the painted face is parallel to the X-rays, but a strong shadow if the rays have to pass through the painted side. The salts of the metals, either solid or in solution, behave generally as the metals themselves.

(3) The preceding experiments lead to the conclusion that the density of the bodies is the property whose variation mainly affects their permeability. At least no other property seems so marked in this connection. But that the density alone does not determine the transparency is shown by an experiment wherein plates of similar thickness of Iceland spar, glass, aluminium, and quartz were employed as screens. Then the Iceland spar showed itself much less transparent than the other bodies, though of approximately the same density. I have not remarked any strong fluorescence of Iceland spar compared with glass (see below, No. 4).

(4) Increasing thickness increases the hindrance offered to the rays by all bodies. A picture has been impressed on a photographic plate of a number of superposed layers of tinfoil, like steps, presenting thus a regularly increasing thickness. This is to be submitted to photometric processes when a suitable instrument is available.

(5) Pieces of platinum, lead, zinc, and aluminium foil were so arranged as to produce the same weakening of the effect. The annexed table shows the relative thickness and density of the equivalent sheets of metal.

	Thickness	Relative thickness	Density
Platinum	0.018 mm	1	21.5
Lead	0.050 mm	3	11.3
Zinc	0.100 mm	6	7.1
Aluminium	3.500 mm	200	2.6

From these values it is clear that in no case can we obtain the transparency of a body from the product of its density and thickness. The transparency increases much more rapidly than the product decreases.

(6) The fluorescence of barium platinocyanide is not the only noticeable action of the X-rays. It is to be observed that other bodies exhibit fluorescence, *e.g.* calcium sulphide, uranium glass, Iceland spar, rock-salt, &c.

水和其他几种液体对于这种射线都是非常透明的。氢气的透明度并没有明显强于空气。铜、银、铅、金和铂质的金属板只有在很薄的时候才能使这种射线透过。0.2 毫米的铂能使这种射线部分透过，银和铜则更透明一些。1.5 毫米厚的铅板基本上是不透明的。将一根边长为 20 毫米的方木棒的一个侧面涂上铅白，当木棒涂有铅白的面与射线平行时，几乎不会产生阴影，但是当射线必须穿过涂有铅白的一面时，就会产生明显的阴影。不论是固态的金属盐还是金属盐溶液，一般都能像金属本身一样阻挡该射线。

(3) 根据上述实验我们可以得出结论：物质的密度是这样一种性质，它的变化主要影响射线在该物质中的透过程度。至少其他性质的影响看起来都不如密度明显。不过，单是密度还不能完全决定物质对该射线的透明度。我用厚度相近的冰洲石板、玻璃板、铝板和石英板作为样品进行的实验表明，尽管这些物质具有近似相同的密度，但冰洲石对该射线的透明度却比其他物质小得多。在用冰洲石进行的实验中，我从来没有观察到像用玻璃进行的实验中出现的那样明显的荧光（见下文，第 4 部分）。

(4) 对于所有物体，增加厚度都会提高其对 X 射线的阻挡程度。我们已经在照相底片上对阶梯状叠放的多层锡箔进行了成像，得到的图像表现出了厚度的这种有规律的增加。如果根据此原理制成适当的仪器，则可以作为光度计使用。

(5) 为了得到对 X 射线相同的减弱效果，我将铂、铅、锌和铝分别制成如下规格的金属片。附表给出了具有相同减弱效果的各种金属片的密度和相对厚度。

	厚度	相对厚度	密度
铂	0.018 毫米	1	21.5
铅	0.050 毫米	3	11.3
锌	0.100 毫米	6	7.1
铝	3.500 毫米	200	2.6

从这些数据中可以清楚地看出，我们不可能根据金属密度与其厚度的乘积来确定其透明度。透明度增加的速度比该乘积减少的速度快很多。

(6) X 射线所产生的显著作用并不是只能使铂氰酸钡发出荧光。可以观测到，X 射线也能使其他一些物质发出荧光，例如硫化钙、铀玻璃、冰洲石和岩盐等。

Of special interest in this connection is the fact that photographic dry plates are sensitive to the X-rays. It is thus possible to exhibit the phenomena so as to exclude the danger of error. I have thus confirmed many observations originally made by eye observation with the fluorescent screen. Here the power of the X-rays to pass through wood or cardboard becomes useful. The photographic plate can be exposed to the action without removal of the shutter of the dark slide or other protecting case, so that the experiment need not be conducted in darkness. Manifestly, unexposed plates must not be left in their box near the vacuum tube.

It seems now questionable whether the impression on the plate is a direct effect of the X-rays, or a secondary result induced by the fluorescence of the material of the plate. Films can receive the impression as well as ordinary dry plates.

I have not been able to show experimentally that the X-rays give rise to any calorific effects. These, however, may be assumed, for the phenomena of fluorescence show that the X-rays are capable of transformation. It is also certain that all the X-rays falling on a body do not leave it as such.

The retina of the eye is quite insensitive to these rays: the eye placed close to the apparatus sees nothing. It is clear from the experiments that this is not due to want of permeability on the part of the structures of the eye.

(7) After my experiments on the transparency of increasing thicknesses of different media, I proceeded to investigate whether the X-rays could be deflected by a prism. Investigations with water and carbon bisulphide in mica prisms of 30° showed no deviation either on the photographic or the fluorescent plate. For comparison, light rays were allowed to fall on the prism as the apparatus was set up for the experiment. They were deviated 10 mm and 20 mm respectively in the case of the two prisms.

With prisms of ebonite and aluminium, I have obtained images on the photographic plate, which point to a possible deviation. It is, however, uncertain, and at most would point to a refractive index 1.05. No deviation can be observed by means of the fluorescent screen. Investigations with the heavier metals have not as yet led to any result, because of their small transparency and the consequent enfeebling of the transmitted rays.

On account of the importance of the question it is desirable to try in other ways whether the X-rays are susceptible of refraction. Finely powdered bodies allow in thick layers but little of the incident light to pass through, in consequence of refraction and reflection. In the case of the X-rays, however, such layers of powder are for equal masses of substance equally transparent with the coherent solid itself. Hence we cannot conclude any regular reflection or refraction of the X-rays. The research was conducted by the aid of finely-powdered rock-salt, fine electrolytic silver powder, and zinc dust already many times employed in chemical work. In all these cases the result, whether by the fluorescent screen

在这方面，特别让人感兴趣的是照相干版对X射线是敏感的。这就使我们可以将实验现象记录下来以避免出现错误。利用照相的方法，我已经确认了很多最初通过肉眼在荧光屏上观测得到的实验结果。X射线穿透木块或纸板的能力很有用处。在对照相干版进行曝光时，可以不用除去遮光板或者其他保护盒，因此实验就不必在暗室中进行。当然，千万不要把装有未曝光照相干版的盒子放在真空管附近。

干版上留下的影像到底是X射线的直接效应，还是由干版材料发出的荧光引起的次级效应，现在看来还是一个令人疑惑的问题。和普通的干版一样，胶片也可以记录到影像。

我还没能通过实验证明X射线是否可以产生热效应，不过我们猜测它可以，因为荧光现象表明X射线可以引起能量转移。而且可以肯定的是，照射在物体上的X射线并没有全部以荧光形式离开物体。

人眼的视网膜对这种射线非常不敏感：即使眼睛离装置很近也看不见任何东西。实验结果清楚地表明，这并不是因为该射线在眼睛这部分结构中的透过程度不够。

(7) 在通过实验研究了不同介质随着厚度增加对该射线的透明度的变化之后，我又研究了X射线是否会被棱镜偏转。在顶角为30°的云母棱镜中分别装入水和二硫化碳进行实验，结果发现照相干版和荧光板上都没有显示出偏移。为了对照，我也用可见光在相同的实验装置上进行了实验，结果发现可见光在穿过上述两种棱镜时分别偏转了10毫米和20毫米。

在使用硬质橡胶和铝制成的棱镜进行实验时，我得到的照相干版上的影像显示射线可能发生了偏转，不过这一点还不能确定，而且偏转所对应的折射率最多也只有1.05。使用荧光屏时则观测不到偏转。用较重金属进行的研究目前还没有任何结果，这是因为它们的透明度都很小，因而透射的射线非常微弱。

考虑到X射线能否被偏转这个问题的重要性，我们就有必要尝试用其他方法来研究X射线能否发生折射。由于反射和折射的原因，微细粉末形成的厚层几乎不能使入射光透过。不过，这种多层粉末对于X射线的透明度与同质量同组成的整块固体是一样的。因此我们不能得出X射线具有常规的反射或折射特性的结论。我又对细粉末状的岩盐、电解得到的细银粉和已经在化学实验中使用了多次的锌粉进行了研究。所有研究结果都表明，不论是用荧光屏还是用照相的方法，粉末与相应的固

or the photographic method, indicated no difference in transparency between the powder and the coherent solid.

It is, hence, obvious that lenses cannot be looked upon as capable of concentrating the X-rays; in effect, both an ebonite and a glass lens of large size prove to be without action. The shadow photograph of a round rod is darker in the middle than at the edge; the image of a cylinder filled with a body more transparent than its walls exhibits the middle brighter than the edge.

(8) The preceding experiments, and others which I pass over, point to the rays being incapable of regular reflection. It is, however, well to detail an observation which at first sight seemed to lead to an opposite conclusion.

I exposed a plate, protected by a black paper sheath, to the X-rays so that the glass side lay next to the vacuum tube. The sensitive film was partly covered with star-shaped pieces of platinum, lead, zinc, and aluminium. On the developed negative the star-shaped impression showed dark under platinum, lead, and, more markedly, under zinc; the aluminium gave no image. It seems, therefore, that these three metals can reflect the X-rays; as, however, another explanation is possible, I repeated the experiment with this only difference, that a film of thin aluminium foil was interposed between the sensitive film and the metal stars. Such an aluminium plate is opaque to ultra-violet rays, but transparent to X-rays. In the result the images appeared as before, this pointing still to the existence of reflection at metal surfaces.

If one considers this observation in connection with others, namely, on the transparency of powders, and on the state of the surface not being effective in altering the passage of the X-rays through a body, it leads to the probable conclusion that regular reflection does not exist, but that bodies behave to the X-rays as turbid media to light.

Since I have obtained no evidence of refraction at the surface of different media, it seems probable that the X-rays move with the same velocity in all bodies, and in a medium which penetrates everything, and in which the molecules of bodies are embedded. The molecules obstruct the X-rays, the more effectively as the density of the body concerned is greater.

(9) It seemed possible that the geometrical arrangement of the molecules might affect the action of a body upon the X-rays, so that, for example, Iceland spar might exhibit different phenomena according to the relation of the surface of the plate to the axis of the crystal. Experiments with quartz and Iceland spar on this point lead to a negative result.

(10) It is known that Lenard, in his investigations on kathode rays, has shown that they belong to the ether, and can pass through all bodies. Concerning the X-rays the same may be said.

体对 X 射线的透明度没有任何差别。

因此，很明显透镜是不能会聚 X 射线的。事实上，大尺寸的玻璃透镜和硬质橡胶透镜对 X 射线都没有会聚作用，这已经得到了证明。圆柱的透视影像显示，中间部分的阴影比边缘更深一些；如果在圆柱内部填入透明度比柱体材料更高的物质，那么在得到的影像中，中间部分会比边缘更亮一些。

(8) 上述实验和另外一些我未提及的实验，都表明这种射线不具备常规的反射能力。不过，我还是要详细介绍一个乍看上去似乎会使人们得出相反结论的实验。

实验中，我将一块用黑纸套保护起来的玻璃干版置于 X 射线中，使其玻璃面靠近真空管，并用铂、铅、锌和铝质的星形金属片部分地遮挡干版的感光膜。在显影后的负片上，铂片和铅片下方出现了黑色的星形影像，锌片下方的影像更加清晰，而铝片并没有产生阴影。由此看来，前三种金属可以反射 X 射线。不过也可能有另外的解释。我又重复了这一实验，这次唯一的不同之处是，我在感光膜和星形金属片之间插入了一块薄薄的铝箔。这块铝箔对紫外线是不透明的，但对 X 射线是透明的。结果出现了和以前一样的影像。这再次表明 X 射线在金属表面发生了反射。

如果综合考虑这一观测结果和其他一些结果，包括关于粉末透明度的结果以及关于表面状态不能有效改变 X 射线穿过物体的路径的结果，我们就会得出这样一个可能的结论：对于 X 射线来说，并不存在普通意义上的反射，物体对于 X 射线的作用，就像混浊介质对于可见光一样。

我还没有得到任何可以表明在不同介质表面 X 射线会发生折射的证据，这样看来，X 射线在所有物质中的传播速度可能都相同，而且在一种渗透一切物质、包容各种物质分子的介质中也是一样的。随着物质密度的增大，其分子对 X 射线的阻挡效果也变得更加明显。

(9) 分子的几何构型看起来可能会影响物质对 X 射线的阻挡作用，例如，对于冰洲石晶体而言，表面与晶轴之间相对取向的不同可能就会导致不同的现象。但是，为此而用石英和冰洲石进行的实验却得到了阴性的结果。

(10) 我们知道，莱纳德在对阴极射线的研究中已经指出，阴极射线属于以太，可以穿透任何物体。估计 X 射线可能也是这样的。

In his latest work, Lenard has investigated the absorption coefficients of various bodies for the kathode rays, including air at atmospheric pressure, which gives 4.10, 3.40, 3.10 for 1 cm, according to the degree of exhaustion of the gas in discharge tube. To judge from the nature of the discharge, I have worked at about the same pressure, but occasionally at greater or smaller pressures. I find, using a Weber's photometer, that the intensity of the fluorescent light varies nearly as the inverse square of the distance between screen and discharge tube. This result is obtained from three very consistent sets of observations at distances of 100 and 200 mm. Hence air absorbs the X-rays much less than the kathode rays. This result is in complete agreement with the previously described result, that the fluorescence of the screen can be still observed at 2 metres from the vacuum tube. In general, other bodies behave like air; they are more transparent for the X-rays than for the kathode rays.

(11) A further distinction, and a noteworthy one, results from the action of a magnet. I have not succeeded in observing any deviation of the X-rays even in very strong magnetic fields.

The deviation of kathode rays by the magnet is one of their peculiar characteristics; it has been observed by Hertz and Lenard, that several kinds of kathode rays exist, which differ by their power of exciting phosphorescence, their susceptibility of absorption, and their deviation by the magnet; but a notable deviation has been observed in all cases which have yet been investigated, and I think that such deviation affords a characteristic not to be set aside lightly.

(12) As the result of many researches, it appears that the place of most brilliant phosphorescence of the walls of the discharge-tube is the chief seat whence the X-rays originate and spread in all directions; that is, the X-rays proceed from the front where the kathode rays strike the glass. If one deviates the kathode rays within the tube by means of a magnet, it is seen that the X-rays proceed from a new point, *i.e.* again from the end of the kathode rays.

Also for this reason the X-rays, which are not deflected by a magnet, cannot be regarded as kathode rays which have passed through the glass, for that passage cannot, according to Lenard, be the cause of the different deflection of the rays. Hence I conclude that the X-rays are not identical with the kathode rays, but are produced from the kathode rays at the glass surface of the tube.

(13) The rays are generated not only in glass. I have obtained them in an apparatus closed by an aluminium plate 2 mm thick. I purpose later to investigate the behaviour of other substances.

(14) The justification of the term "rays", applied to the phenomena, lies partly in the regular shadow pictures produced by the interposition of a more or less permeable body between the source and a photographic plate or fluorescent screen.

莱纳德在最近的工作中研究了各种物体对阴极射线的吸收系数，比如一个大气压下的空气的吸收系数，根据放电管抽真空程度的不同，每一厘米对应的吸收系数分别是 4.10、3.40 和 3.10。为了根据放电的本质来作出判断，我在基本相同的压强下进行了研究，不过偶尔也会用更高一点或更低一点的压强。利用韦伯光度计，我发现荧光的强度近似与屏幕到放电管距离的平方成反比。这个结论是根据三组非常一致的观测结果得到的，其观测距离分别为 100 毫米和 200 毫米。因此，空气对 X 射线的吸收比对阴极射线的吸收低很多。这一结果与前述的在距真空管 2 米处的屏幕上仍会出现荧光的结果是完全一致的。大体上，其他物质的性质与空气类似，它们对 X 射线比对阴极射线更加透明。

(11) 另一个更明显也更值得关注的区别是磁场的作用。即使是在非常强的磁场中，我也没有观测到 X 射线的任何偏转。

阴极射线在磁场作用下会发生偏转，这是它的独特性质之一。赫兹和莱纳德曾经观测到存在好几种阴极射线，它们的区别在于激发磷光的能力不同、被吸收的容易程度不同以及在磁场作用下的偏转不同。但是对于所有已经被研究过的阴极射线，人们都观测到了显著的偏转，我认为这种偏转代表了阴极射线的一种绝不该被忽视的特性。

(12) 很多研究结果表明，放电管管壁上磷光最强的位置是在 X 射线产生并向四周各个方向发散的那个源头处，也就是说，X 射线产生于阴极射线轰击玻璃的前沿位置。如果利用磁场使管中的阴极射线偏转，就会看到 X 射线从另一个位置上产生，但仍然是在阴极射线的终端位置。

基于这一原因，我们不能把在磁场作用下并不偏转的 X 射线看作是已经穿透玻璃的阴极射线，因为按照莱纳德的说法，这条通道不可能是由阴极射线的不同偏转造成的。由此我断定，X 射线与阴极射线是不同的，它是阴极射线作用于真空管的玻璃表面而产生的。

(13) 并不是只有用玻璃才能产生 X 射线。我曾利用一种被 2 毫米厚的铝板包裹起来的装置得到了 X 射线。以后我将研究其他物质是否也能产生 X 射线。

(14) 在描述这种现象时我使用了“射线”这个词，这在一定程度上是因为，将不太透明的物体插入到源和照相干版或荧光屏之间时会产生规则的阴影。

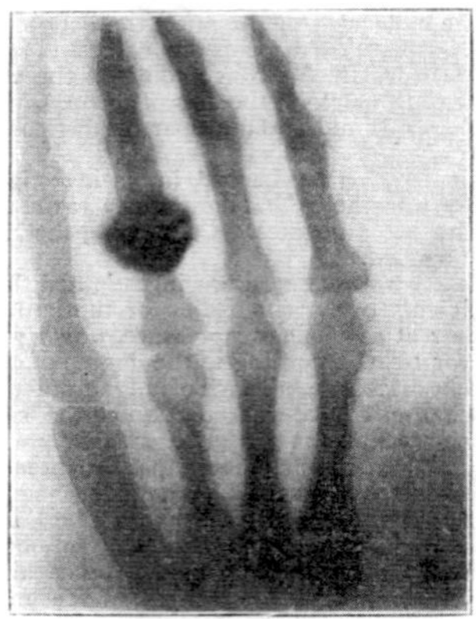

Fig. 1. Photograph of the bones in the fingers of a living human hand.
The third finger has a ring upon it.

I have observed and photographed many such shadow pictures. Thus, I have an outline of part of a door covered with lead paint; the image was produced by placing the discharge-tube on one side of the door, and the sensitive plate on the other. I have also a shadow of the bones of the hand (Fig. 1), of a wire wound upon a bobbin, of a set of weights in a box, of a compass card and needle completely enclosed in a metal case (Fig. 2), of a piece of metal where the X-rays show the want of homogeneity, and of other things.

Fig. 2. Photograph of a compass card and needle completely enclosed in a metal case

For the rectilinear propagation of the rays, I have a pin-hole photograph of the discharge apparatus covered with black paper. It is faint but unmistakable.

(15) I have sought for interference effects of the X-rays, but possibly, in consequence of their small intensity, without result.

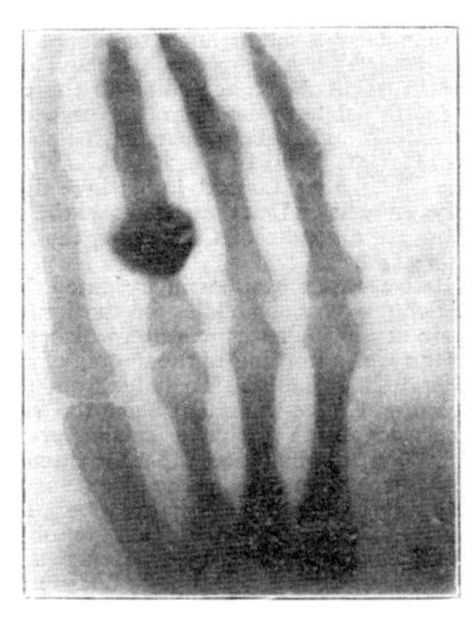

图 1. 活人手指（第三指上戴着一枚戒指）骨骼的影像

我已经观察并用照相记录了很多这样的阴影。由此，我记录下了门的局部轮廓。我是用含铅涂料刷了门的轮廓，然后把放电管放置在门的一侧，而把光敏照相干版放置在另一侧，这样就得到了门的局部轮廓的影像。我还记录了其他许多物体的阴影，这包括手掌骨骼（图 1）、缠在绕线筒上的导线、一套装在盒子里的砝码、完全密封于金属盒子中的罗经刻度盘和指针（图 2）、一块在 X 射线下显示出具有不均匀缺陷的金属片，以及其他一些物品。

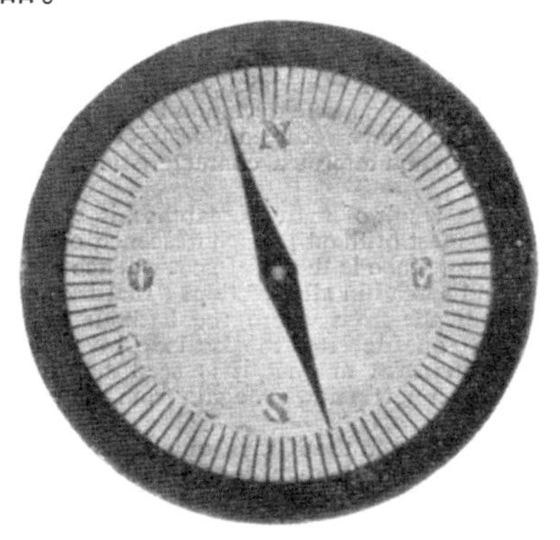

图 2. 完全密封于金属盒子中的罗经刻度盘和指针的影像

为了说明射线的直线传播，我用针孔照相的方法拍摄了用黑纸覆盖的放电装置。照片虽然有些模糊，但却可以明白无误地分辨出装置。

(15) 我曾经试图寻找 X 射线的干涉效应，但并没有检测到，这可能是由于强度太低的缘故。

(16) Researches to investigate whether electrostatic forces act on the X-rays are begun but not yet concluded.

(17) If one asks, what then are these X-rays; since they are not kathode rays, one might suppose, from their power of exciting fluorescence and chemical action, them to be due to ultra-violet light. In opposition to this view a weighty set of considerations presents itself. If X-rays be indeed ultra-violet light, then that light must possess the following properties.

(*a*) It is not refracted in passing from air into water, carbon bisulphide, aluminium, rock-salt, glass or zinc.
(*b*) It is incapable of regular reflection at the surfaces of the above bodies.
(*c*) It cannot be polarised by any ordinary polarising media.
(*d*) The absorption by various bodies must depend chiefly on their density.

That is to say, these ultra-violet rays must behave quite differently from the visible, infra-red, and hitherto known ultra-violet rays.

These things appear so unlikely that I have sought for another hypothesis.

A kind of relationship between the new rays and light rays appears to exist; at least the formation of shadows, fluorescence, and the production of chemical action point in this direction. Now it has been known for a long time, that besides the transverse vibrations which account for the phenomena of light, it is possible that longitudinal vibrations should exist in the ether, and, according to the view of some physicists, must exist. It is granted that their existence has not yet been made clear, and their properties are not experimentally demonstrated. Should not the new rays be ascribed to longitudinal waves in the ether?

I must confess that I have in the course of this research made myself more and more familiar with this thought, and venture to put the opinion forward, while I am quite conscious that the hypothesis advanced still requires a more solid foundation.

(**53**, 274-276; 1896)

(16) 关于静电力对 X 射线是否有作用的研究工作正在进行，但目前尚无结论。

(17) 也许有人会问 X 射线到底是什么。既然这种射线不是阴极射线，有人可能就会根据其激发荧光和引发化学反应的能力猜想它是紫外光。然而，一系列认真的思考都是反对这种观点的。如果 X 射线真是一种紫外光，那么这种紫外光就必须具有如下性质：

(*a*) 它在由空气进入水、二硫化碳、铝、岩盐、玻璃或锌时，不会发生折射。

(*b*) 它在上述物质的表面不会发生常规的反射。

(*c*) 任何普通的偏振介质都不能使它偏振。

(*d*) 不同物质对它的吸收主要取决于该物质的密度。

也就是说，这种紫外线必须具有与可见光、红外线以及迄今为止已知的紫外线都十分不同的性质。

看起来这些是很难成立的，因此我想到了另一种假说。

这种新型的射线与普通光之间看起来应该存在着某种关联，至少在形成阴影、激发荧光以及引发化学反应这些方面都是相似的。长期以来我们都知道，除了能够解释光现象的横向振动外，在以太中可能存在纵向振动，某些物理学家甚至认为纵向振动是必定存在的。尽管目前人们还不完全清楚纵向振动是否存在，也没有通过实验论证这种纵向振动的性质，但是，难道我们就不能认为这种新型的射线属于以太中的纵波吗？

我必须要承认的是，在研究过程中我越来越倾向于这一观点。另外我也十分清楚这一新假说还需要更为可靠的证据，我承认在目前的情况下抛出这一观点是比较冒昧的。

（王耀杨 翻译；江丕栋 审稿）

Professor Röntgen's Discovery

A. A. C. Swinton

Editor's Note

One of the most sobering things about this verification of Röntgen's discovery of X-rays, less than a month after they were first reported, is that it shows how tepid the reception of great discoveries can be among scientific peers. Campbell-Swinton hints that the newspapers have been getting excited over a phenomenon that is not "entirely novel". But that is because he somewhat misinterprets Röntgen's results. Swinton insists on regarding the X-rays as "some portion of the kathode radiations", and points out that cathode rays are already known to produce photographic images—missing Röntgen's claim that his X-rays are not cathode rays at all.

THE newspaper reports of Prof. Röntgen's experiments have, during the past few days, excited considerable interest. The discovery does not appear, however, to be entirely novel, as it was noted by Hertz that metallic films are transparent to the kathode rays from a Crookes or Hittorf tube, and in Lenard's researches, published about two years ago, it is distinctly pointed out that such rays will produce photographic impressions. Indeed, Lenard, employing a tube with an aluminium window, through which the kathode rays passed out with comparative ease, obtained photographic shadow images almost identical with those of Röntgen, through pieces of cardboard and aluminium interposed between the window and the photographic plate.

Prof. Röntgen has, however, shown that this aluminium window is unnecessary, as some portion of the kathode radiations that are photographically active will pass through the glass walls of the tube. Further, he has extended the results obtained by Lenard in a manner that has impressed the popular imagination, while, perhaps most important of all, he has discovered the exceedingly curious fact that bone is so much less transparent to these radiations than flesh and muscle, that if a living human hand be interposed between a Crookes tube and a photographic plate, a shadow photograph can be obtained which shows all the outlines and joints of the bones most distinctly.

Working upon the lines indicated in the telegrams from Vienna, recently published in the daily papers, I have, with the assistance of Mr. J. C. M. Stanton, repeated many of Prof. Röntgen's experiments with entire success. According to one of our first experiments, an ordinary gelatinous bromide dry photographic plate was placed in an ordinary camera back. The wooden shutter of the back was kept closed, and upon it were placed miscellaneous articles such as coins, pieces of wood, carbon, ebonite, vulcanised fibre, aluminium, &c., all being quite opaque to ordinary light. Above was supported a Crookes tube, which was excited for some minutes. On development, shadows of all the articles

伦琴教授的发现

斯温顿

编者按

距离伦琴最初宣布发现X射线还不到一个月，就有了这篇对伦琴的发现的查证，这是一个应该引起人们警醒的事例，它表明科学界同行对重大发现的态度也会是冷淡的。坎贝尔–斯温顿含蓄地指出，报业为之感到兴奋的现象实际上并不是一个“全新的”现象。但他之所以这样说是因为他在某种程度上误会了伦琴得到的结论。斯温顿坚持把X射线看作是“阴极射线的一部分”，并指出人们早就知道阴极射线能够用于拍摄照片，可他没有注意到伦琴所称的X射线根本就不是阴极射线。

前一段时间，关于伦琴教授的实验的新闻报道引起了相当广泛的关注。不过，这一发现似乎并不是全新的，因为赫兹就曾注意到从克鲁克斯管或希托夫管中发射出来的阴极射线能够穿透金属薄片，而大约两年前莱纳德就在其发表的研究报告中明确地指出这种射线可以产生影像。莱纳德使用了一个带铝窗的管子，阴极射线可以比较容易地从此窗中穿出，伦琴的实验中则是射线穿过了插在管窗与照相干版之间的纸板和铝片。实际上，莱纳德得到了与伦琴的结果几乎完全一样的阴影图像。

不过，伦琴教授已经阐明这个铝窗并不是必需的，因为一部分能够引起成像的阴极辐射是从玻璃管壁中穿出的。此外，他还以一种能给人们留下深刻印象的方式推广了莱纳德得到的结果，而也许最为重要的是，他发现了一个极为新奇的现象，即这种辐射穿透骨骼的能力比穿透肌肉的能力差很多，如果将活人的手置于克鲁克斯管与照相干版之间，就能得到一张非常清晰地显示出骨骼关节轮廓的阴影图像。

最近，许多日报都刊载了来自维也纳的电报，根据其中提供的线索，在斯坦顿先生的协助下，我已经完全成功地重复了伦琴教授的很多实验。在最初的一次实验中，我们将一张普通的凝胶溴化物照相干版放置在普通相机后面。背面的木质快门始终保持关闭状态，并紧接着放置各种物品，诸如硬币、木块、炭、硬质橡胶、硬化纤维和铝等，所有这些物品对于普通的可见光都是完全不透明的。在这些物品上方固定一个已经激发了几分钟的克鲁克斯管。显影后，放置的所有物品的阴影都清晰可

placed on the slide were clearly visible, some being more opaque than others. Further experiments were tried with thin plates of aluminium or of black vulcanised fibre interposed between the objects to be photographed and the sensitive surface, this thin plate being used in place of the wood of the camera back. In this manner sharper shadow pictures were obtained. While most thick metal sheets appear to be entirely opaque to the radiations, aluminium appears to be relatively transparent. Ebonite, vulcanised fibre, carbon, wood, cardboard, leather and slate are all very transparent, while, on the other hand, glass is exceedingly opaque. Thin metal foils are moderately opaque, but not altogether so.

As tending to the view that the radiations are more akin to ultraviolet than to infra-red light, it may be mentioned that a solution of alum in water is distinctly more transparent to them than a solution of iodine in bisulphide of carbon.

So far as our own experiments go, it appears that, at any rate without very long exposures, a sufficiently active excitation of the Crookes tube is not obtained by direct connection to an ordinary Rhumkorff induction coil, even of a large size. So-called high frequency currents, however, appear to give good results, and our own experiments have been made with the tube excited by current obtained from the secondary circuit of a Tesla oil coil, through the primary of which were continuously discharged twelve half-gallon Leyden jars, charged by an alternating current of about 20,000 volts pressure, produced by a transformer with a spark-gap across its high-pressure terminals.

For obtaining shadow photographs of inanimate objects, and for testing the relative transparency of different substances, the particular form of Crookes tube employed does not appear to greatly signify, though some forms are, we find, better than others. When, however, the human hand is to be photographed, and it is important to obtain sharp shadows of the bones, the particular form of tube used and its position relative to the hand and sensitive plate appear to be of great importance. So far, owing to the frequent destruction of the tubes, due to overheating of the terminals, we have not been able to ascertain exactly the best form and arrangement for this purpose, except that it appears desirable that the electrodes in the tube should consist of flat and not curved plates, and that these plates should be of small dimensions.

The accompanying photograph of a living human hand (Fig. 1) was exposed for twenty minutes through an aluminium sheet 0.0075 in thickness, the Crookes tube, which was one of the kind containing some white phosphorescent material (probably sulphide of barium), being held vertically upside down, with its lowest point about two inches above the centre of the hand.

见，其中一些物品的阴影比其他的更加明显。在后来的实验中，我们尝试着将薄铝板或黑色硬化纤维薄板插入待成像的物体与感光表面之间，这一薄板用来代替相机后的木质快门。用这种方式我们得到了更加清晰的阴影图像。大部分厚金属板对于这种辐射似乎都是完全不透明的，而铝板则似乎比较透明。硬质橡胶、硬化纤维、炭、木块、纸板、皮革以及石板都是非常透明的，相反，玻璃则是非常不透明的。薄金属板是中等透明的，不过也不全是这样。

为了支持该辐射更类似于紫外线而不是红外线的观点，我们要说明一下，与碘的二硫化碳溶液相比，明矾的水溶液对该辐射的透明度明显好得多。

就我们的实验情况来说，如果不进行很长时间的曝光，单靠将一个普通的拉姆科夫感应线圈与克鲁克斯管直接相连，即使是用大号的线圈，看起来似乎也无法使克鲁克斯管产生足够引起成像活性的激发辐射。不过，我们常说的高频电流看来能够给出好的结果。我们在实验中使用的克鲁克斯管，是由特斯拉油线圈的次级电路产生的电流来激发的，通过其初级电路的是连续放电的 12 个半加仑莱顿瓶，这些莱顿瓶通过电压约为 20,000 伏特的交流电进行充电，此交流电由一高压端带有放电间隙的变压器产生。

要获得无生命物体的阴影或检验不同物质对此种辐射的相对透明度，使用哪种结构的克鲁克斯管看起来并不是非常要紧，尽管我们发现某些结构的管子比另外一些好一点。但是，在获取人手的影像时，重要的是得到骨骼的清晰阴影，那么所使用的管子的特殊结构以及管子相对于人手和感光干版的位置就显得至关重要了。到目前为止，由于管子经常因其末端过热而毁坏，我们还没能弄清楚对于上述目的来说什么样的管子结构和摆放位置是最好的，不过能够确定的是，管子中的电极应该采用平板电极而不是曲面电极，并且应该用尺寸较小的电极板。

本文所附的活人手掌的影像（图 1）是该辐射穿过厚度为 0.0075 的铝片持续曝光 20 分钟而得到的，实验中使用的克鲁克斯管中包含某种白色磷光物质（可能是硫化钡），管子颠倒后垂直放置，其最低点位于掌心上方大约 2 英寸处。

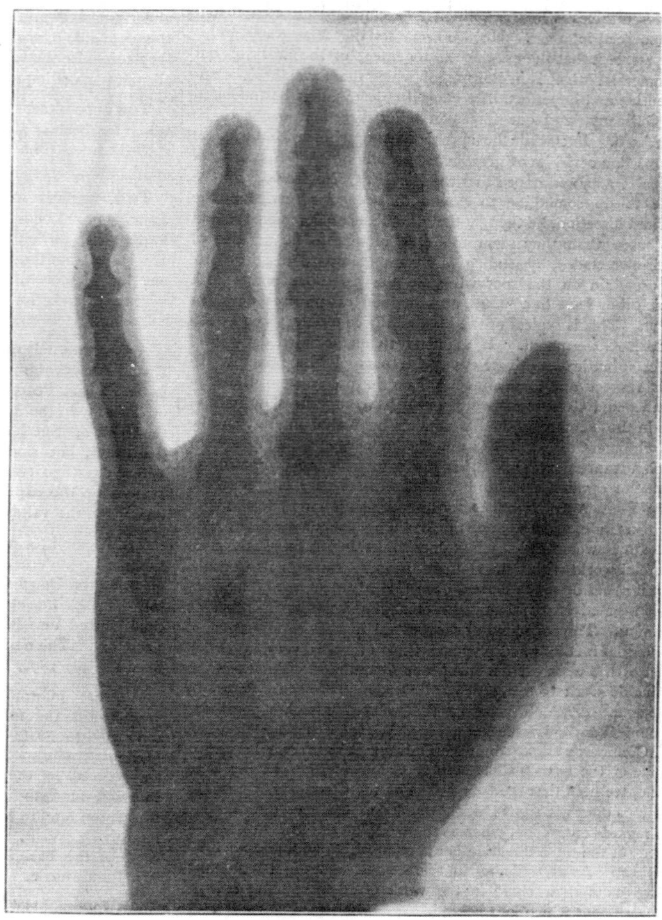

Fig. 1. Photograph of a living human hand

By substituting a thin sheet of black vulcanised fibre for the aluminium plate, we have since been able to reduce the exposure required to four minutes. Indeed with the aluminium plate, the twenty minutes' exposure appears to have been longer than was necessary. Further, having regard to the great opacity of glass, it seems probable that where ordinary Crookes tubes are employed, a large proportion of the active radiations must be absorbed by the glass of the tube itself. If this is so, by the employment of a tube partly constructed of aluminium, as used by Lenard, the necessary length of exposure could be much reduced.

(**53**, 276-277; 1896)

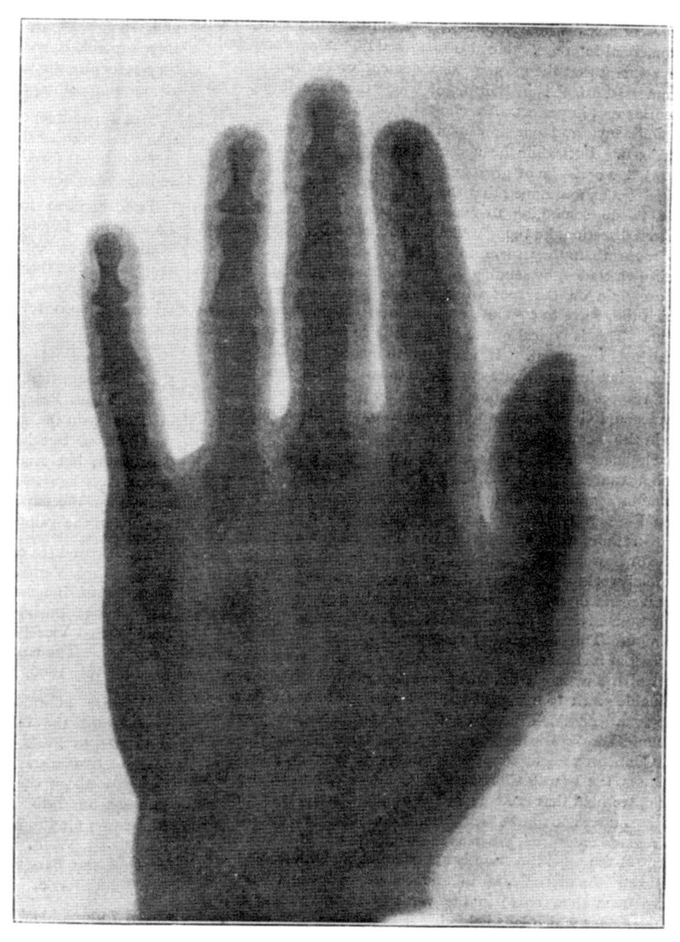

图 1. 活人手掌的影像

用一块黑色硬化纤维薄板代替铝板后，我们发现可以把曝光时间缩短到 4 分钟。实际上，在使用铝板的情况下，20 分钟的曝光时间似乎也比必需的曝光时间长一些。此外，考虑到玻璃很显著的不透明性，看来在使用普通克鲁克斯管时，大部分能够形成影像的辐射一定被管子自身的玻璃壁吸收了。如果确实如此，那么使用部分为铝质材料的管子（如同莱纳德所用的那样）的话，必需的曝光时间应该会大大缩短。

（王耀杨 翻译；江丕栋 审稿）

New Experiments on the Kathode Rays*

M. J. Perrin

Editor's Note

Physicists were puzzled by cathode rays, which carried energy from a negative electrode (cathode) toward a positive electrode inside a vacuum tube. Experimenters had determined that they originated at the cathode, but could not say what they were. In this classic paper the French physicist Jean Perrin reports experiments that helped to clarify the mystery. He placed into a cathode ray tube a metal cylinder linked to an electroscope, which would measure any charge deposited into it. When the cathode rays were directed into the tube, Perrin detected a significant negative charge. Perrin speculated that the charge carriers were negative ions created near the cathode. In fact they were electrons, as J. J. Thomson discovered one year later.

(1) Two hypotheses have been propounded to explain the properties of the kathode rays.

Some physicists think with Goldstein, Hertz, and Lenard, that this phenomenon is like light, due to vibrations of the ether †, or even that it is light of short wavelength. It is easily understood that such rays may have a rectilinear path, excite phosphorescence, and affect photographic plates.

Others think, with Crookes and J. J. Thomson, that these rays are formed by matter which is negatively charged and moving with great velocity, and on this hypothesis their mechanical properties, as well as the manner in which they become curved in a magnetic field, are readily explicable.

This latter hypothesis has suggested to me some experiments which I will now briefly describe, without for the moment pausing to inquire whether the hypothesis suffices to explain all the facts at present known, and whether it is the only hypothesis that can do so. Its adherents suppose that the kathode rays are negatively charged; so far as I know, this electrification has not been established, and I first attempted to determine whether it exists or not.

(2) For that purpose I had recourse to the laws of induction, by means of which it is possible to detect the introduction of electric charges into the interior of a closed electric conductor, and to measure them. I therefore caused the kathode rays to pass into a

---

* Translation of a paper by M. Jean Perrin, read before the Paris Academy of Sciences on December 30, 1895.

† These vibrations might be something different from light; recently M. Jaumann, whose hypotheses have since been criticised by M. H. Poincaré, supposed them to be longitudinal.

关于阴极射线的新实验*

佩兰

编者按

物理学家们对在真空管中把能量从负极（阴极）带到正极的阴极射线感到迷惑不解。实验可以证明它们来自阴极，但不能说明它们到底是什么。法国物理学家让·佩兰在这篇经典论文中用实验揭开了阴极射线的神秘面纱。他在阴极射线管内放置了一个与验电器相连的金属圆柱体，验电器可以测量进入其中的电荷。当阴极射线进入真空管时，佩兰检测到了大量的负电荷。佩兰推测这些载流子是在阴极附近产生的负离子。实际上它们就是一年之后汤姆逊发现的电子。

（1）现在可以解释阴极射线性质的假说有两种。

一部分物理学家和戈尔德施泰因、赫兹、莱纳德的意见一致，认为这种现象和光一样，是由以太的振动引起的†，或者它就是一种短波长的光。很容易就可以理解，这种射线可能是沿直线传播的，能激发磷光，而且可以使照相干版感光。

另一部分人则与克鲁克斯、汤姆逊持相同的观点，认为这种射线是由带负电的物质组成，并以极快的速度运动。用这种假说可以解释它们的力学性质，也可以很容易地说明为什么它们在磁场中的路径会弯曲。

后一种假说启发我进行了一些实验，我将在这里简单地描述这些实验，暂时先不管这个假说是否能解释目前所有已知的现象，或者是否只有这一种假说可以解释这些现象。它的支持者认为阴极射线是带负电的，而据我所知，这种带电性还没有被确认，我首先要确定它是否带电。

（2）为了达到这个目的，我将借助电磁感应定律，用这个定律，我们可以检测引入闭合导电体内部的电量，并且进行定量测量。因此，我让阴极射线通过法拉第

---

* 这篇文章翻译自让·佩兰在 1895 年 12 月 30 日向巴黎科学院宣读的论文。

† 这种振动可能与光有些不同。最近，尧曼（其猜想曾经被普安卡雷批判过）提出这种振动可能是纵向的。

Faraday's cylinder. For this purpose I employed the vacuum tube represented in Fig. 1. A B C D is a tube with an opening α in the centre of the face B C. It is this tube which plays the part of a Faraday's cylinder. A metal thread soldered at S to the wall of the tube connects this cylinder with an electroscope.

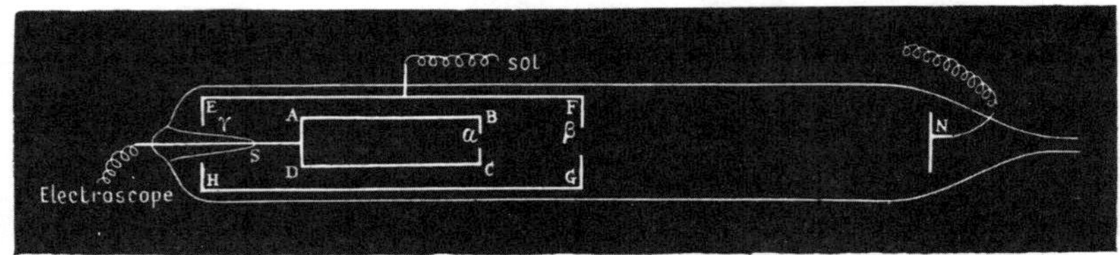

Fig. 1

E F G H is a second cylinder in permanent communication with the earth, and pierced by two small openings at β and γ; it protects the Faraday's cylinder from all external influence. Finally, at a distance of about 0.10 m in front of F G , was placed an electrode N. The electrode N served as kathode; the anode was formed by the protecting cylinder E F G H; thus a pencil of kathode rays passed into the Faraday's cylinder. This cylinder invariably became charged with negative electricity.

The vacuum tube could be placed between the poles of an electro-magnet. When this was excited, the kathode rays, becoming deflected, no longer passed into the Faraday's cylinder, and this cylinder was then not charged; it, however, became charged immediately the electromagnet ceased to be excited.

In short, the Faraday's cylinder became negatively charged when the kathode rays entered it, and only when they entered it; *the kathode rays are then charged with negative electricity*.

The quantity of electricity which these rays carry can be measured. I have not finished this investigation, but I shall give an idea of the order of magnitude of the charges obtained when I say that for one of my tubes, at a pressure of 20 microns of mercury, and for a single interruption of the primary of the coil, the Faraday's cylinder received a charge of electricity sufficient to raise a capacity of 600 C. G. S. units to 300 volts.

(3) The kathode rays being negatively charged, the principle of the conservation of electricity drives us to seek somewhere the corresponding positive charges. I believe that I have found them in the very region where the kathode rays are formed, and that I have established the fact that they travel in the opposite direction, and fall upon the kathode. In order to verify this hypothesis, it is sufficient to use a hollow kathode pierced with a small opening by which a portion of the attracted positive electricity might enter. This electricity could then act upon a Faraday's cylinder inside the kathode.

圆筒。为此我设计了如图 1 所示的真空管。A B C D 是一根管子，在 B C 面中心的 α 处有一个小孔。正是这根管子起到了法拉第圆筒的作用。一根焊接在管壁 S 处的金属线将圆筒和外部的验电器连接起来。

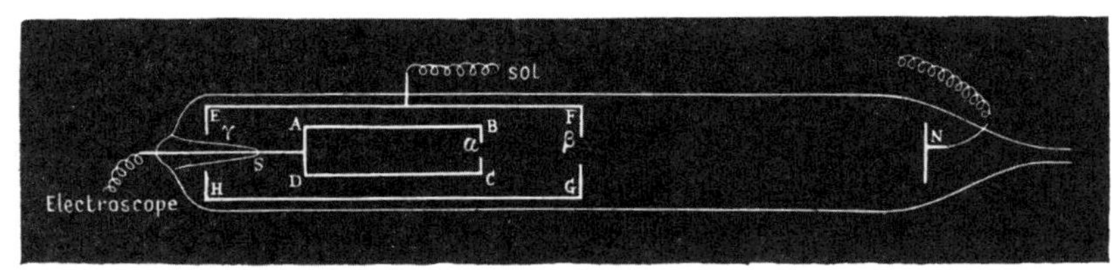

图 1

E F G H 是另一个圆筒，永久接地，并且在 β、γ 处穿有两个小孔；这个圆筒可以屏蔽外界对法拉第圆筒的干扰。最后，在 FG 前面大约 0.1 米的地方有一个电极 N。电极 N 作为阴极，屏蔽圆筒 E F G H 作为阳极。在这样的条件下，将一束阴极射线通入法拉第圆筒，这个圆筒将一直带负电。

真空管可以放置在电磁铁的两极之间。当电磁铁通电时，阴极射线将发生偏转，不能再通入法拉第圆筒，这个圆筒也将不再带电，而当电磁铁断电之后，法拉第圆筒马上又带电了。

简单地说就是，当且仅当有阴极射线进入时，法拉第圆筒带负电，**所以阴极射线一定带负电**。

射线所带的电量可以被测量出来。我还没有完成这项研究，但是我可以给出一个有关所获电量的数量级的概念，对于一个压力为 20 微米汞柱的真空管，将初级线圈截断，法拉第圆筒接收的电量足以使 600 单位（厘米克秒制）的电容器的电势差提高到 300 伏特。

（3）阴极射线带负电，根据电荷守恒定律，我们应该能在某处找到相应的正电荷。我确信我已经在阴极射线产生的地方找到了正电荷，我认为它们向相反的方向运动，而后撞在了阴极上。为了证明这种说法，只要用一个中空的阴极就行，在阴极上穿一个小孔，以使一部分被吸引过来的正电荷可以由此通过。进入阴极的正电荷会影响阴极内部的法拉第圆筒。

The protecting cylinder E F G H with its opening β fulfilled these conditions, and this time I therefore employed it as the kathode, the electrode N being the anode. The Faraday's cylinder is then invariably charged with *positive electricity*. The positive charges were of the order of magnitude of the negative charges previously obtained.

Thus, at the same time as negative electricity is *radiated* from the kathode, positive electricity travels towards that kathode.

I endeavoured to determine whether this positive flux formed a second system of rays absolutely symmetrical to the first.

(4) For that purpose I constructed a tube (Fig. 2) similar to the preceding, except that between the Faraday's cylinder and the opening β was placed a metal diaphragm pierced with an opening β', so that the positive electricity which entered by β could only affect the Faraday's cylinder if it also traversed the diaphragm β'. Then I repeated the preceding experiments.

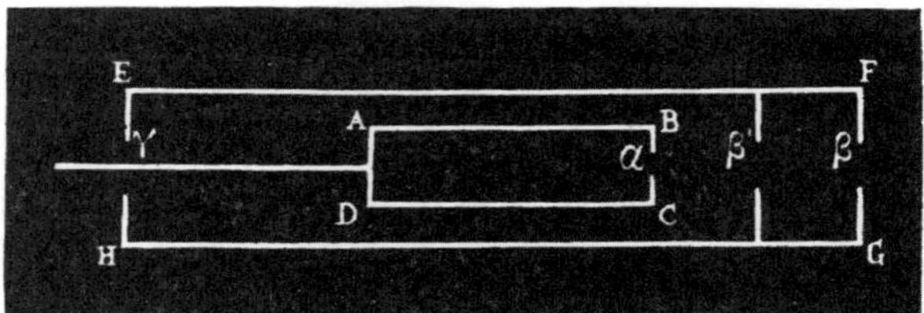

Fig. 2

When N was the kathode, the rays emitted from the kathode passed through the two openings β and β' without difficulty, and caused a strong divergence of the leaves of the electroscope. But when the protecting cylinder was the kathode, the positive flux, which, according to the preceding experiment, entered at β, did not succeed in separating the gold leaves except at very low pressures. When an electrometer was substituted for the electroscope, it was found that the action of the positive flux was real but very feeble, and increased as the pressure decreased. In a series of experiments at a pressure of 20 microns, it raised a capacity of 2,000 C. G .S. units to 10 volts; and at a pressure of 3 microns, during the same time, it raised the potential to 60 volts.*

By means of a magnet this action could be entirely suppressed.

(5) These results as a whole do not appear capable of being easily reconciled with the theory which regards the kathode rays as an ultra-violet light. On the other hand, they agree well with the theory which regards them as a material radiation, and which, as it appears to me, might be thus enunciated.

---

* The breaking of the tube has temporarily prevented me from studying the phenomenon at lower pressures.

带有小孔 β 的屏蔽圆筒 E F G H 满足以上条件，因此这一回我用 E F G H 作阴极，电极 N 作阳极。这样，法拉第圆筒就会一直带**正电**。其所带正电荷和前面所测的负电荷的数量级相同。

这说明，在阴极**发射**负电荷的同时，正电荷也在向阴极运动。

我下决心要确定这种正电流是否能形成另一个和阴极射线完全对称的射线系统。

(4) 为此我构造了一个和前面类似的管子（见图 2），与前面管子唯一的不同之处是，在法拉第圆筒和小孔 β 之间放置了一个金属膜片，膜片上有一个小孔 β'，这样，从 β 进入的正电荷只有在也通过 β' 的情况下才能作用于法拉第圆筒。我用这个装置重复了上面的实验。

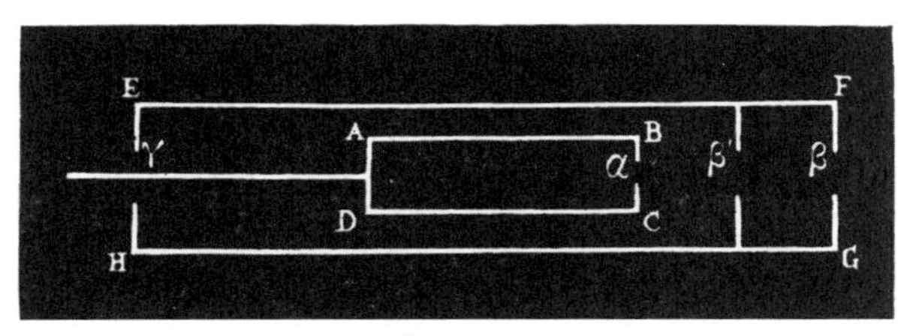

图 2

当 N 作为阴极时，由阴极发射的射线可以顺利地通过 β 和 β' 两个孔，使验电器的两个叶片张得很大。但是当屏蔽圆筒作为阴极时，根据前述的实验，通过 β 进入的正电流并没有使验电器的金箔张开，除非是在压力很低的情况下。当用静电计代替验电器时，可以看到正电流的确存在但非常微弱，并随着压力的减小而增大。在 20 微米汞柱条件下进行的一系列实验中，这一正电流可以把 2,000 单位（厘米克秒制）的电容器的电势差提高到 10 伏特；在压力为 3 微米汞柱时，同样时间内电势差被提高到了 60 伏特。*

利用磁铁可以完全地抑制这种作用。

(5) 总的来说，把阴极射线看作是紫外线的假说似乎不太容易解释这些实验结果。另一方面，这些实验结果与认为阴极射线是一种物质辐射的假说符合得很好，在我看来，这些实验结果恐怕只能这么解释。

* 真空管的爆裂使我暂时无法再在低压条件下研究这一现象。

In the neighbourhood of the kathode, the electric field is sufficiently intense to break into pieces (*into ions*) certain of the molecules of the residual gas. The negative ions move towards the region where the potential is increasing, acquire a considerable speed, and form the kathode rays; their electric charge, and consequently their mass (at the rate of one valence-gramme for 100,000 Coulombs) is easily measurable. The positive ions move in the opposite direction; they form a diffused brush, sensitive to the magnet, and not a radiation in the correct sense of the word.*

(**53**, 298-299; 1896)

* This work has been carried out in the laboratory of the Normal School, and in that of M. Pellat at the Sorbonne.

在阴极附近，电场强度强到足以把一定量的残余气体分子打成碎片（**变成离子**）。负离子向着电势增加的方向运动，速度很大，这形成了阴极射线，它们的电量很容易被测定，从而其质量（100,000 库仑对应 1 克当量）也很容易得到。正离子向相反方向运动，形成一个发散的尾巴，对磁场非常敏感，准确地说这就不是辐射了。*

（王锋 翻译；江丕栋 审稿）

* 这项研究是在师范学院的实验室和索邦大学佩拉的实验室中进行的。

The Effect of Magnetisation on the Nature of Light Emitted by a Substance*

P. Zeeman

Editor's Note

Does a magnetic field influence the light emitted by an atom? Here Pieter Zeeman reports the first evidence that it does. Zeeman heated sulphur in a ceramic chamber with transparent ends, and placed the chamber in a magnetic field. With the light of an arc lamp, he then measured the absorption spectrum and found a broadening of certain lines, attributing this to a change in the frequency of the absorbed light. Zeeman noted that the polarization of light emitted in the presence of a field behaves as predicted by Lorentz, owing to the circular motion of charged particles within the atom. He estimates the charge/mass ratio for these particles as being about 10^7.

IN consequence of my measurements of Kerr's magneto-optical phenomena, the thought occurred to me whether the period of the light emitted by a flame might be altered when the flame was acted upon by magnetic force. It has turned out that such an action really occurs. I introduced into an oxyhydrogen flame, placed between the poles of a Ruhmkorff's electromagnet, a filament of asbestos soaked in common salt. The light of the flame was examined with a Rowland's grating. Whenever the circuit was closed both D lines were seen to widen.

Since one might attribute the widening to the known effects of the magnetic field upon the flame, which would cause an alteration in the density and temperature of the sodium vapour, I had resort to a method of experimentation which is much more free from objection.

Sodium was strongly heated in a tube of biscuit porcelain, such as Pringsheim used in his interesting investigations upon the radiations of gases. The tube was closed at both ends by plane parallel glass plates, whose effective area was 1 cm. The tube was placed horizontally between the poles, at right angles to the lines of force. The light of an arc lamp was sent through. The absorption spectrum showed both D lines. The tube was continuously rotated round its axis to avoid temperature variations. Excitation of the magnet caused immediate widening of the lines. It thus appears very probable that the period of sodium light is altered in the magnetic field. It is remarkable that Faraday, as early as 1862, had made the first recorded experiment in this direction, with the incomplete resources of that period, but with a negative result (Maxwell, "Collected Works", vol. II, p. 790).

---

* Translated by Arthur Stanton from the *Proceedings of the Physical Society of Berlin*.

磁化对物质发射的光的性质的影响*

塞曼

编者按

磁场会影响原子发射的光吗？彼得·塞曼在这篇报告中首次证明这种效应是存在的。塞曼在两端透明的陶瓷真空室中加热硫黄，并把这个真空室放入磁场中。在弧光灯的照射下，他测量了吸收光谱并发现某些特定的谱线出现了加宽的现象，他把这归因于被吸收光线的频率的改变。塞曼特别提到，有场存在时发射出来的光的偏振与洛伦兹预言的一样，是由原子内带电粒子的圆周运动产生的。他估计这些粒子的荷质比约为 10^7。

我在对克尔磁光效应进行测量时突然产生了这样的想法：当磁力作用于火焰时，火焰发射出的光的周期是否会发生变化。结果证实这样的作用确实存在。我把浸泡在普通食盐中的石棉丝放在置于鲁姆科夫电磁体两极之间的氢氧焰中。火焰光用罗兰光栅检验。每当电路接通时都能看到两条 D 线的加宽。

鉴于也许有人会将谱线加宽归因于磁场对火焰的某种已知作用使钠蒸气的密度和温度发生了变化，我已采用了更加没有异议的实验方法进行了确证。

我们在素瓷管（与普林斯海姆在他著名的气体辐射实验中所用的一样）内对钠进行高温加热。管的两端用两块相互平行的平玻璃板密封，其有效区域为 1 厘米。该管被水平地置于两极之间，与磁力线垂直。弧光灯的光穿过其中，吸收光谱中显示出两条 D 线。管子不停地绕着它的轴自转以保持各处温度均衡，磁作用使谱线迅速加宽。很可能是因为钠光的周期在磁场中发生了变化。值得注意的是，这方面第一个有记录的实验是法拉第早在 1862 年进行的，那时的资源并不完备，得到的是阴性的结果（麦克斯韦，《文集》，第 2 卷，第 790 页）。

* 由阿瑟·斯坦顿翻译自《柏林物理学会会刊》。

It has been already stated what, in general, was the origin of my own research on the magnetisation of the lines in the spectrum. The possibility of an alteration of period was first suggested to me by the consideration of the accelerating and retarding forces between the atoms and Maxwell's molecular vortices; later came an example suggested by Lord Kelvin, of the combination of a quickly rotating system and a double pendulum. However, a true explanation appears to me to be afforded by the theory of electric phenomena propounded by Prof. Lorentz.

In this theory, it is considered that, in all bodies, there occur small molecular elements charged with electricity, and that all electrical processes are to be referred to the equilibrium or motion of these "ions". It seems to me that in the magnetic field the forces directly acting on the ions suffice for the explanation of the phenomena.

Prof. Lorentz, to whom I communicated my idea, was good enough to show me how the motion of the ions might be calculated, and further suggested that if my application of the theory be correct there would follow these further consequences: that the light from the edges of the widened lines should be circularly polarised when the direction of vision lay along the lines of force; further, that the magnitude of the effect would lead to the determination of the ratio of the electric charge the ion bears to its mass. We may designate the ratio e/m. I have since found by means of a quarter-wave length plate and an analyser, that the edges of the magnetically-widened lines are really circularly polarised when the line of sight coincides in direction with the lines of force. An altogether rough measurement gives 10^7 as the order of magnitude of the ratio e/m when e is expressed in electromagnetic units.

On the contrary, if one looks at the flame in a direction at right angles to the lines of force, then the edges of the broadened sodium lines appear plane polarised, in accordance with theory. Thus there is here direct evidence of the existence of ions.

This investigation was conducted in the Physical Institute of Leyden University, and will shortly appear in the "Communications of the Leyden University".

I return my best thanks to Prof. K. Onnes for the interest he has shown in my work.

(**55**, 347; 1897)

P. Zeeman: Amsterdam.

前面已经介绍了我对谱线磁化进行研究的起因。周期变化的可能性使我首先想到的是原子与麦克斯韦分子涡旋之间的加速和减速作用力；然后想到的是开尔文勋爵提出的一个快速旋转体系与双摆复合体的例子。然而，使我受到启发并最终得出正确结论的是洛伦兹教授提出的关于电现象的理论。

这个理论认为：在所有物体中，都存在小的、带电的分子单元，所有电的过程都与这些“离子”的平衡或运动有关。在我看来，只要认为在磁场中力直接作用于这些离子上，就足以解释这些现象。

我向洛伦兹教授阐述了我的观点，他友好地告诉我离子如何运动也许是可以计算的，并进一步建议说，如果我对该理论的应用是正确的，那么就会出现以下结果：当沿磁力线方向观察时，从加宽谱线边缘发出的光应该是圆偏振光；此外，这个效应的大小将能使人们测定离子所带电荷与其质量的比值。我们可以用 e/m 表示这个比值。后来我用四分之一波片和检偏器测量发现，当观测方向与磁力线一致时，磁场加宽谱线的边缘果然是圆偏振的。粗略的测定表明，如果用 e 来表示电磁单位，e/m 这一比值的数量级大约为 10^7。

反之，如果观察火焰的方向与磁力线垂直，加宽的钠线边缘出现的是平面偏振光，这与理论相符。这些都是离子存在的直接证据。

这项研究是在莱顿大学物理研究所进行的，不久之后研究报告将刊登在《莱顿大学学报》上。

非常感谢昂内斯教授对我的工作的重视。

（沈乃澂 翻译；赵见高 审稿）

An Undiscovered Gas

W. Ramsay

Editor's Note

Sir William Ramsay was professor of chemistry at University College London and was awarded a Nobel prize in 1904 for his discovery of four of the rare gases, helium (in concert with Lord Rayleigh), neon, argon and xenon. Ramsay's address to the Chemistry Section of the British Association for the Advancement of Science in 1897 was principally about his reasons for believing that neon would exist, but is also interesting because it illustrates how tentative were ideas about the periodic table of the elements.

A sectional address to members of the British Association falls under one of three heads. It may be historical, or actual, or prophetic; it may refer to the past, the present, or the future. In many cases, indeed in all, this classification overlaps. Your former presidents have given sometimes a historical introduction, followed by an account of the actual state of some branch of our science, and, though rarely, concluding with prophetic remarks. To those who have an affection for the past, the historical side appeals forcibly; to the practical man, and to the investigator engaged in research, the actual, perhaps, presents more charm; while to the general public, to whom novelty is often more of an attraction than truth, the prophetic aspect excites most interest. In this address I must endeavour to tickle all palates; and perhaps I may be excused if I take this opportunity of indulging in the dangerous luxury of prophecy, a luxury which the managers of scientific journals do not often permit their readers to taste.

The subject of my remarks today is a new gas. I shall describe to you later its curious properties; but it would be unfair not to put you at once in possession of the knowledge of its most remarkable property—it has not yet been discovered. As it is still unborn, it has not yet been named. The naming of a new element is no easy matter. For there are only twenty-six letters in our alphabet, and there are already over seventy elements. To select a name expressible by a symbol which has not already been claimed for one of the known elements is difficult, and the difficulty is enhanced when it is at the same time required to select a name which shall be descriptive of the properties (or want of properties) of the element.

It is now my task to bring before you the evidence for the existence of this undiscovered element.

It was noticed by Döbereiner, as long ago as 1817, that certain elements could be arranged in groups of three. The choice of the elements selected to form these triads was made on account of their analogous properties, and on the sequence of their atomic

一种尚未发现的气体

拉姆齐

编者按

威廉·拉姆齐爵士是伦敦大学学院的化学教授，他曾因发现了氦（与瑞利勋爵合作）、氖、氪和氙这4种稀有气体而荣获1904年的诺贝尔奖。拉姆齐于1897年向英国科学促进会化学分部作了报告，主要阐述了他认为氖存在的观点及理由，报告之所以吸引人还在于它表明了当时关于元素周期表的许多观点还是非常不确定的。

向英国科学促进会成员所作的部门性演讲，大体可归入以下三类中的一种。它可以是历史性的，现实性的或是预见性的；它可能会涉及过去、现在或将来。在很多情况下，实际上是在所有情况下，这些分类间都是有交集的。前任主席有时会先给出一段历史介绍，随后是一段关于某些分支学科现状的说明，接着，虽然很少见，还是会以预见性的评述作为结语。对那些热衷于过去的听众来说，历史性的一面似乎更吸引人；对于实干家以及研究人员来说，可能现实性的介绍更有吸引力；而对于广大民众来说，新奇事物比科学真理更有诱惑力，预言性的内容将引起更多关注。在这段演讲中我将竭尽全力满足所有人的需求；如果我有凭借诸位给予我的宽容而作出危险的狂妄预言，敬请大家原谅，这种狂妄是科学杂志的管理者们不常让读者们体验的。

今天我要谈的主题是一种新的气体。随后我将向各位介绍它的奇妙性质；但是，为了公平起见，我要让你们马上了解到它最不同寻常的性质——它至今尚未被发现。由于尚未问世，它也还没有名字。一种新元素的命名可不是件简单的事。因为在我们的字母表中只有26个字母，而目前已有70多种元素。选择一个从未被其他已知元素用过的符号作为名字是很困难的，同时还要使所选择的名字能描述该元素性质（或者是希望它所具备的性质），那就更困难了。

现在，我要将这种尚未被发现的元素存在的证据呈现给你们。

早在1817年，德贝赖纳就注意到，某些元素可以按照三个一组的方式进行排列。选择元素构成这样的三元素组，是以它们具有的相似的性质，以及它们的原子量大

weights, which had at that time only recently been discovered. Thus calcium, strontium, and barium formed such a group; their oxides, lime, strontia, and baryta are all easily slaked, combining with water to form soluble lime-water, strontia-water, and baryta-water. Their sulphates are all sparingly soluble, and resemblance had been noticed between their respective chlorides and between their nitrates. Regularity was also displayed by their atomic weights. The numbers then accepted were 20, 42.5, and 65; and the atomic weight of strontium, 42.5, is the arithmetical mean of those of the other two elements, for $(65+20)/2=42.5$. The existence of other similar groups of three was pointed out by Döbereiner, and such groups became known as "Döbereiner's triads".

Another method of classifying the elements, also depending on their atomic weights, was suggested by Pettenkofer, and afterwards elaborated by Kremers, Gladstone, and Cooke. It consisted in seeking for some expression which would represent the differences between the atomic weights of certain allied elements. Thus, the difference between the atomic weight of lithium, 7, and sodium, 23, is 16; and between that of sodium and of potassium, 39, is also 16. The regularity is not always so conspicuous; Dumas, in 1857, contrived a somewhat complicated expression which, to some extent, exhibited regularity in the atomic weights of fluorine, chlorine, bromine, and iodine; and also of nitrogen, phosphorus, arsenic, antimony and bismuth.

The upshot of these efforts to discover regularity was that, in 1864, Mr. John Newlands, having arranged the elements in eight groups, found that when placed in the order of their atomic weights, "the eighth element, starting from a given one, is a kind of repetition of the first, like the eighth note of an octave in music." To this regularity he gave the name "The Law of Octaves".

The development of this idea, as all chemists know, was due to the late Prof. Lothar Meyer, of Tübingen, and to Prof. Mendeléeff, of St. Petersburg. It is generally known as the "Periodic Law". One of the simplest methods of showing this arrangement is by means of a cylinder divided into eight segments by lines drawn parallel to its axis; a spiral line is then traced round the cylinder, which will, of course, be cut by these lines eight times at each revolution. Holding the cylinder vertically, the name and atomic weight of an element is written at each intersection of the spiral with a vertical line, following the numerical order of the atomic weights. It will be found, according to Lothar Meyer and Mendeléeff, that the elements grouped down each of the vertical lines form a natural class; they possess similar properties, form similar compounds, and exhibit a graded relationship between their densities, melting-points, and many of their other properties. One of these vertical columns, however, differs from the others, inasmuch as on it there are three groups, each consisting of three elements with approximately equal atomic weights. The elements in question are iron, cobalt, and nickel; palladium, rhodium, and ruthenium; and platinum, iridium, and osmium. There is apparently room for a fourth group of three elements in this column, and it may be a fifth. And the discovery of such a group is not unlikely, for when this table was first drawn up Prof. Mendeléeff drew attention to certain gaps, which have since been filled up by the discovery of gallium, germanium, and others.

小顺序（当时刚刚发现的）为依据的。据此，钙、锶和钡形成一个三元素组；它们的氧化物，石灰、锶土和重土都容易熟化，即与水结合形成可溶的石灰水、锶土水和重土水。它们的硫酸盐的溶解度都很小，而且它们各自的氯化物以及硝酸盐也都具有相似之处。它们的原子量也体现出了规律性。当时公认的数值分别为 20、42.5 和 65；而锶的原子量 42.5 是另外两种元素原子量的算术平均值，因为 (65+20)/2=42.5。德贝赖纳还指出了其他一些类似的三元素组的存在，它们后来被称为“德贝赖纳三元素组”。

另外一种根据原子量进行元素分类的方法是由佩滕科费尔提出的，后来克雷默斯、格拉德斯通和库克进行了详细阐述。这种方法是寻找某种表达式以描述某些相关元素的原子量之差。比如，锂的原子量是 7，它和钠的原子量 23 之间的差是 16；而钠和钾(原子量 39)的原子量之差也是 16。规律性并非总是这样显而易见；1857 年，杜马设计出一个稍显复杂的表达式，从某种程度上体现了氟、氯、溴和碘以及氮、磷、砷、锑和铋的原子量之间的规律性。

这些寻找规律性的努力的最终结果是，在 1864 年约翰 · 纽兰兹先生将元素分为 8 组，他发现当元素按照原子量的顺序排列时，“以某一特定元素为起始的 8 个元素形成一种周而复始的循环，犹如音乐中八度音阶中的 8 个音符。”他将这种规律性称为“八音律”。

正如所有化学家都知道的那样，这一观点的发展应当归功于图宾根大学已故的洛塔尔·迈耶尔教授和圣彼得堡大学的门捷列夫教授。这种规律通常被称为“周期律”。展示这种排列的最简单的方法之一，是利用一个被若干条与轴平行的直线分割为 8 部分的圆柱面；一条螺旋线环绕柱面，当然，它会在每一次环绕中被这些直线切割 8 次。将柱面垂直放置，元素的名称和原子量就写在螺旋线与垂线的每个交点上，前面还写着按原子量排列的序数。根据洛塔尔 · 迈耶尔和门捷列夫的观点，每条垂线上的元素组成一个自然类；它们表现出相似的性质，形成类似的化合物，并在密度、熔点和许多其他性质上体现出递变性。不过，在这些垂直列中有一个是与众不同的，因为其中包含 3 个组，每一组都由原子量几乎相同的元素组成。这些元素是铁、钴和镍；钯、铑和钌；铂、铱和锇。在这一纵列中显然还为第 4 个三元素组留有余地，可能还会有第 5 个。发现这样一个组并不是不可能的，因为早在这张表刚刚被草拟出来时，门捷列夫教授就注意到其中特定位置的空缺，这些空缺逐渐被后来发现的镓、锗等其他元素填充。

The discovery of argon at once raised the curiosity of Lord Rayleigh and myself as to its position in this table. With a density of nearly 20, if a diatomic gas, like oxygen and nitrogen, it would follow fluorine in the periodic table; and our first idea was that argon was probably a mixture of three gases, all of which possessed nearly the same atomic weights, like iron, cobalt, and nickel. Indeed, their names were suggested, on this supposition, with patriotic bias, as Anglium, Scotium, and Hibernium! But when the ratio of its specific heats had, at least in our opinion, unmistakably shown that it was molecularly monatomic, and not diatomic, as at first conjectured, it was necessary to believe that its atomic weight was 40, and not 20, and that it followed chlorine in the atomic table, and not fluorine. But here arises a difficulty. The atomic weight of chlorine is 35.5, and that of potassium, the next element in order in the table, is 39.1; and that of argon, 40, follows, and does not precede, that of potassium, as it might be expected to do. It still remains possible that argon, instead of consisting wholly of monatomic molecules, may contain a small percentage of diatomic molecules; but the evidence in favour of this supposition is, in my opinion, far from strong. Another possibility is that argon, as at first conjectured, may consist of a mixture of more than one element; but, unless the atomic weight of one of the elements in the supposed mixture is very high, say 82, the case is not bettered, for one of the elements in the supposed trio would still have a higher atomic weight than potassium. And very careful experiments, carried out by Dr. Norman Collie and myself, on the fractional diffusion of argon, have disproved the existence of any such element with high atomic weight in argon, and, indeed, have practically demonstrated that argon is a simple substance, and not a mixture.

The discovery of helium has thrown a new light on this subject. Helium, it will be remembered, is evolved on heating certain minerals, notably those containing uranium; although it appears to be contained in others in which uranium is not present, except in traces. Among these minerals are clèveite, monazite, fergusonite, and a host of similar complex mixtures, all containing rare elements, such as niobium, tantalum, yttrium, cerium, &c. The spectrum of helium is characterised by a remarkably brilliant yellow line, which had been observed as long ago as 1868 by Profs. Frankland and Lockyer in the spectrum of the sun's chromosphere, and named "helium" at that early date.

The density of helium proved to be very close to 2.0, and, like argon, the ratio of its specific heat showed that it, too, was a monatomic gas. Its atomic weight therefore is identical with its molecular weight, viz. 4.0, and its place in the periodic table is between hydrogen and lithium, the atomic weight of which is 7.0.

The difference between the atomic weights of helium and argon is thus 36, or 40 – 4. Now there are several cases of such a difference. For instance, in the group the first member of which is fluorine we have—

Fluorine	19	16.5
Chlorine	35.5	
Manganese	55	19.5

氩的发现立即激起了瑞利勋爵和我本人对它在周期表中位置的好奇心。它的密度大约是 20，如果是像氧气和氮气一样的双原子气体，那么在周期表中它应该是紧随氟元素之后；我们最初的想法是，氩很可能是 3 种气体的混合物，它们具有几乎相同的原子量，就像铁、钴和镍一样。实际上，基于这种假定，按照具有爱国主义倾向的方式来命名，应该把它们分别称作 Anglium，Scotiun 和 Hibernium。但是，当比热比准确无误地表明它是单原子分子而非最初所猜测的双原子分子时，我们只能相信它的原子量是 40 而不是 20，而它在周期表中的位置也应是位于氯之后，而不是氟之后——至少在我们看来是这样。不过这又产生了新的困难。氯的原子量是 35.5，周期表中下一个元素钾的原子量是 39.1；而氩的原子量是 40，并不像我们期待的那样比钾的小，而是比它大。还有可能氩并不完全是由单原子分子组成，而是包含一小部分的双原子分子；不过在我看来，支持这一假设的证据非常不足。另外一种可能性就是氩可能是多种元素的混合物，正如最初猜测的那样；但是，除非这个假定的混合物中某一种元素的原子量非常高，比如说有 82，否则情况也并不理想，因为在这个假定的三元素组中要有一种元素的原子量比钾的高。由诺曼 · 科利和我本人所做的非常严谨的关于氩分馏扩散的实验已经证明，在氩中不存在任何高原子量的元素，事实上，氩是一种纯净物而不是混合物。

氦的发现给这一问题带来了转机。我们应该记得，加热某些矿物，特别是那些含铀的矿物时，会释放出氦；尽管氦似乎也存在于另外一些并不含铀的矿物中，但是含量非常低。这些矿物包括钇铀矿、独居石、褐钇铌（钽）矿以及很多类似的复杂混合物，它们都含有诸如铌、钽、钇和铈等稀有元素。氦光谱的特征是有一条显著而明亮的黄线，弗兰克兰和洛克耶教授早在 1868 年就在太阳色球的光谱中观察到了这一点，并从那时起称它为“氦”。

经证实，氦的密度非常接近 2.0，而且，与氩一样，它的比热比表明它也是一种单原子气体。因此，它的原子量与分子量是一样的，也就是 4.0，而它在周期表中的位置则介于氢和原子量为 7.0 的锂之间。

因此，氦与氩的原子量之差为 40 − 4，也就是 36。这里还有几个差值与此相同的例子。例如，在以氟作为第一个成员的那组元素中，我们看到——

氟	19	16.5
氯	35.5	
锰	55	19.5

In the oxygen group—

Oxygen	16	16
Sulphur	32	
Chromium	52.3	20.3

In the nitrogen group—

Nitrogen	14	17
Phosphorus	31	
Vanadium	51.4	20.4

And in the carbon group—

Carbon	12	16.3
Silicon	28.3	
Titanium	48.1	19.8

These instances suffice to show that approximately the differences are 16 and 20 between consecutive members of the corresponding groups of elements. The total differences between the extreme members of the short series mentioned are—

Manganses – Fluorine	36
Chromium – Oxygen	36.3
Vanadium – Nitrogen	37.4
Titanium – Carbon	36.1

This is approximately the difference between the atomic weights of helium and argon, 36.

There should, therefore, be an undiscovered element between helium and argon, with an atomic weight 16 units higher than than of helium, and 20 units lower than that of argon, namely 20. And if this unknown element, like helium and argon, should prove to consist of monatomic molecules, then its density should be half its atomic weight, 10. And pushing the analogy still further, it is to be expected that this element should be as indifferent to union with other elements as the two allied elements.

My assistant, Mr. Morris Travers, has indefatigably aided me in a search for this unknown gas. There is a proverb about looking for a needle in a haystack; modern science, with the aid of suitable magnetic appliances, would, if the reward were sufficient, make short work of that proverbial needle. But here is a supposed unknown gas, endowed no doubt with negative properties, and the whole world to find it in. Still, the attempt had to be made.

We first directed our attention to the sources of helium—minerals. Almost every mineral which we could obtain was heated in a vacuum, and the gas which was evolved examined. The results are interesting. Most minerals give off gas when heated, and the gas contains, as a rule, a considerable amount of hydrogen, mixed with carbonic acid, questionable traces of nitrogen, and carbonic oxide. Many of the minerals, in addition, gave helium, which proved to be widely distributed, though only in minute proportion. One mineral—malacone—gave appreciable quantities of argon; and it is noteworthy that argon was not

在氧这一组中——

氧	16	16
硫	32	
铬	52.3	20.3

在氮这一组中——

氮	14	17
磷	31	
钒	51.4	20.4

而在碳这一组中——

碳	12	16.3
硅	28.3	
钛	48.1	19.8

这些实例足以表明，各组元素的连续成员之间的原子量之差大约都是 16 和 20。以上各组中两端元素的总的原子量之差为——

锰－氟	36
铬－氧	36.3
钒－氮	37.4
钛－碳	36.1

这与氦和氩的原子量之差 36 也是近似一致的。

可见，在氦与氩之间应该有一种尚未发现的元素，它的原子量比氦的原子量大 16 而比氩的原子量小 20，也就是 20。而且，如果这种未知元素像氦和氩一样被证实由单原子分子构成，那么它的密度应该是其原子量的一半——10。进一步类比，可以预期这种元素像它的两种同类元素一样不易与其他元素结合。

我的助手莫里斯 · 特拉弗斯先生一直毫不厌倦地帮助我研究这种未知气体。有句俗语叫大海捞针；如果报酬足够丰厚，那么在有合适的磁力仪器帮助下的现代科学将大大简化大海捞针的工作。但现在是要在全世界寻找一种假想的未知气体，而且无疑它还具有不利的性质。所以，仍然需要努力尝试。

我们首先将注意力集中在氦的来源——矿物上。我们几乎将每一种可能得到的矿物都在真空中加热，并检验释放出来的气体。结果很有趣，大多数矿物在加热时释放出气体，通常有大量的氢气，另外还混杂着碳酸，不太确定的痕量氮气，还有一氧化碳。此外，有很多种矿物释放出氦气，这证明它的分布很广，不过只占很小的比例。一种叫做水锆石的矿物能释放出大量的氩气；值得注意的是，除了这种矿石和一份陨铁标本以外，在其他物质中并没有发现氩，而且奇怪的是，在水锆石中

found except in it (and, curiously, in much larger amount than helium), and in a specimen of meteoric iron. Other specimens of meteoric iron were examined, but were found to contain mainly hydrogen, with no trace of either argon or helium. It is probable that the sources of meteorites might be traced in this manner, and that each could be relegated to its particular swarm.

Among the minerals examined was one to which our attention had been directed by Prof. Lockyer, named eliasite, from which he said that he had extracted a gas in which he had observed spectrum lines foreign to helium. He was kind enough to furnish us with a specimen of this mineral, which is exceedingly rare, but the sample which we tested contained nothing but undoubted helium.

During a trip to Iceland in 1895, I collected some gas from the boiling springs there; it consisted, for the most part, of air, but contained somewhat more argon than is usually dissolved when air is shaken with water. In the spring of 1896 Mr. Travers and I made a trip to the Pyrenees to collect gas from the mineral springs of Cauterets, to which our attention had been directed by Dr. Bouchard, who pointed out that these gases are rich in helium. We examined a number of samples from the various springs, and confirmed Dr. Bouchard's results, but there was no sign of any unknown lines in the spectrum of these gases. Our quest was in vain.

We must now turn to another aspect of the subject. Shortly after the discovery of helium, its spectrum was very carefully examined by Profs. Runge and Paschen, the renowned spectroscopists. The spectrum was photographed, special attention being paid to the invisible portions, termed the "ultra-violet" and "infra-red". The lines thus registered were found to have a harmonic relation to each other. They admitted of division into two sets, each complete in itself. Now, a similar process had been applied to the spectrum of lithium and to that of sodium, and the spectra of these elements gave only one series each. Hence, Profs. Runge and Paschen concluded that the gas, to which the provisional name of helium had been given, was, in reality, a mixture of two gases, closely resembling each other in properties. As we know no other elements with atomic weights between those of hydrogen and lithium, there is no chemical evidence either for or against this supposition. Prof. Runge supposed that he had obtained evidence of the separation of these imagined elements from each other by means of diffusion; but Mr. Travers and I pointed out that the same alteration of spectrum, which was apparently produced by diffusion, could also be caused by altering the pressure of the gas in the vacuum tube; and shortly after Prof. Runge acknowledged his mistake.

These considerations, however, made it desirable to subject helium to systematic diffusion, in the same way as argon had been tried. The experiments were carried out in the summer of 1896 by Dr. Collie and myself. The result was encouraging. It was found possible to separate helium into two portions of different rates of diffusion, and consequently of different density by this means. The limits of separation, however, were not very great. On the one hand, we obtained gas of a density close on 2.0; and on the other, a sample of

氩含量比氦高很多。在检验其他陨铁标本时，却发现它们主要含氢，没有一丁点氩或氦。也许利用这种方式能够追溯陨星的来源，并将每一个都归入其所在的特定门类中。

我们的注意力被洛克耶教授拉到一种被检验过的名叫脂铅铀矿的矿物上，洛克耶教授说他已经从中提取出一种具有和氦不同的光谱线的气体。他还非常友好地为我们提供了这种极为稀有的矿物的一份样本，但是我们检验的样本中只含有氦。

1895 年在冰岛旅行期间，我从那里的沸泉中收集了一些气体；这些气体主要就是空气，但是其中所含的氩却比通常把水放在空气中振荡时水中所溶解的氩多一些。1896 年春，特拉弗斯先生和我前往比利牛斯山收集科特雷矿泉水中的气体；这一次我们接受了布沙尔博士的指点，他指出这种气体中富含氦气。我们检验了来自不同泉水的大量样本，证实了布沙尔博士的结论，但是这些气体的光谱中没有任何未知谱线的迹象。我们的探索仍然一无所获。

现在我们必须转向问题的另一个方面。在发现氦之后不久，著名光谱学家龙格教授和帕邢教授就非常细致地检验了它的光谱。他们对光谱照了相，并特别注意到其中的不可见部分，人们称之为“紫外区”和“红外区”。他们发现这样记录的谱线彼此间具有谐波关系。它们可以被分为两个子集，各自是一个完整的体系。现在，对锂和钠的光谱也进行了类似的研究，但这两种元素的谱图中每一种只出现了一个谱线系列。由此，龙格教授和帕邢教授得出结论认为，被我们称为氦的这种气体实际上是两种性质极为相似的气体的混合物。正如我们所知，没有其他元素的原子量介于氢和锂之间，因而没有化学上的证据支持或反对这一假设。龙格教授认为他已经得到了通过扩散方法分离这些假想元素的证据；但是特拉弗斯先生和我认为，看似由扩散产生的谱图变化，也可能是由真空管中气体压强的变化引起的；不久之后龙格教授承认了他的错误。

不过，这些研究使人们想到可以用系统扩散的方法研究氦，就像曾经研究氩时所尝试过的一样。1896 年夏天，科利博士和我进行了实验，结果是令人鼓舞的。我们发现,这种方法可以将氦分离为扩散速率不同从而具有不同密度的两个部分。不过，分离的极限还不是很大。我们一方面得到了一种密度接近于 2.0 的气体，另一方面

density 2.4 or thereabouts. The difficulty was increased by the curious behaviour, which we have often had occasion to confirm, that helium possesses a rate of diffusion too rapid for its density. Thus, the density of the lightest portion of the diffused gas, calculated from its rate of diffusion, was 1.874; but this corresponds to a real density of about 2.0. After our paper, giving an account of these experiments, had been published, a German investigator, Herr A. Hagenbach, repeated our work and confirmed our results.

The two samples of gas of different density differ also in other properties. Different transparent substances differ in the rate at which they allow light to pass through them. Thus, light travels through water at a much slower rate than through air, and at a slower rate through air than through hydrogen. Now Lord Rayleigh found that helium offers less opposition to the passage of light than any other substance does, and the heavier of the two portions into which helium had been split offered more opposition than the lighter portion. And the retardation of the light, unlike what has usually been observed, was nearly proportional to the densities of the samples. The spectrum of these two samples did not differ in the minutest particular; therefore it did not appear quite out of the question to hazard the speculation that the process of diffusion was instrumental, not necessarily in separating two kinds of gas from each other, but actually in removing light molecules of the same kind from heavy molecules. This idea is not new. It had been advanced by Prof. Schützenberger (whose recent death all chemists have to deplore), and later, by Mr. Crookes, that what we term the atomic weight of an element is a mean; that when we say the atomic weight of oxygen is 16, we merely state that the average atomic weight is 16; and it is not inconceivable that a certain number of molecules have a weight somewhat higher than 32, while a certain number have a lower weight.

We therefore thought it necessary to test this question by direct experiment with some known gas; and we chose nitrogen, as a good material with which to test the point. A much larger and more convenient apparatus for diffusing gases was built by Mr. Travers and myself, and a set of systematic diffusions of nitrogen was carried out. After thirty rounds, corresponding to 180 diffusions, the density of the nitrogen was unaltered, and that of the portion which should have diffused most slowly, had there been any difference in rate, was identical with that of the most quickly diffusing portion—*i.e.* with that of the portion which passed first through the porous plug. This attempt, therefore, was unsuccessful; but it was worth carrying out, for it is now certain that it is not possible to separate a gas of undoubted chemical unity into portions of different density by diffusion. And these experiments rendered it exceedingly improbable that the difference in density of the two fractions of helium was due to separation of light molecules of helium from heavy molecules.

The apparatus used for diffusion had a capacity of about two litres. It was filled with helium, and the operation of diffusion was carried through thirty times. There were six reservoirs each full of gas, and each was separated into two by diffusion. To the heavier portion of one lot, the lighter portion of the next was added, and in this manner all six reservoirs were successfully passed through the diffusion apparatus. This process was

还得到一种密度在2.4左右的样品。我们常常发现氦的扩散速率相对于它的密度来说实在太快了，这种奇怪的现象也增加了研究的难度。由此，利用扩散速率计算出扩散所得气体中最轻部分的密度为1.874，但对应的真实密度却大约是2.0。在我们介绍上述实验的论文发表之后，一位德国研究者，哈根巴赫先生重复了我们的工作并确认了这一结果。

这两种密度不同的气体样品在其他性质上也不相同。光穿过不同透明物体的速率是不同的。因此，光在水中的传播速率比在空气中低很多，而在空气中的传播速率又比在氢气中低。瑞利勋爵发现，氦对光透过的阻碍比其他任何物质都小，而通过分离氦所得到的两个组分中，较重的部分对光透过的阻碍比较轻的部分大一些。与通常所观测到的不同，对光的阻碍几乎正比于样品的密度。两种样品的光谱在最微小的细节上也毫无区别；由此，看起来我们可以大胆地认为，扩散过程是起作用的，但不一定有助于将两种气体彼此分开，而实际上是对将同一种气体中较轻的分子从较重的分子中分离出去有帮助。这并不是一个全新的观点。最先由舒岑贝热教授（最近他的去世令所有化学家感到痛惜），还有后来的克鲁克斯先生提出，我们所谓的某种元素的原子量是平均值；当我们说氧的原子量是16时，我们只是指出氧的平均原子量是16；不难想象，有一定数量分子的重量比32稍高，同时还有一定数量分子的重量略低一些。

由此，我们认为，必须用某些已知气体直接进行实验来验证这一问题；我们选择了氮气这种合适的物质来检验这一观点。特拉弗斯和我建立了更大、更方便的用于气体扩散的装置，并对氮气进行了一组系统扩散。经过30轮相当于180次扩散之后，氮气的密度没有变化，如果扩散速率确实有所不同的话，那么扩散速率最慢的组分与扩散速率最快的组分——即最先通过多孔塞的部分——密度是一样的。可见，这次的努力是不成功的；不过值得尝试，因为现在可以确定，不可能通过扩散将化学上确实均一的一种气体分离为几个密度不同的部分。这些实验证明，氦的两个组分的密度差异不可能是由较轻的氦分子与较重的分子分离引起的。

用于扩散实验的装置的容积大约是2升。向装置中充满氦气，进行30次扩散操作。共有6个充满气体的储气槽，每个储气槽中的气体通过扩散分离为两部分。一个储气槽中较轻的气体被加入到前一个储气槽中较重的气体中，全部6个储气槽中的气体以这种方式连续地通过扩散装置。将这一过程进行30次，每一次每个储气槽中的

carried out thirty times, each of the six reservoirs having had its gas diffused each time, thus involving 180 diffusions. After this process, the density of the more quickly diffusing gas was reduced to 2.02, while that of the less quickly diffusing had increased to 2.27. The light portion on re-diffusion hardly altered in density, while the heavier portion, when divided into three portions by diffusion, showed a considerable difference in density between the first third and the last third. A similar set of operations was carried out with a fresh quantity of helium, in order to accumulate enough gas to obtain a sufficient quantity for a second series of diffusions. The more quickly diffusing portions of both gases were mixed and re-diffused. The density of the lightest portion of these gases was 1.98; and after other 15 diffusions, the density of the lightest portion had not decreased. The end had been reached; it was not possible to obtain a lighter portion by diffusion. The density of the main body of this gas is therefore 1.98; and its refractivity, air being taken as unity, is 0.1245. The spectrum of this portion does not differ in any respect from the usual spectrum of helium.

As re-diffusion does not alter the density or the refractivity of this gas, it is right to suppose that either one definite element has now been isolated; or that if there are more elements than one present, they possess the same, or very nearly the same, density and refractivity. There may be a group of elements, say three, like iron, cobalt, and nickel; but there is no proof that this idea is correct, and the simplicity of the spectrum would be an argument against such a supposition. This substance, forming by far the larger part of the whole amount of the gas, must, in the present state of our knowledge, be regarded as pure helium.

On the other hand, the heavier residue is easily altered in density by re-diffusion, and this would imply that it consists of a small quantity of a heavy gas mixed with a large quantity of the light gas. Repeated re-diffusion convinced us that there was only a very small amount of the heavy gas present in the mixture. The portion which contained the largest amount of heavy gas was found to have the density 2.275, and its refractive index was found to be 0.1333. On re-diffusing this portion of gas until only a trace sufficient to fill a Plücker's tube was left, and then examining the spectrum, no unknown lines could be detected, but, on interposing a jar and spark gap, the well-known blue lines of argon became visible; and even without the jar the red lines of argon, and the two green groups were distinctly visible. The amount of argon present, calculated from the density, was 1.64 percent, and from the refractivity 1.14 percent. The conclusion had therefore to be drawn that the heavy constituent of helium, as it comes off the minerals containing it, is nothing new, but, so far as can be made out, merely a small amount of argon.

If, then, there is a new gas in what is generally termed helium, it is mixed with argon, and it must be present in extremely minute traces. As neither helium nor argon has been induced to from compounds, there does not appear to be any method, other than diffusion, for isolating such a gas, if it exists, and that method has failed in our hands to give any evidence of the existence of such a gas. It by no means follows that the gas does not exist; the only conclusion to be drawn is that we have not yet stumbled on the material

气体都进行扩散，一共进行 180 次扩散。经过上述过程后，扩散速率较快的气体密度下降到 2.02，而扩散较慢的则升高到 2.27。重复扩散所得的较轻组分密度很难改变，而较重组分通过扩散分成了三部分，第一部分与第三部分的密度表现出相当大的差异。为了收集足够多的气体以进行第二系列的扩散，又对一些新制的氦气进行了一系列类似的操作。最后，两批气体中扩散较快的组分被混在一起并进行了重复扩散。其中最轻组分的气体密度为 1.98；再进行 15 次扩散后，最轻组分的气体密度也没有减小。这已经达到了极限；不可能通过扩散得到更轻的组分了。因此这种气体的主要部分的密度就是 1.98；而若将气体视为均一的话，它的折射率就是 0.1245。这个组分的光谱与通常的氦光谱从任何方面看都没有差别。

重复扩散并没有改变这种气体的密度和折射率，由此可知，不管是现在已分离出一种确定的元素，还是其中含有更多的元素，它们都有相同或近乎相同的密度和折射率。也许会有一组元素，比如像铁、钴和镍这样的三种元素；但是没有证据表明这种观点是正确的，而光谱的简单性也不支持这种假设。从我们目前的认识水平来看，占全部气体中一大部分的这种物质应该是纯净的氦气。

另一方面，这种较重的残余物很容易通过重复扩散改变其密度，这意味着它是由少量的重气体和大量的轻气体组成的混合物。不断重复扩散后得到的结果使我们确信混合物中只存在极少量的重气体。含有最多重气体部分的密度是 2.275，其折射率为 0.1333。将这部分气体继续进行重复扩散，直到剩余的量只够充满一个普吕克管为止，接着检验其光谱，没有检测到未知谱线，不过将其装入广口瓶中再放入放电器，就可以看到众所周知的氩特有的蓝色谱线；即使没有装入广口瓶，氩的红色谱线以及两组绿色谱线也是明显可见的。由密度计算出的氩的含量为 1.64%，而由折射率计算得到的是 1.14%。由此可以得出结论，在我们目前的认识水平下，从含氦矿物中释放出来的氦气中的较重组分并不是什么新物质，而只是少量的氩气。

那么，如果在我们通常所称的氦气中存在新的气体，那它应该是混在氩气中的，而且是痕量的。如果这种气体存在的话，由于尚不能使氦气或氩气形成化合物，看起来除了扩散之外也没有任何其他方法能够分离出这种气体，而扩散的方法又无法给出任何关于这种气体存在的证据。但这决不能说明该气体不存在；唯一可以得出

which contains it. In fact, the haystack is too large and the needle too inconspicuous. Reference to the periodic table will show that between the elements aluminium and indium there occurs gallium, a substance occurring only in the minutest amount on the earth's surface; and following silicon, and preceding tin, appears the element germanium, a body which has as yet been recognised only in one of the rarest of minerals, argyrodite. Now, the amount of helium in fergusonite, one of the minerals which yields it in reasonable quantity, is only 33 parts by weight in 100,000 of the mineral; and it is not improbable that some other mineral may contain the new gas in even more minute proportion. If, however, it is accompanied in its still undiscovered source by argon and helium, it will be a work of extreme difficulty to effect a separation from these gases.

In these remarks it has been assumed that the new gas will resemble argon and helium in being indifferent to the action of reagents, and in not forming compounds. This supposition is worth examining. In considering it, the analogy with other elements is all that we have to guide us.

We have already paid some attention to several triads of elements. We have seen that the differences in atomic weights between the elements fluorine and manganese, oxygen and chromium, nitrogen and vanadium, carbon and titanium, is in each case approximately the same as that between helium and argon, viz. 36. If elements further back in the periodic table be examined, it is to be noticed that the differences grow less, the smaller the atomic weights. Thus, between boron and scandium, the difference is 33; between beryllium (glucinum) and calcium, 31; and between lithium and potassium, 32. At the same time, we may remark that the elements grow liker each other, the lower the atomic weights. Now, helium and argon are very like each other in physical properties. It may be fairly concluded, I think, that in so far they justify their position. Moreover, the pair of elements which show the smallest difference between their atomic weights is beryllium and calcium; there is a somewhat greater difference between lithium and potassium. And it is in accordance with this fragment of regularity that helium and argon show a greater difference. Then again, sodium, the middle element of the lithium triad, is very similar in properties both to lithium and potassium; and we might, therefore, expect that the unknown element of the helium series should closely resemble both helium and argon.

Leaving now the consideration of the new element, let us turn our attention to the more general question of the atomic weight of argon, and its anomalous position in the periodic scheme of the elements. The apparent difficulty is this: The atomic weight of argon is 40; it has no power to form compounds, and thus possesses no valency; it must follow chlorine in the periodic table, and precede potassium; but its atomic weight is greater than that of potassium, whereas it is generally contended that the elements should follow each other in the order of their atomic weights. If this contention is correct, argon should have an atomic weight smaller than 40.

Let us examine this contention. Taking the first row of elements, we have:

的结论是，我们还没有碰到过包含它的物质。事实上，大海太宽广了，而针则太不显眼了。周期表显示在铝和铟元素之间还应该存在镓，它是一种在地壳中含量极少的物质；而位于硅元素之后锡元素之前的是锗——一种至今只在硫银锗矿（最稀有的矿物之一）中发现过的元素。褐钇铌（钽）矿是可以产生适量氦气的矿物之一，当这种矿物的总重量为100,000时，氦只占其中的33；其他一些矿物中所含新气体的比例可能更低。然而，如果在尚未发现的来源中新气体也是与氩和氦共存的话，那么把它从这些气体中分离出来将是一项极其艰巨的任务。

在上面的评述中我们假定，新气体与氩和氦类似，对于试剂反应呈惰性，不形成化合物。这一假定尚有待检验。在考虑这一问题时，与其他元素的类比是我们唯一的指导思想。

我们已经关注了若干个三元素组。我们看到，氟与锰、氧与铬、氮与钒、碳与钛这些元素之间的原子量差值和氦与氩之间的差值是近似相等的，大约都是36。如果进一步检验周期表中靠后的元素，可以看到原子量越小，其差值越小。硼与钪之间的原子量差值为33；铍与钙之间的差值为31；而锂与钾之间的差值则是32。同时，我们也可以说，元素之间越相似，其原子量的差值越小。在物理性质上，氦和氩非常相似。我认为，在此范围内可以适当地得出结论，它们的性质证实了它们位置的正确性。此外，原子量差值最小的元素对是铍和钙；锂和钾之间的差值则更大一些。与这一规律相吻合的是，氦和氩之间存在更大的原子量差值。而且，锂所在的三元素组中的中间元素钠的性质与锂和钾颇为相似；由此，我们就可以预期，位于氦所在系列中的这一未知元素的性质与氦和氩应该非常相似。

暂时不考虑新元素的问题，让我们将注意力转向更一般性的问题，即氩的原子量以及它在元素周期表中的反常位置。我们面临的困难显然是这样的：氩的原子量是40；它没有形成化合物的能力，由此表现为没有化合价；在周期表中它必须紧随氯元素之后而位于钾元素之前；但是它的原子量比钾的原子量大，而一般认为元素应该按照其原子量的顺序依次排列。如果这一观点是正确的，氩的原子量就应该小于40。

让我们检验一下这种观点。考察第一行元素，我们得到：

Li = 7, Be = 9.8, B = 11, C = 12, N = 14, O = 16, F = 19, ? = 20.

The differences are:

2.8, 1.2, 1.0, 2.0, 2.0, 3.0, 1.0.

It is obvious that they are irregular. The next row shows similar irregularities. Thus:

(? = 20), Na = 23, Mg = 24.3, Al = 27, Si = 28, P = 31, S = 32,
Cl = 35.5, A = 40.

And the differences:

3.0, 1.3, 2.7, 1.0, 3.0, 1.0, 3.5, 4.5.

The same irregularity might be illustrated by a consideration of each succeeding row. Between argon and the next in order, potassium, there is a difference of −0.9; that is to say, argon has a higher atomic weight than potassium by 0.9 unit; whereas it might be expected to have a lower one, seeing that potassium follows argon in the table. Further on in the table there is a similar discrepancy. The row is as follows:

Ag = 108, Cd = 112, In = 114, Sn = 119, Sb = 120.5,
Te = 127.7, I = 127.

The differences are:

4.0, 2.0, 5.0, 1.5, 7.2, −0.7.

Here, again, there is a negative difference between tellurium and iodine. And this apparent discrepancy has led to many and careful redeterminations of the atomic weight of tellurium. Prof. Brauner, indeed, has submitted tellurium to methodical fractionation, with no positive results. All the recent determinations of its atomic weight give practically the same number, 127.7.

Again, there have been almost innumerable attempts to reduce the differences between the atomic weights to regularity, by contriving some formula which will express the numbers which represent the atomic weights, with all their irregularities. Needless to say, such attempts have in no case been successful. Apparent success is always attained at the expense of accuracy, and the numbers reproduced are not those accepted as the true atomic weights. Such attempts, in my opinion, are futile. Still, the human mind does not rest contented in merely chronicling such an irregularity; it strives to understand why such an irregularity should exist. And, in connection with this, there are two matters which call for our consideration. These are: Does some circumstance modify these "combining proportions" which we term "atomic weights"? And is there any reason to suppose that we can modify them at our will? Are they true "constants of nature", unchangeable, and once for all determined? Or are they constant merely so long as other circumstances, a change in which would modify them, remain unchanged?

Li = 7，Be = 9.8，B = 11，C = 12，N = 14，O = 16，F = 19，？ = 20。

其差值为：

2.8，1.2，1.0，2.0，2.0，3.0，1.0。

很显然，这是不规则的。下一行显示出类似的不规则性。它们是：

（？ = 20），Na = 23，Mg = 24.3，Al = 27，Si = 28，P = 31，S = 32，
Cl = 35.5，A = 40。

而其差值为：

3.0，1.3，2.7，1.0，3.0，1.0，3.5，4.5。

考察后面每一行都可以发现同样的不规则性。氩与紧随其后的钾之间的原子量差值为 –0.9；也就是说，氩的原子量比钾大 0.9 个单位；但鉴于在周期表中钾位于氩之后，氩的原子量应该低于钾的原子量。在表中还有类似的矛盾。如此行所示：

Ag = 108，Cd = 112，In = 114，Sn = 119，Sb = 120.5，
Te = 127.7，I = 127。

其差值为：

4.0，2.0，5.0，1.5，7.2，–0.7。

这里，在碲与碘之间再次出现了负的原子量差值。这一明显的矛盾引发了很多谨慎地重测碲原子量的工作。确实，布劳纳教授已经对碲进行了系统的分馏，但是没有得到预期的结果。近来所有对碲的原子量进行测定的工作都得出了同样的结果——127.7。

另外，为了归纳出原子量差值所具有的规律性，人们进行了几乎无数次的尝试工作，设计某种可以反映原子量数值的不规律性的公式。不用说，这些尝试都没有成功。一些表面上的成功总是通过牺牲精确性来获得的，公式中得出的数值并不是我们公认的原子量的真实值。在我看来，这些努力是徒劳的。当然，人类的头脑不会仅仅满足于记述这一不规律性,而是要努力理解这种不规律性为什么会存在。并且，有两件与此相关的事需要考虑，即，是否有某种条件可以改变被我们称为“原子量”的“组合比例”？以及是否有理由使我们相信我们可以按照自己的意愿改变它们？它们是被一次性确定了的真正不可改变的“自然常量”，还是只在其他条件不变时才保持恒定，而其中任一条件的改变都将使它们发生变化呢？

In order to understand the real scope of such questions, it is necessary to consider the relation of the "atomic weights" to other magnitudes, and especially to the important quantity termed "energy".

It is known that energy manifests itself under different forms, and that one form of energy is quantitatively convertible into another form, without loss. It is also known that each form of energy is expressible as the product of two factors, one of which has been termed the "intensity factor", and the other the "capacity factor". Prof. Ostwald, in the last edition of his "Allgemeine Chemie", classifies some of these forms of energy as follows:

Kinetic energy is the product of			Mass into the square of velocity.
Linear	...	...	Length into force.
Surface	...	...	Surface into surface tension.
Volume	...	...	Volume into pressure.
Heat	...	...	Heat capacity (entropy) into temperature.
Electrical	...	...	Electrical capacity into potential.
Chemical	...	...	"Atomic weight" into affinity.

In each statement of factors, the "capacity factor" is placed first, and the "intensity factor" second.

In considering the "capacity factors", it is noticeable that they may be divided into two classes. The two first kinds of energy, kinetic and linear, are *independent of the nature of the material* which is subject to the energy. A mass of lead offers as much resistance to a given force, or, in other words, possesses as great inertia as an equal mass of hydrogen. A mass of iridium, the densest solid, counterbalances an equal mass of lithium, the lightest known solid. On the other hand, surface energy deals with molecules, and not with masses. So does volume energy. The volume energy of two grammes of hydrogen, contained in a vessel of one litre capacity, is equal to that of thirty-two grammes of oxygen at the same temperature, and contained in a vessel of equal size. Equal masses of tin and lead have not equal capacity for heat; but 119 grammes of tin has the same capacity as 207 grammes of lead; that is, equal atomic masses have the same heat capacity. The quantity of electricity conveyed through an electrolyte under equal difference of potential is proportional, not to the mass of the dissolved body, but to its equivalent; that is, to some simple fraction of its atomic weight. And the capacity factor of chemical energy is the atomic weight of the substance subjected to the energy. We see, therefore, that while mass or inertia are important adjuncts of kinetic and linear energies, all other kinds of energy are connected with atomic weights, either directly or indirectly.

Such considerations draw attention to the fact that quantity of matter (assuming that there exists such a carrier of properties as we term "matter") need not necessarily be measured by its inertia, or by gravitational attraction. In fact the word "mass" has two totally distinct significations. Because we adopt the convention to measure quantity of matter by its mass, the word "mass" has come to denote "quantity of matter." But it is open to any one to

为了理解这些问题的真正内涵，有必要考虑“原子量”与其他参量之间的关联，尤其是与我们称为“能量”的这个重要参量之间的关系。

我们知道能量以不同形式存在，并且一种形式的能量可以毫无损耗地转化为另一种形式的能量。我们还知道，每一种形式的能量都可以表示为两个因子的乘积，其中一个称为“强度因子”，另一个称为“容量因子”。奥斯特瓦尔德教授在他最新版的《普通化学》中，对几种形式的能量进行了如下分类：

动能	是	质量与速率平方的	乘积。
线性能	是	长度与力的	乘积。
表面能	是	表面与表面张力的	乘积。
体积能	是	体积与压力的	乘积。
热能	是	热容量（熵）与温度的	乘积。
电能	是	电容与电势的	乘积。
化学能	是	“原子量”与亲合力的	乘积。

在对各因子的每一组表述中，“容量因子”放在前面，而“强度因子”放在后面。

考察一下“容量因子”，不难看出它们可以分为两类。前两种能量，动能和线性能，**不依赖于物质的属性**（它是受能量支配的）。一定质量的铅与等质量的氢对给定的力产生相同的反作用力，换言之，它们具有同样大的惯性。一定质量的铱（密度最大的固体）与等质量的锂（已知最轻的固体）也能相平衡。另一方面，表面能所涉及的是分子而不是质量，体积能也是如此。放置在一个容量为 1 升的容器中的 2 克氢，与相同温度下放置在相同体积的容器中的 32 克氧具有相同的体积能。等质量的锡和铅的热容量并不相同；但 119 克锡与 207 克铅的热容量是相同的；也就是说，相同的原子数量对应相同的热容量。同一电势差，在电解质溶液中传输的电量正比于溶质的当量而不是其质量；也就是正比于其原子量的某个简单百分比。化学能的容量因子则是该能量所支配物质的原子量。由此我们看到，因为质量或惯性是动能与线性能的重要附属条件，所以其他各种类型的能量都直接或间接地与原子量相关联。

这些考虑使我们注意到物质的量（假定对于我们所说的“物质”的性质确实存在这样一个载体）不一定非要通过其惯性或重力来测定。实际上，“质量”这个词具有两方面完全不同的含义。由于我们习惯于通过质量测量物质的多少，“质量”这个词逐渐用以表示“物质的量”。但是，任何其他的能量因子也都可以用来测定物质

measure a quantity of matter by any other of its energy factors. I may, if I choose, state that those quantities of matter which possess equal capacities for heat are equal; or that "equal numbers of atoms" represent equal quantities of matter. Indeed, we regard the value of material as due rather to what it can do, than to its mass; and we buy food, in the main, on an atomic, or perhaps, a molecular basis, according to its content of albumen. And most articles depend for their value on the amount of food required by the producer or the manufacturer.

The various forms of energy may therefore be classified as those which can be referred to an "atomic" factor, and those which possess a "mass" factor. The former are in the majority. And the periodic law is the bridge between them; and yet, an imperfect connection. For the atomic factors, arranged in the order of their masses, display only a partial regularity. It is undoubtedly one of the main problems of physics and chemistry to solve this mystery. What the solution will be is beyond my power of prophecy; whether it is to be found in the influence of some circumstance on the atomic weights, hitherto regarded as among the most certain "constants of nature"; or whether it will turn out that mass and gravitational attraction are influenced by temperature, or by electrical charge, I cannot tell. But that some means will ultimately be found of reconciling these apparent discrepancies, I firmly believe. Such a reconciliation is necessary, whatever view be taken of the nature of the universe and of its mode of action; whatever units we may choose to regard as fundamental among those which lie at our disposal.

In this address I have endeavoured to fulfil my promise to combine a little history, a little actuality, and a little prophecy. The history belongs to the Old World; I have endeavoured to share passing events with the New; and I will ask you to join with me in the hope that much of the prophecy may meet with its fulfilment on this side of the ocean.

(**56**, 378-382; 1897)

的量。如果让我选择，我会说，那些具有相同热容量的物质的量是相同的；或者说，“相同数量的原子”代表相同的物质的量。实际上，我们衡量物质的价值是看它的能力而不是它的质量；我们购买食物时，大体上在原子或者也许是分子的基础上来讲，是根据其中的蛋白质含量。而大多数商品的价格都取决于生产者或制造商所需食物的量。

由此，各种形式的能量可以分为两类，即具有可以称之为“原子”因子的能量，和具有“质量”因子的能量。前者是占大多数的。周期律则是两者之间的桥梁；只是目前尚不够完善。因为按照其质量顺序排列的原子因子只表现出部分规律性。如何揭开这个秘密无疑是物理和化学领域的主要难题之一。我没有能力预言答案会是什么。是否会发现某些条件可以影响迄今仍被视为最为确定的“自然常量”之一的原子量，或者是否会发现质量和重力作用受温度或者电荷的影响，我都无法预言。但是我绝对相信，我们终将找到调和这些表面矛盾的方法。无论对宇宙的本质及其运动模式有何看法，无论我们将选择什么作为基本单位，这样一种调和都是必需的。

在这篇报告中，我努力实践了我开始的承诺——将一些历史、现实和预言结合起来。历史属于过去的世界；我一直努力与新世界分享过去的事；让我们共同期望，很多预言将会在大洋此岸得到证实。

（王耀杨 翻译；汪长征 审稿）

Distant Electric Vision

A. A. Campbell-Swinton

Editor's Note

Here the electrical engineer Alan Archibald Campbell-Swinton comments on the recent discussion of a Mr. Bidwell on the technical obstacles to transmitting visual signals electrically over long distances—that is, to television. The need for 160,000 synchronized operations per second, he notes, might be achieved with separate cathode rays at the transmitting and receiving ends, which could be swept over a display surface in less than the one-tenth of a second necessary for visual persistence. The more demanding challenge lay in finding a means for transmitting the high-frequency signals required for visual fields over long distances. Campbell-Swinton is today seen as the first man to have clearly envisaged how television might work.

REFERRING to Mr. Shelford Bidwell's illuminating communication on this subject published in *Nature* of June 4, may I point out that though, as stated by Mr. Bidwell, it is wildly impracticable to effect even 160,000 synchronised operations per second by ordinary mechanical means, this part of the problem of obtaining distant electric vision can probably be solved by the employment of two beams of kathode rays (one at the transmitting and one at the receiving station) synchronously deflected by the varying fields of two electromagnets placed at right angles to one another and energised by two alternating electric currents of widely different frequencies, so that the moving extremities of the two beams are caused to sweep synchronously over the whole of the required surfaces within the one-tenth of a second necessary to take advantage of visual persistence.

Indeed, so far as the receiving apparatus is concerned, the moving kathode beam has only to be arranged to impinge on a sufficiently sensitive fluorescent screen, and given suitable variations in its intensity, to obtain the desired result.

The real difficulties lie in devising an efficient transmitter which, under the influence of light and shade, shall sufficiently vary the transmitted electric current so as to produce the necessary alterations in the intensity of the kathode beam of the receiver, and further in making this transmitter sufficiently rapid in its action to respond to the 160,000 variations per second that are necessary as a minimum.

Possibly no photoelectric phenomenon at present known will provide what is required in this respect, but should something suitable be discovered, distant electric vision will, I think, come within the region of possibility.

(**78**, 151; 1908)

A. A. Campbell-Swinton: 66 Victoria Street, London, S.W., June 12.

远程电视系统

坎贝尔–斯温顿

编者按

电气工程师艾伦·阿奇博尔德·坎贝尔–斯温顿在这篇文章中评价了比德韦尔先生最近关于视觉信号在通过电力远距离传输（即电视）时遇到技术障碍的论述。他特别提到，每秒所需的160,000次同步操作也许可以通过分别位于发送端和接收端的阴极射线来完成，这样才能在显示区域内以低于1/10秒的时间进行扫描以满足视觉暂留的需要。找到一种传输高频信号的手段以满足远程视场的需要对我们来说是一个更高层次的挑战。在今天看来，坎贝尔–斯温顿是第一个明确提出电视工作原理的人。

谢尔福德·比德韦尔先生在6月4日的《自然》上就远程电视系统这一主题发表了一篇颇具启发性的通讯文章，他在该文章中指出，通过普通的力学方法想要在每秒内有效执行160,000次同步操作是完全不可能实现的，但我想说的是，对于实现远程电视系统所面临的这部分问题，通过引入两束阴极射线（一束在发送站，一束在接收站）就很有可能得到解决。用两个互相垂直放置的电磁铁产生的交变磁场来使这两束射线发生同步偏转，用两个频率迥异的交流电流来驱动电磁铁，这样就能使两束射线变化中的极限状态在1/10秒的时间内同步地扫过整个屏幕表面，而在1/10秒内完成这些是达到视觉暂留效果所必需的。

对于接收装置来说，实际上只要使变化的阴极射线在撞击足够敏感的荧光屏后能在其上形成影像，并使不同强度射线产生的影像有合适的差异即可，这样就能获得期望的结果了。

真正的困难在于发射装置的设计。首先，这个发射装置要有能力在光线和阴影的作用下发射出变化细节足够丰富的电流，以便能使接收端阴极射线的强度产生必要的变化。其次，这个发射装置的速度要足够快，每秒最少要能对160,000次变化作出响应。

可能现在还没有任何光电仪器能够达到这些要求，但人们肯定能发明出满足要求的东西。我相信远程电视系统最终可以实现。

（刘东亮 翻译；李军刚 审稿）

Intra-Atomic Charge

F. Soddy

Editor's Note

What was the internal structure of an atom? While Ernest Rutherford's experiments in 1911 had convinced him that the atom contained a dense, positively charged nucleus, others were not so sure. Here Rutherford's sometime collaborator Frederick Soddy suggested that the nucleus must also contain negative charges, expelled during so-called radioactive beta decay. Soddy introduces the term "isotope": atoms essentially identical in their chemical properties but with differing nuclei. For any given nuclear charge, he asserted, an atom may have any number of electrons in an "outer ring system". Changes in this number are a consequence of chemical action, with no effect on the nucleus. Clarification of this view awaited the discovery of the proton and neutron.

THAT the intra-atomic charge of an element is determined by its place in the periodic table rather than by its atomic weight, as concluded by A. van der Broek (*Nature*, November 27, p. 372), is strongly supported by the recent generalisation as to the radio-elements and the periodic law. The successive expulsion of one α and two β particles in three radio-active changes in any order brings the intra-atomic charge of the element back to its initial value, and the element back to its original place in the table, though its atomic mass is reduced by four units. We have recently obtained something like a direct proof of van der Broek's view that the intra-atomic charge of the nucleus of an atom is not a purely positive charge, as on Rutherford's tentative theory, but is the difference between a positive and a smaller negative charge.

Fajans, in his paper on the periodic law generalisation (*Physikal. Zeitsch.*, 1913, vol. XIV, p. 131), directed attention to the fact that the changes of chemical nature consequent upon the expulsion of α and β particles are precisely of the same kind as in ordinary electrochemical changes of valency. He drew from this the conclusion that radio-active changes must occur in the same region of atomic structure as ordinary chemical changes, rather than with a distinct inner region of structure, or "nucleus", as hitherto supposed. In my paper on the same generalisation, published immediately after that of Fajans (*Chem. News*, February 28), I laid stress on the absolute identity of chemical properties of different elements occupying the same place in the periodic table.

A simple deduction from this view supplied me with a means of testing the correctness of Fajans's conclusion that radio-changes and chemical changes are concerned with the same region of atomic structure. On my view his conclusion would involve nothing else than that, for example, uranium in its tetravalent uranous compounds must be chemically

原子内的电荷

索迪

编者按

原子的内部结构是怎样的？1911年，当欧内斯特·卢瑟福用实验验证了原子包含一个致密的带正电的核时，其他人并没有表示十分肯定。在这篇文章中，曾经与卢瑟福合作过的弗雷德里克·索迪提出原子核中必须同时也包含负电荷，这些负电荷会在放射性 β 衰变时发射出去。索迪引入了“同位素”的概念，即原子核结构不同但化学性质基本一致的原子。他认为，对于任意给定的核电荷，原子“外层系统”可以排布任意的电子数量。化学反应能引起外层电子数的变化，但对原子核没有影响。后来质子和中子的发现证实了他的观点。

范德布鲁克断言（《自然》，11月27日，第372页），一种元素原子内的电荷是由它在周期表中的位置而不是它的原子量确定的，这一论断受到最近一些关于放射性元素与周期律的结论的强力支持。如果在3次放射性变化中相继发射出1个 α 粒子和2个 β 粒子，那么不管这3次衰变的次序如何，都会使该元素的原子内电荷回到初始数值，元素也回到了它在周期表中的原始位置，但是它的原子质量却减少了4个单位。最近我们获得了一些证据，能够直接支持范德布鲁克的观点：就像卢瑟福的初步理论所指出的，原子核内的电荷并不是单纯的正电荷，而是正电荷与较小的负电荷的差值。

法扬斯在他那篇关于周期律的一般法则的论文(《物理学杂志》,1913年,第14卷,第131页）中，特别指出了如下事实：由于 α 粒子和 β 粒子的发射而引起的元素化学性质的改变，与普通的会发生价态变动的由电化学变化引起的物质化学性质的改变完全属于同一类型。他由此得出的结论是，放射性变化必定与普通的化学变化一样发生在原子结构的同一区域，而不是像我们目前所假设的——放射性变化发生在被称为“核”的一个完全不同的内部区域。在法扬斯的论文发表后不久，我也针对同一主题发表了一篇论文（《化学新闻》，2月28日），文中我将重点放在了周期表中处于同一位置的不同元素具有完全相同的化学性质这一点上。

此观点的一个简单推论为我提供了一种检验法扬斯认为的放射性变化与化学变化发生于原子结构中同一区域这一观点是否正确的方法。我认为，他的结论其实就是下面的意思：举例来说，铀的四价化合物中的铀元素必定与钍化合物中的钍元

identical with and non-separable from thorium compounds. For uranium X, formed from uranium I by expulsion of an α particle, is chemically identical with thorium, as also is ionium formed in the same way from uranium II. Uranium X loses two β particles and passes back into uranium II, chemically identical with uranium. Uranous salts also lose two electrons and pass into the more common hexavalent uranyl compounds. If these electrons come from the same region of the atom uranous salts should be chemically non-separable from thorium salts. But they are not.

There is a strong resemblance in chemical character between uranous and thorium salts, and I asked Mr. Fleck to examine whether they could be separated by chemical methods when mixed, the uranium being kept unchanged throughout in the uranous or tetravalent condition. Mr. Fleck will publish the experiments separately, and I am indebted to him for the result that the two classes of compounds can readily be separated by fractionation methods.

This, I think, amounts to a proof that the electrons expelled as β rays come from a nucleus not capable of supplying electrons to or withdrawing them from the ring, though this ring is capable of gaining or losing electrons from the exterior during ordinary electro-chemical changes of valency.

I regard van der Broek's view, that the number representing the net positive charge of the nucleus is the number of the place which the element occupies in the periodic table when all the possible places from hydrogen to uranium are arranged in sequence, as practically proved so far as the relative value of the charge for the members of the end of the sequence, from thallium to uranium, is concerned. We are left uncertain as to the absolute value of the charge, because of the doubt regarding the exact number of rare-earth elements that exist. If we assume that all of these are known, the value for the positive charge of the nucleus of the uranium atom is about 90. Whereas if we make the more doubtful assumption that the periodic table runs regularly, as regards numbers of places, through the rare-earth group, and that between barium and radium, for example, two complete long periods exist, the number is 96. In either case it is appreciably less than 120, the number were the charge equal to one-half the atomic weight, as it would be if the nucleus were made out of α particles only. Six nuclear electrons are known to exist in the uranium atom, which expels in its changes six β rays. Were the nucleus made up of α particles there must be thirty or twenty-four respectively nuclear electrons, compared with ninety-six or 102 respectively in the ring. If, as has been suggested, hydrogen is a second component of atomic structure, there must be more than this. But there can be no doubt that there must be some, and that the central charge of the atom on Rutherford's theory cannot be a pure positive charge, but must contain electrons, as van der Broek concludes.

素具有完全相同的化学性质且不可区分。因为，由铀 I 发射 1 个 α 粒子而形成的铀 X 与钍，以及由铀 II 以同样方式形成的“锾”，在化学性质上是一致的。铀 X 发射 2 个 β 粒子就又回到铀 II，铀 II 与铀的化学性质是一致的。亚铀盐也可以失去 2 个电子并形成更为常见的六价铀的化合物。如果这些电子来自原子中的同一区域，那么亚铀盐和钍盐应该具有相同的化学性质并且不可区分。然而事实并非如此。

亚铀盐与钍盐的化学性质非常相似，我已经请弗莱克先生研究是否可以在始终保持铀元素的四价或六价状态不发生改变的前提下用化学方法将铀盐与钍盐的混合物中的铀和钍分离开来。弗莱克先生将独立发表他的实验，我很感激能够使用他取得的结果，即通过分馏的方法可以很顺利地分离这两类化合物。

我认为这足以证明，以 β 射线形式发射出的电子来自核，尽管核无法为核外圈层提供电子或从中取走电子，但核外圈层可以在普通的会发生价态变动的电化学变化中从外部获取电子或失去电子。

在研究从铊到铀这些处在周期表序列尾部的成员的电荷相对值时，我把范德布鲁克的观点当作是已经被证实的，即从氢到铀按顺序排好周期表中所有可能的位置后，代表核所具有的净正电荷的数值就正好是元素在周期表中所处位置对应的数值。不过我们仍旧不能确定电荷的绝对值，因为对到底存在多少种稀土元素还存在疑问。如果我们假定这些都是已知的，那么铀原子核中正电荷的数值大约是 90。如果我们采取另一个更不确定的假设，即从周期表中位置对应的数值的角度来看，包括全部稀土元素以及介于钡与镭之间的元素在内的周期表是规则排布的，整个周期表有两个完整的长周期，这样得到的数值就会是 96。不管是哪种情况，该数值都明显小于 120，即便假设其电荷数值就是 120，那也只是其原子量的一半，如果真是这样，那核可能就只有 α 粒子了。已知铀原子中存在 6 个核电子，在放射性变化中这些核电子以 6 次 β 射线的形式发射出来。如果核是由 α 粒子构成的话，那么与核外圈层中具有 96 个或者 102 个电子相对应，就必须有 30 个或者 24 个核电子。另外已经有人提出氢是原子结构的另一种构件，如果考虑上这一点的话，那么其核电子数就不止于此了。但是有一点毋庸置疑，那就是必定存在一些核电子，而且卢瑟福理论中原子的中心电荷也不是单纯的正电荷，而是必定包含负电子，就像范德布鲁克断定的那样。

So far as I personally am concerned, this has resulted in a great clarification of my ideas, and it may be helpful to others, though no doubt there is little originality in it. The same algebraic sum of the positive and negative charges in the nucleus, when the arithmetical sum is different, gives what I call "isotopes" or "isotopic elements", because they occupy the same place in the periodic table. They are chemically identical, and save only as regards the relatively few physical properties which depend upon atomic mass directly, physically identical also. Unit changes of this nuclear charge, so reckoned algebraically, give the successive places in the periodic table. For any one "place," or any one nuclear charge, more than one number of electrons in the outer-ring system may exist, and in such a case the element exhibits variable valency. But such changes of number, or of valency, concern only the ring and its external environment. There is no in- and out-going of electrons between ring and nucleus.

(**92**, 399-400; 1913)

Frederick Soddy: Physical Chemistry Laboratory, University of Glasgow.

就目前我个人的思考结果来说，以上是对我的观点的一个很明晰的解释。虽然很明显其中并无多少创见，但也许会对其他人有所帮助吧。根据核内正负电荷的代数和相同而算数和不同的现象，我提出了“同位素”或“同位置元素”的概念，因为它们在周期表中处于相同的位置。它们在化学性质上是完全相同的，除了很有限的一些直接决定于原子量的物理性质外，其他大部分物理性质也是完全相同的。从数学角度来看，核电荷数会单位递增地发生变化，这使得周期表中的位置也连续地变化。对于周期表中任意一个位置，或者说任意一个确定的核电荷数，可以存在不止一种外层电子数，在这种情况下元素就表现出了不同的价态。不过，这种电子数或者价态的变化只是考虑了核外圈层及其外部环境，并没有把核外圈层与核之间的电子进出过程考虑进去。

（王耀杨 翻译； 汪长征 审稿）

The Structure of the Atom

E. Rutherford

Editor's Note

Responding to a comment by Frederick Soddy, Ernest Rutherford here clarifies his view on the structure of the atomic nucleus. Soddy had suggested that Rutherford believed the nucleus to contain positive charges only. On the contrary, Rutherford insists, he believes only that the atomic nucleus is small, dense, and positively charged overall. Moreover, he thinks that two of the key products of radioactive decay—alpha and beta particles—might both originate from the nucleus. Rutherford supports a recent suggestion that the charge on the atomic nucleus is equal to the atomic number, and not to half the atomic weight. This observation prefigured the revelation that the atomic number is the equal to the number of protons in the nucleus.

IN a letter to this journal last week, Mr. Soddy has discussed the bearing of my theory of the nucleus atom on radio-active phenomena, and seems to be under the impression that I hold the view that the nucleus must consist entirely of positive electricity. As a matter of fact, I have not discussed in any detail the question of the constitution of the nucleus beyond the statement that it must have a resultant positive charge. There appears to me no doubt that the α particle does arise from the nucleus, and I have thought for some time that the evidence points to the conclusion that the β particle has a similar origin. This point has been discussed in some detail in a recent paper by Bohr (*Phil. Mag.*, September, 1913). The strongest evidence in support of this view is, to my mind, (1) that the β ray, like the α ray, transformations are independent of physical and chemical conditions, and (2) that the energy emitted in the form of β and γ rays by the transformation of an atom of radium C is much greater than could be expected to be stored up in the external electronic system. At the same time, I think it very likely that a considerable fraction of the β rays which are expelled from radio-active substances arise from the external electrons. This, however, is probably a secondary effect resulting from the primary expulsion of a β particle from the nucleus.

The original suggestion of van der Broek that the charge on the nucleus is equal to the atomic number and not to half the atomic weight seems to me very promising. This idea has already been used by Bohr in his theory of the constitution of atoms. The strongest and most convincing evidence in support of this hypothesis will be found in a paper by Moseley in *The Philosophical Magazine* of this month. He there shows that the frequency of the X radiations from a number of elements can be simply explained if the number of unit charges on the nucleus is equal to the atomic number. It would appear that the charge on the nucleus is the fundamental constant which determines the physical and chemical properties of the atom, while the atomic weight, although it approximately follows the order of the nucleus charge, is probably a complicated function of the latter depending on the detailed structure of the nucleus.

(**92**, 423; 1913)

E. Rutherford: Manchester, December 6, 1913.

原子结构

卢瑟福

编者按

为了回应弗雷德里克·索迪的意见，欧内斯特·卢瑟福在这篇文章中进一步解释了他对原子核结构的观点。索迪曾指出卢瑟福认为原子核只包含正电荷，而卢瑟福却强调说他只承认原子核很小、很致密，以及整体带正电。此外，卢瑟福认为，放射性衰变的两个主要产物，即 α 粒子和 β 粒子，可能都源自原子核。卢瑟福同意最近有人提出的关于核电荷等于原子序数而非原子量的一半的观点。这个结果预示了原子序数与原子核中的质子数相等这一关系。

在上周致贵刊的一封信中，索迪先生对我关于放射性现象中原子核的理论进行了相关讨论，他似乎以为，我认定原子核必须完全由带正电荷的粒子构成。事实上，我只是认为原子核必定具有总和为正的电荷，而对它的具体构成并未发表看法。在我看来，α 粒子无疑是产生于核的，而且经过一段时间的思考，我认为有证据表明 β 粒子也源自核。玻尔在最近的一篇论文（《哲学杂志》，1913 年 9 月）中对这一点进行了较为详细的讨论。支持这一观点的最强有力的证据是：(1) 与 α 射线一样，β 射线的衰变也是与物理和化学条件无关的；(2) 镭原子发生衰变时以 β 射线或 γ 射线的形式放出的能量 C，比预想的外部电子系统所存储的能量大得多。不过，放射性物质发射出的 β 射线可能有相当一部分来源于外部电子。这也许是原子核中的 β 粒子的初级辐射引起的一种次级效应。

范德布鲁克最先提出，核所带的基本电荷数应等于原子序数而不是原子量的一半，我认为这是非常有可能的。玻尔在他的原子结构理论中已经应用了这种观点。在莫塞莱本月发表于《哲学杂志》上的文章中，可以找到支持这个假说的最强有力也最令人信服的证据。文中表明，如果核所带的基本电荷数等于原子序数，就能方便地对多种元素所发出的 X 辐射的频率进行解释。看起来，核所带的基本电荷数是确定原子物理和化学性质的基本常数，尽管原子量的顺序与相应原子核电荷数的顺序基本一致，但原子量并非如人们以前所设想的那样是其核电荷数的两倍，可能是核电荷数的复杂函数，其函数关系与核的具体结构有关。

（王耀杨 翻译；鲍重光 审稿）

The Reflection of X-Rays

Editor's Note

The German physicist Max von Laue demonstrated the phenomenon of X-ray diffraction in 1912. This high-energy form of light, with wavelengths comparable to the spacing between molecules in crystalline solids, could be used to reveal crystal structures. Other physicists had deduced the relationship between the lattice spacing and the angles at which bright diffraction spots should occur. Here Maurice de Broglie, the brother of physicist Louis de Broglie, introduces what came to be known as the rotating-crystal method for recording X-ray diffraction from a single crystal. The technique detects X-rays reflected along the surface of a series of so-called "Laue cones". It became the standard method of X-ray diffraction for many years.

IN view of the great interest of Prof. Bragg's and Messrs. Moseley and Darwin's researches on the distribution of the intensity of the primary radiation from X-ray tubes, it may be of interest to describe an alternate method which I have found very convenient (*Comptes rendus*, November 17, 1913).

As we know, the wavelength of the reflected ray is defined by the equation $n\lambda = 2d\sin\theta$, where n is a whole number, d the distance of two parallel planes, and θ the glancing angle. If one mounts a crystal with one face in the axis of an instrument that turns slowly and regularly, such as, for instance, a registering barometer, the angle changes gradually and continuously.

If, therefore, one lets a pencil of X-rays, emerging from a slit, be reflected from this face on to a photographic plate, one finds the true spectrum of the X-rays on the plate, supposing intensity of the primary beam to have remained constant. (This can be tested by moving another plate slowly before the primary beam during the exposure.)

The spectra thus obtained are exactly analogous to those obtained with a diffraction grating, and remind one strongly of the usual visual spectra containing continuous parts, bands, and lines.

So far I have only identified the doublet, 11°17′ and 11°38′, described by Messrs. Moseley and Darwin. The spectra contain also a number of bright lines about two octaves shorter than these, and the continuous spectrum is contained within about the same limits. These numbers may be used in the interpretations of diffraction Röntgen patterns, as they were obtained with tubes of the same hardness as those used for producing these latter.

X 射线的反射

编者按

1912 年，德国物理学家马克斯·冯·劳厄证明了 X 射线的衍射现象。这种形式的光能量很高，其波长与晶格中分子之间的距离相近，因而可以用于研究晶体的结构。其他物理学家推算出了晶格间距与预计会出现明亮衍射斑点的角度之间的关系。在这篇文章中，物理学家路易斯·德布罗意的哥哥莫里斯·德布罗意介绍了记录单晶 X 射线衍射的方法，后来被称作旋转晶体法。该技术探测沿着一组被称作“劳厄锥”的晶面反射的 X 射线。许多年来这种方法一直是人们研究 X 射线衍射的标准方法。

鉴于大家对布拉格教授、莫塞莱先生以及达尔文先生关于从 X 射线管发射的初级辐射强度分布的研究有极大的兴趣，我已发现的另一种很方便的方法可能也会引起大家的兴趣（《法国科学院院刊》，1913 年 11 月 17 日）。

如我们所知，反射射线的波长由方程 $n\lambda = 2d\sin\theta$ 确定，式中 n 是一个整数，d 是两个平行平面间的距离，θ 是掠射角。如果我们将一块晶体放在一台缓慢而有规律旋转的仪器（例如，记录式气压计）上，使晶体的一个表面沿仪器的轴向，那么反射的角度将连续不断地发生变化。

如果我们使一束 X 射线从狭缝中射出，并被此晶面反射到照相干版上，假定初级光束的强度保持不变，即可在干版上得到 X 射线的真实光谱。（可以通过曝光时在初级光束前方缓慢移动另一块照相干版来检验光束强度是否保持不变。）

这样得到的光谱与用衍射光栅得到的光谱非常类似，而且很容易使人想起那些含有连续区、谱带和谱线的普通可见光谱。

至今，我仅确认了莫塞莱先生和达尔文先生所描述的双线：11° 17′和 11° 38′。这个光谱还包含比双线短大约两个倍频程的许多亮线，还有在大致相同范围内的连续谱。这些光谱线也许可用于解释伦琴的衍射图样，因为产生伦琴的衍射图样时使用的管子与产生这些谱线时使用的管子具有相同的硬度。

The arrangement described above enables us to distinguish easily the spectra of different orders, as the interposition of an absorbing layer cuts out the soft rays, but does not weaken appreciably the hard rays of the second and higher orders.

It is convenient also for absorption experiments; thus a piece of platinum foil of 0.2 mm thickness showed transparent bands. The exact measurements will be published shortly, as well as the result of some experiments I am engaged upon at present upon the effect of changing the temperature of the crystal.

Maurice de Broglie

* * *

As W. L. Bragg first showed, when a beam of soft X-rays is incident on a cleavage plane of mica, a well-defined proportion of the beam suffers a reflection strictly in accordance with optical laws. In addition to this generally reflected beam, Bragg has shown that for certain angles of incidence, there occurs a kind of selective reflection due to reinforcement between beams incident at these angles on successive parallel layers of atoms.

Experiments I am completing seem to show that a generally reflected beam of rays on incidence at a second crystal surface again suffers optical reflection; but the degree of reflection is dependent on the orientation of this second reflector relative to the first.

The method is a photographic one. The second reflector is mounted on a suitably adapted goniometer, and the photographic plate is mounted immediately behind the crystal. The beam is a pencil 1.5 mm in diameter. When the two reflectors are parallel the impression on the plate, due to the two reflections, is clear. But as the second reflector is rotated about an axis given by the reflected beam from the first and fixed reflector, the optically reflected radiation from the second reflector—other conditions remaining constant—diminishes very appreciably. As the angle between the reflectors is increased from 0° to 90°, the impression recorded on the photographic plate diminishes in intensity. For an angle of 20° it is still clear; for angles in the neighbourhood of 50° it is not always detectable; and for an angle of 90° it is very rarely detectable in the first stages of developing, and is then so faint that it never appears on the finished print.

These results, then, would show that the generally reflected beam of X-rays is appreciably polarised in a way exactly analogous to that of ordinary light. Owing to the rapidity with which the intensity of the generally reflected beam falls off with the angle of incidence of the primary beam, it has not been possible to work with any definiteness with angles of incidence greater than about 78°, and this is unfortunately a considerably larger angle than the probable polarising angle. Experiments with incidence in the neighbourhood of 45° should prove peculiarly decisive, for whereas ordinary light cannot as a rule be completely polarised by reflection, the reflection of X-rays, which occurs at planes of atoms, is

上述装置能使我们很容易地区别不同级的谱线，因为插入吸收层可以截断软射线，但并不会太明显地减弱次级和更高级的硬射线。

进行吸收实验也是很方便的。用这样的方法，一片 0.2 毫米厚的铂箔会显示出透射带。至于确切的测量方法以及目前我在改变晶体温度方面所做的实验得到的结果，都将在我即将发表的文章中进行介绍。

莫里斯 · 德布罗意

* * *

正如布拉格首先指出的，当一束软 X 射线入射到云母的一个解理面上时，光束反射部分所占的比例严格遵照光学定律。除了这种普通的反射光束，布拉格指出，对于特定的入射角，由于以这些角度入射到原子中连续的平行层上的光束彼此相互增强，因而会产生一种选择性的反射。

我正在进行的实验似乎表明，当一个普通的反射束入射到下一个晶体表面上时，还会再次产生光学反射，但是，反射的程度取决于第二个反射面与第一个反射面之间的夹角。

我用的是照相记录法。第二个反射面被安装在调整好的测角仪上，照相干版紧贴在晶体后面。光束直径为 1.5 毫米。当两个反射面平行时，两次反射在照相干版上产生的影像是清晰的。但是，当以第一个固定反射面上反射的射线为轴旋转第二个反射面时，假如其他条件不变，则从第二个反射面上反射的光学辐射明显减弱。当反射面之间的夹角从 0° 增加到 90° 时，照相干版上记录的影像的清晰度不断降低。在角度为 20° 时，影像仍然很清晰；角度为 50° 左右时，经常检测不到影像；角度为 90° 时，由于影像非常微弱，在第一步显影阶段已经很难看到，而在最终冲洗出的照片上则从来没有出现过。

以上结果表明，X 射线普通反射束的偏振特性在某种程度上完全类似于普通光的偏振。由于普通反射束的强度随着初级光束入射角的增加而迅速下降，当入射角超过 78° 后就很难得到确定无疑的影像了。遗憾的是，这个角度比可能的偏振角大很多。入射角在 45° 附近的实验尤其具有决定意义，虽然通常不能通过反射使普通光完全偏振，但在原子平面上 X 射线的反射并不受被辐照的晶体表面上任何污染的影响，一旦出现偏振，那么偏振角处被反射的辐射将是完全偏振的。被选择性反射

independent of any contamination of the exposed crystal surface, and polarisation, once established, should prove complete for radiation reflected at the polarising angle. The selectively reflected X-rays seem to show the same effects as does the generally reflected beam. Selectively reflected radiation is always detectable after the second reflection, but this seems due to the selectively reflected radiation produced at the second reflector by the unpolarised portion of the beam generally reflected at the first reflector.

The application of a theory of polarisation to explain the above results is interestingly supported by the fact that in the case of two reflections by parallel reflectors, the proportion of X-rays reflected at the second reflector is invariably greater than the proportion of rays reflected at the first; that is, the ratio of reflected radiation to incident radiation at the second reflector is always greater than the same ratio at the first reflector. This might be expected if vibrations perpendicular to the plane of incidence are to be reflected to a greater extent than those in the plane of incidence. The proportion of such vibrations is larger in the beam incident on the second reflector than in the original beam, and a greater proportion of radiation would be reflected at the second reflector than could be at the first. For the case of parallel reflectors and incidence of a primary beam on the first at the polarising angle, the reflection at the second should be complete.

E. Jacot

(**92**, 423-424; 1913)

Maurice de Broglie: 29, Rue Chateaubriand, Paris, December 1.
E. Jacot: South African College, Cape Town, November 14.

的 X 射线与普通的反射有相同的效应。在第二次反射后总能检测到被选择性反射的辐射，但这似乎是由第一个反射面处被反射的那部分非偏振辐射在第二个反射面上发生选择性反射而造成的。

偏振理论可以用来解释上述结果，这得到了以下事实的强烈支持，在由两个平行反射面引起的两次反射中，第二个反射面上被反射的 X 射线总是多于第一个反射面上被反射的 X 射线；即在第二个反射面上被反射的辐射相对于入射辐射的比例，总是大于第一个反射面。这就可以预期，垂直于入射平面的振动被反射的量比在入射平面内的振动更大。与原光束相比，入射到第二个反射面上的光束中垂直于入射平面的振动所占比例更大一些，因此在第二个反射面上被反射的辐射的比例大于第一个反射面。在两个反射面相互平行的情况下，如果初始光束以偏振角入射到第一个反射面上，则在第二个反射面上它将被完全反射。

贾科

（沈乃澂 翻译；江丕栋 审稿）

The Constitution of the Elements

F. W. Aston

Editor's Note

Francis Aston was a physicist at the Cavendish Laboratory of the University of Cambridge who invented the instrument now called the mass spectrometer, which could in principle measure the masses of individual atoms. The principle was to electrify the atoms by using an electric field to remove one or more of their electrons, accelerate them in the same electric field, and deflect them by means of a magnetic field. A beam of atoms thus ionised would travel in a curved path onto a screen and the mass of the atom related to some standard could then be inferred from the deflection of its path. At the beginning of what proved to be his life's work, Aston had come to the conclusion that several familiar atoms consisted of isotopes with different masses but the same overall electric charge. In this letter, he drew attention to the fact that the measured masses of atoms appeared to be very near to integral multiples of the mass of the hydrogen atom, and suggested that this may "do much to elucidate the ultimate structure of matter".

IT will doubtless interest readers of *Nature* to know that other elements besides neon (see *Nature* for November 27, p. 334) have now been analysed in the positive-ray spectrograph with remarkable results. So far oxygen, methane, carbon monoxide, carbon dioxide, neon, hydrochloric acid, and phosgene have been admitted to the bulb, in which, in addition, there are usually present other hydrocarbons (from wax, etc,) and mercury.

Of the elements involved hydrogen has yet to be investigated; carbon and oxygen appear, to use the terms suggested by Paneth, perfectly "pure"; neon, chlorine, and mercury are unquestionably "mixed". Neon, as has been already pointed out, consists of isotopic elements of atomic weights 20 and 22. The mass-spectra obtained when chlorine is present cannot be treated in detail here, but they appear to prove conclusively that this element consists of at least two isotopes of atomic weights 35 and 37. Their elemental nature is confirmed by lines corresponding to double charges at 17.50 and 18.50, and further supported by lines corresponding to two compounds HCl at 36 and 38, and in the case of phosgene to two compounds COCl at 63 and 65. In each of these pairs the line corresponding to the smaller mass has three or four times the greater intensity.

Mercury, the parabola of which was used as a standard of mass in the earlier experiments, now proves to be a mixture of at least three or four isotopes grouped in the region corresponding to 200. Several, if not all, of these are capable of carrying three, four, five, or even more charges. Accurate values of their atomic weights cannot yet be given.

元素的组成

阿斯顿

编者按

剑桥大学卡文迪什实验室的物理学家弗郎西斯·阿斯顿发明了我们今天称为质谱仪的仪器，这种仪器原则上可以测定单个原子的质量。其原理是，利用电场移除原子中的一个或多个电子而将原子电离，使原子在该电场中加速，再利用磁场使之偏转。电离了的原子束经由弯曲轨迹到达接收屏，通过其轨迹的偏转程度可以推断此原子相对于某种标准的质量。在这项堪称终身成就工作的早期，阿斯顿断言几种常见元素是由质量不同但总电荷数相同的同位素组成。在这封快报中，他注意到已测得的原子质量似乎非常接近于氢原子质量的整数倍这一事实，并指出这将“大大有助于阐明物质的基本结构”。

《自然》的读者们一定会有兴趣知道，用阳极射线谱仪分析除氖之外的其他一些元素（参见《自然》，11 月 27 日，第 334 页）后得到了不同寻常的结果。目前已经用此仪器对氧、甲烷、一氧化碳、二氧化碳、氖、氢氯酸和光气进行了分析，不过仪器中通常还存在其他烃类（来自蜡等）和汞。

相关元素中，氢还需要被研究。按照帕内特的说法，碳和氧似乎是绝对“纯的”。而氖、氯和汞则无疑是“混合的”。如同已经指出的那样，氖由原子量分别为 20 和 22 的同位素组成。这里无法详细讨论有氯存在时得到的质谱，但似乎可以确认这种元素至少由原子量为 35 和 37 的两种同位素组成。根据位于 17.50 和 18.50 的与双电荷微粒有关的谱线，可以证实它们的元素性质。与两种 HCl 相对应的位于 36 和 38 处的两条谱线以及在研究光气时由两种 COCl 所形成的位于 63 和 65 处的两条谱线进一步支持了这样的元素性质。在上述的每组谱线对中，较小质量对应的谱线强度是较大质量对应谱线强度的 3~4 倍。

在早期实验中汞的抛物线曾被用作质量标准，现在证实汞至少是由 3~4 种原子量集中在 200 左右的同位素混合而成。其中几种，如果不是全部的话，可以携带 3 个、4 个、5 个甚至更多个电荷。目前尚无法给出它们的原子量的准确数值。

A fact of the greatest theoretical interest appears to underlie these results, namely, that of more than forty different values of atomic and molecular mass so far measured, all, without a single exception, fall on whole numbers, carbon and oxygen being taken as 12 and 16 exactly, and due allowance being made for multiple charges.

Should this integer relation prove general, it should do much to elucidate the ultimate structure of matter. On the other hand, it seems likely to make a satisfactory distinction between the different atomic and molecular particles which may give rise to the same line on a mass-spectrum a matter of considerable difficulty.

(**104**, 393; 1919)

F. W. Aston: Cavendish Laboratory, December 6.

这些结果之下似乎存在一个极具理论价值的事实，即目前已测得的超过 40 个原子或分子的质量数据，无一例外都是整数，这里以碳原子量和氧原子量正好为 12 和 16 为准确值，并适当考虑带多个电荷造成的影响。

如果这一整数关系被证明是普遍存在的，它将大大有助于阐明物质的基本结构。另一方面，要对质谱中给出相同谱线的不同原子和分子进行令人满意的区分，看来还是一件相当困难的事情。

（王耀杨 翻译；李芝芬 审稿）

Einstein's Relativity Theory of Gravitation

E. Cunningham

Editor's Note

Arthur Eddington had recently announced his measurements of the deflection of starlight during a solar eclipse, in apparent agreement with Einstein's general theory of relativity. Here Ebenezer Cunningham surveys Einstein's ideas. Einstein's 1905 work had established a link between inertia and energy, and his new work pursued the question of whether gravity too might be linked to energy. What emerges from the theory, Cunningham argues, is a view in which there is no ultimate criterion for the equality of space or time intervals, but only the equivalence of an infinite number of ways of mapping out physical events. All this has been made possible, he notes, by Einstein adopting mathematics already developed by Riemann, Levi-Civita and others.

I

THE results of the Solar Eclipse Expeditions announced at the joint meeting of the Royal Society and Royal Astronomical Society on November 6 brought for the first time to the notice of the general public the consummation of Einstein's new theory of gravitation. The theory was already in being before the war; it is one of the few pieces of pure scientific knowledge which have not been set aside in the emergency; preparations for this expedition were in progress before the war had ceased.

Before attempting to understand the theory which, if we are to believe the daily Press, has dimmed the fame of Newton, it may be worth while to recall what it was that he did. It was not so much that he, first among men, used the differential calculus. That claim was disputed by Leibniz. Nor did he first conceive the exact relations of inertia and force. Of these, Galileo certainly had an inkling. Kepler, long before, had a vague suspicion of a universal gravitation, and the law of the inverse square had, at any rate, been mooted by Hooke before the "Principia" saw the light. The outstanding feature of Newton's work was that it drew together so many loose threads. It unified phenomena so diverse as the planetary motions, exactly described by Kepler, the everyday facts of falling bodies, the rise and fall of the tides, the top-like motion of the earth's axis, besides many minor irregularities in lunar and planetary motions. With all these drawn into such a simple scheme as the three laws of motion combined with the compact law of the inverse square, it is no wonder that flights of speculation ceased for a time. The universe seemed simple and satisfying. For a century at least there was little to do but formal development of Newton's dynamics. In the mid-eighteenth century Maupertuis hinted at a new physical doctrine. He was not content to think of the universe as a great clock the wheels of which turned inevitably and irrevocably according to a fixed rule. Surely there must be some purpose, some divine economy in all its motions. So he propounded a principle of least action. But it soon appeared that this was only Newton's laws in a new guise; and so the eighteenth century closed.

爱因斯坦关于万有引力的相对论

坎宁安

编者按

阿瑟·爱丁顿最近宣布，他在日食期间对星光偏转的测量明确地证实了爱因斯坦的广义相对论。在这篇文章中，埃比尼泽·坎宁安简单描述了爱因斯坦的理论。爱因斯坦在1905年就已经建立了惯性和能量之间的联系，他现在的工作主要是考查重力是否也有可能与能量相关联。坎宁安指出，这个理论说明了这样一个观点：虽然不存在衡量空间间隔和时间间隔完全相等的绝对标准，但可以设计无限多种方式以保证多个物理事件的等同性。他说，爱因斯坦利用黎曼、列维齐维塔和其他人的数学理论已经验证了这些结论。

I

在11月6日皇家学会和皇家天文学会共同举办的会议上宣布的日食观测结果，使得爱因斯坦关于万有引力的新理论受到了公众的广泛关注。这个理论在战争之前就已经被提出来了；它是在战争中极少数没有被丢到一边的纯科学工作之一；日食观测的准备工作在战争结束之前就已经着手进行了。

如果在试图理解这个新理论之前，我们就已经像日报上说的那样，认为牛顿在其面前也会黯然失色，那么最好首先回顾一下牛顿所做的工作。并不能说牛顿是第一个使用微分计算的人。有些议论认为这是莱布尼茨的首创。牛顿也不是第一个认识到惯性和力之间关系的人。这方面，肯定是伽利略首先对此进行了初步的设想。开普勒在很久以前就对万有引力有过模糊的猜想，但是，引力与距离平方成反比的规律是在该“原理”建立之前，由胡克首先发现的。牛顿所做工作的杰出之处在于，他把这么多松散的线索整合到了一起。行星运动（已被开普勒精确描述过）、日常生活中的落体运动、潮汐的涨落、地轴的进动以及在月球和行星运动中出现的许多小的不规则性——牛顿将所有这些现象都统一了起来。牛顿把所有这些都归入了一个包括力学三大定律和简洁的平方反比关系的基本框架中，也难怪在后来很长一段时间内，科学上的思索都停滞了。整个宇宙都看似简单而圆满。在后来至少一个世纪的时间里，除了在形式上发展一下牛顿力学，没有其他工作可做。在18世纪中叶，莫佩尔蒂暗示了一种新的物理学说。他不满足于认为宇宙是一个在某种确定的法则下永不停止、永不倒退的钟表。宇宙的运动一定是有目的的，一定有一种神圣的力量在支配着它的运动。因此，他提出了最小作用原理。但是人们很快发现这只不过是牛顿定律的一种新的外在形式而已，然后18世纪就这样结束了。

The nineteenth saw great changes. When it closed, the age of electricity had come. Men were peering into the secrets of the atom. Space was no longer a mighty vacuum in the cold emptiness of which rolled the planets. It was filled in every part with restless energy. Ether, not matter, was the last reality. Mass and matter were electrical at bottom. A great problem was set for the present generation: to reconcile one with the other the new laws of electricity and the classical dynamics of Newton. At this point the principle of least action began to assume greater importance; for the old and the new schemes of the universe had this in common, that in each of them the time average of the difference between the kinetic and the potential energies appears to be a minimum.

One of the main difficulties encountered by the electrical theory of matter has been the obstinate refusal of gravitation to come within its scope. Quietly obeying the law of the inverse square, it heeded not the bustle and excitement of the new physics of the atom, but remained, independent and inevitable, a constant challenge to rash claimants to the key of the universe. The electrical theory seemed on the way to explain every property of matter yet known, except the one most universal of them all. It could trace to its origins the difference between copper and glass, but not the common fact of their weight; and now the ether began silently to steal away.

One matter that has seriously troubled men in Newton's picture of the universe is its failure to accord with the philosophic doctrine of the relativity of space and time. The vital quantity in dynamics is the acceleration, the change of motion of a body. This does not mean that Newton assumed the existence of some ultimate framework in space relative to which the actual velocity of a body can be uniquely specified, for no difference is made to his laws if any arbitrary constant velocity is added to the velocity of every particle of matter at all time. The serious matter is that the laws cannot possibly have the same simplicity of form relative to two frameworks of which one is in rotation or non-uniform motion relative to the other. It seems, for instance, that if Newton were right, the term "fixed direction" in space means something, but "fixed position" means nothing. It seems as if the two must stand or fall together. And yet the physical relations certainly make a distinction. Why this should be so has not yet been made known to us. Whatever new theory we adopt must take account of the fact.

It was with some feeling of relief that men hailed the advent of the ether as a substitute for empty space, though we may note in passing that some philosophers—Comte, for example—have held that the concept of an ether, infinite and intangible, is as illogical as that of an absolute space. But, jumping at the notion, physicists proposed to measure all velocities and rotations relative to it. Alas! the ether refused to disclose the measurements. Explanations were soon forthcoming to account for its reluctance; but these were so far-reaching that they explained away the ether itself in the sense in which it was commonly understood. At any rate, they proved that this creature of the scientific imagination was not one, but many. It quite failed to satisfy the cravings for a permanent standard against which motion might be measured. The problem was left exactly where it was before. This was prewar relativity, summarised by Einstein in 1905. The physicists complained loudly that he was taking away their ether.

19 世纪发生了重大的变化。在这个世纪末，电的时代到来了。人类要揭示原子中的秘密。太空也不再是有行星在其中运行的寒冷而空洞的广袤真空。宇宙中的每个部分都注满了运动着的能量。以太，而不是物质，才是最终的存在形式。物质实际上都是带电的。一个重大的问题摆在了当代人的面前：如何将新的电学理论和经典的牛顿力学原理统一起来。这样，最小作用原理就变得重要起来；因为它是新旧两种宇宙观相通的部分，在两者中，动能和势能之差的时间平均都应该取最小值。

物质的电学理论遇到的一个最主要的问题是，无法把万有引力引入到这个理论框架之中。在电学理论遵从平方反比定律的情况下，新兴的原子物理的蓬勃发展却被忽视掉了，电学独自向破解宇宙之谜的目标发起了挑战。电学理论力图说明物质的所有其他性质，而唯独将人们最为熟知的属性排除在外。它可以说清铜和玻璃之间存在差别的原因，但是不能解释它们共同的特性：重量。至此，以太学说也开始默默地销声匿迹了。

在牛顿的宇宙框架中最令人困扰的一点是：它无法与时空相对性的哲学学说达成一致。加速度在力学中是一个关键的量，它反映了物体运动状态的变化。但这并不意味着牛顿假设了空间中存在一个终极的参考系，一个物体的实际速度相对于这个参考系是唯一确定的，因为在他的理论中，任意一个质点的运动速度总可以加上一个常数速度。这种处理方法的严重缺陷是，当处理两个参考系的问题时，如果其中一个参考系相对于另一个做转动或者非匀速运动，那么前面所说的法则就不可能保持如此简单的形式。这样看来，如果牛顿是正确的，那么在空间中，“固定方向”是有意义的，但“固定位置”却没有任何意义。两者看似应该同时成立或者同时不成立，但是物理上的关系显然是有区别的。为什么会这样呢？我们还不知道。无论我们采用什么样的新理论，都要考虑到上面的问题。

当我们引入了以太的概念，用它来代替空无一物的空间时，问题看似得到了解决，尽管我们也许注意到，在传播这一概念的时候，一些哲学家，比如孔德，认为无限且无形的以太和绝对空间一样不合逻辑。但让我们先把这些看法抛在一边，物理学家们提出要测量所有相对于以太的平动和转动。唉！可惜以太却拒绝我们的测量。很快就出现了对这一难题的解释，但是这与能够用一种大家可以接受的方式来为以太辩解还有非常遥远的距离。无论如何，它们证明了这种科学想象的创造并不是唯一的，而是有很多种。我们渴望找到一种可以用来测量运动的永久标准，而以太的概念是非常失败的。问题和从前一样没有得到解决。还是战争之前由爱因斯坦于 1905 年总结出来的相对论解决了这个问题。物理学家们则强烈地抱怨爱因斯坦摒弃了他们的以太。

Let it not be thought, however, that the results of the hypothesis then advanced were purely negative. They showed quite clearly that many current ideas must be modified, and in what direction this must be done. Most notably it emphasised the fact that inertia is not a fundamental and invariable property of matter; rather it must be supposed that it is consequent upon the property of energy. And, again, energy is a relative term. One absolute quantity alone remained; one only stood independent of the taste or fancy of the observer, and that was "action". While the ether and the associated system of measurement could be selected as any one of a legion, the principle of least action was satisfied in each of them, and the magnitude of the action was the same in all.

But, still, gravitation had to be left out; and the question from which Einstein began the great advance now consummated in success was this. If energy and inertia are inseparable, may not gravitation, too, be rooted in energy? If the energy in a beam of light has momentum, may it not also have weight?

The mere thought was revolutionary, crude though it be. For if at all possible it means reconsidering the hypothesis of the constancy and universality of the velocity of light. This hypothesis was essential to the yet infant principle of relativity. But if called in question, if the velocity of light is only approximately constant because of our ordinary ways of measuring, the principle of relativity, general as it is, becomes itself an approximation. But to what? It can only be to something more general still. Is it possible to maintain anything at all of the principle with that essential limitation removed?

Here was exactly the point at which philosophers had criticised the original work of Einstein. For the physicist it did too much. For the philosopher it was not nearly drastic enough. He asked for an out-and-out relativity of space and time. He would have it that there is no ultimate criterion of the equality of space intervals or time intervals, save complete coincidence. All that is asked is that the order in which an observer perceives occurrences to happen and objects to be arranged shall not be disturbed. Subject to this, any way of measuring will do. The globe may be mapped on a Mercator projection, a gnomonic, a stereographic, or any other projection; but no one can say that one is a truer map than another. Each is a safe guide to the mariner or the aviator. So there are many ways of mapping out the sequences of events in space and time, all of which are equally true pictures and equally faithful servants.

This, then, was the mathematical problem presented to Einstein and solved. The pure mathematics required was already in existence. An absolute differential calculus, the theory of differential invariants, was already known. In pages of pure mathematics that the majority must always take as read, Riemann, Christoffel, Ricci, and Levi-Civita supplied him with the necessary machinery. It remained out of their equations and expressions to select some which had the nearest kinship to those of mathematical physics and to see what could be done with them.

(**104**, 354-356; 1919)

但是，我们不要以为后来提出的假说都是不正确的。这些假说明确地表明现有的许多观点需要修正，并且说明了修正的方向应该在哪里。尤其是它强调了一个事实：惯性不是物质的一个不会发生变化的基本属性，而应该被看作是随着能量的变化而变化的。我们要再次强调，能量是一个相对量。一个绝对量是独立不变的；它不依赖于观察者的体验和想象，这就是“作用”。当以太及与它相关的测量系统被选择作为大量作用中的任意一员时，它们都满足最小作用原理，并且作用量的总和不变。

但是，万有引力仍然没有被考虑进去，爱因斯坦正是从这个问题出发，现在已成功地获得了巨大的进展。如果能量和惯性是不可分割的，那么重力难道就不能建立在能量的基础之上吗？如果能量是一束具有动量的光束，那么它为什么不能有重量呢？

这样的想法是革命性的，尽管它还不够成熟。因为，如果这是可能的，那么它就意味着需要重新考虑光速不变性和普适性的假说。这个假说是尚不成熟的相对论的基本原则。但是如果它被质疑，如果光速只是因为我们通常的测量方法不够精确才大致不变，那么，被大家普遍接受的相对论法则就只是一种近似，就像它本身的系统一样。但这是对什么的近似呢？只能是对一种更加普遍的原理的近似。当消除了那种根本上的限制以后，原来的法则中还有没有什么东西可以保留下来呢？

就是在这个问题上，哲学家们批判了爱因斯坦早期的工作。对物理学家来说，这样的批评太偏激了。对哲学家来说，这种批评还远算不上严厉。爱因斯坦开始寻找一种彻底的时空相对论。他认为，除了完全重合之外，不存在衡量空间间隔和时间间隔完全相等的绝对标准。他只要求观察者观察到的事件的发生顺序和物体摆放的顺序不被打乱。在这个前提下，任何测量方法都将是可行的。这就好像我们可以用墨卡托投影法、心射切面投影法、立体投影法或者任何一种其他的方法去画地球，而没有人会说其中哪一种地图较之其他地图更准确。飞行员和海员使用任何一种地图都是安全可靠的。因此，也有很多方法可以标定时空中事件发生的顺序，每一种都描述了真实的情况，每一种都同样可信。

这样，爱因斯坦接下来就只需要解决那些数学上的问题了。他所需要的纯数学方法已经存在。绝对微分、微分不变量理论，这些都是已知的。大多数人经常研读的纯数学著作，如黎曼、克里斯托弗尔、里奇和列维齐维塔的著作，都为爱因斯坦提供了必要的数学工具。剩下的工作就是从那些方程和表达式中选出最接近数学物理的部分，并想办法把它们解出来。

II. The Nature of the Theory

In the first article an attempt was made to show the roads which led to Einstein's adventure of thought. On the physical side briefly it was this. Newton associated gravitation definitely with mass. Electromagnetic theory showed that the mass of a body is not a definite and invariable quantity inherent in matter alone. The energy of light and heat certainly has inertia. Is it, then, also susceptible to gravitation, and, if so, exactly in what manner? The very precise experiments of Eötvös rather indicated that the mass of a body, as indicated by its inertia, is the same as that which is affected by gravitation.

Also, how must the expression of Newton's law of gravitation be modified to meet the new view of mass? How, also, must the electromagnetic theory and the related pre-war relativity be adapted to allow of the effect of gravitation? With the relaxation of the stipulation that the velocity of light shall be constant, will the principle of relativity become more general and acceptable to the philosophic doctrine of relativity, or will it, on the other hand, become completely impossible?

One point arises immediately. The out-and-out relativist will not admit an absolute measure of acceleration any more than of velocity. The effect, however, of an accelerated motion is to produce an apparent change in gravitation; the measure of gravitation at any place must therefore be a relative quantity depending upon the choice which the observer makes as to the way in which he will measure velocities and accelerations. This is one of Einstein's fundamental points. It has been customary in expositions of mechanics to distinguish between so-called "centrifugal force" and "gravitational force". The former is said to be fictitious, being simply a manifestation of the desire of a body to travel uniformly in a straight line. On the other hand, gravitation has been called a real force because associated with a cause external to the body on which it acts.

Einstein asks us to consider the result of supposing that the distinction is not essential. This was his so-called "principle of equivalence". It led at once to the idea of a ray of light being deviated as it passes through a field of gravitational force. An observer near the surface of the earth notes objects falling away from him towards the earth. Ordinarily, he attributes this to the earth's attraction. If he falls with them, his sense of gravitation is lost. His watch ceases to press on the bottom of his pocket; his feet no longer press on his boots. To this falling observer there is no gravitation. If he had time to think or make observations of the propagation of light, according to the principle of equivalence he would now find nothing gravitational to disturb the rectilinear motion of light. In other words, a ray of light propagated horizontally would share in his vertical motion. To an observer not falling, and, therefore, cognisant of a gravitational field, the path of the ray would therefore be bending downward towards the earth.

The systematic working out of this idea requires, as has been remarked, considerable mathematics. All that can be attempted here is to give a faint indication of the line of attack, mainly by way of analogy.

II. 理论的本质

第一篇文章旨在说明爱因斯坦的思考方法，从物理学的角度来看，简要的说明就是这样。牛顿明确地将万有引力和质量联系起来。电磁学理论表明，一个物体的质量并不是物质确定不变的内在属性。光能和热能当然都具有惯性，那么，它也会受到万有引力的影响吗？如果确实如此，确切的作用方式又是怎样的呢？厄缶的精确实验更表明了由其惯性所表示的物体质量同样受到万有引力的影响。

牛顿万有引力定律的描述要怎样修正才能符合关于质量的新观点呢？电磁学理论和战争前提出的相对论要怎样调整才能允许万有引力效应的存在呢？在解除了光速不变的约束之后，相对论原理会不会成为一种更加普遍且被相对性的哲学理论所接受的法则呢？或者说，另一方面，它会不会被证明完全不可行呢？

这里马上就引出了一个观点。彻底的相对论者只能对速度进行绝对测量，却不能对加速度进行绝对测量。然而，加速运动在万有引力场中会发生明显的变化；因此在任意地点，万有引力的测量值肯定都是相对的，它取决于测量者所选择的测量速度和加速度的方式。这是爱因斯坦的主要观点之一。在力学上，对所谓“离心力”和“万有引力”的解释通常是有区别的。前者是一个虚拟的力，仅仅表现了物体要做匀速直线运动的趋势。另一方面，万有引力被认为是一个真实的力，因为它和作用于物体的外界因素有关。

爱因斯坦让我们考虑，如果这种区别不是本质的，结果会怎样。这就是他所说的“等效原理”。它立刻就引出了这样的设想：一束光穿过万有引力场时会发生弯曲。一个在地球表面附近的观察者会看到物体远离自己落向地球。一般来说，他会把这归结为地球的引力。如果他和物体一起下落，那么他对万有引力的感觉就会消失。他的怀表不再压在衣袋底部，他的脚也不再压在靴子上。对于这位正在下落的观察者来说，他是观察不到万有引力存在的。如果他有时间观察和思考光的传播，那么按照等效原理，他将看不到光线的直线运动被万有引力所干扰。换句话说，一束沿水平方向传播的光线，将与观察者一起同时做垂直运动。因此，一个没有下落的观察者可以感觉到万有引力的存在，所以光线在向地球运动的过程中会发生弯曲。

就像前面提到的那样，要把这样一个想法系统地求解出来，需要大量的数学运算。这里我们所能做的只是用模糊的示意来说明这个原理，主要通过类比法。

It is no new discovery to speak of time as a fourth dimension. Every human mind has the power in some degree of looking upon a period of the history of the world as a whole. In doing this, little difference is made between intervals of time and intervals of space. The whole is laid out before him to comprehend in one glance. He can at the same time contemplate a succession of events in time, and the spatial relations of those events. He can, for instance, think simultaneously of the growth of the British Empire chronologically and territorially. He can, so to speak, draw a map, a four-dimensional map, incapable of being drawn on paper, but none the less a picture of a domain of events.

Let us pursue the map analogy in the familiar two-dimensional sense. Imagine that a map of some region of the globe is drawn on some material capable of extension and distortion without physical restriction save that of the preservation of its continuity. No matter what distortion takes place, a continuous line marking a sequence of places remains continuous, and the places remain in the same order along that line. The map ceases to be any good as a record of distance travelled, but it invariably records certain facts, as, for example, that a place called London is in a region called England, and that another place called Paris cannot be reached from London without crossing a region of water. But the common characteristic of maps of correctly recording the shape of any small area is lost.

The shortest path from any place on the earth's surface to any other place is along a great circle; on all the common maps, one series of great circles, the meridians, is mapped as a series of straight lines. It might seem at first sight that our extensible map might be so strained that all great circles on the earth's surface might be represented by straight lines. But, as a matter of fact, this is not so. We might represent the meridians and the great circles through a second diameter of the earth as two sets of straight lines, but then every other great circle would be represented as a curve.

The extension of this to four dimensions gives a fair idea of Einstein's basic conception. In a world free from gravitation we ordinarily conceive of free particles as being permanently at rest or moving uniformly in straight lines. We may imagine a four-dimensional map in which the history of such a particle is recorded as a straight line. If the particle is at rest, the straight line is parallel to the time axis; otherwise it is inclined to it. Now if this map be strained in any manner, the paths of particles are no longer represented as straight lines. Any person who accepts the strained map as a picture of the facts may interpret the bent paths as evidence of a "gravitational field", but this field can be explained right away as due to his particular representation, for the paths can all be made straight.

But our two-dimensional analogy shows that we may conceive of cases where no amount of straining will make all the lines that record the history of free particles simultaneously straight; pure mathematics can show the precise geometrical significance of this, and can write down expressions which may serve as a measure of the deviations that cannot be removed. The necessary calculus we owe to the genius of Riemann and Christoffel.

把时间作为第四维并不是一个新的发现。每一个人在某种程度上都会把世界历史的一段时期当作一个整体来看待。在这样做的时候，时间段和空间段没有什么区别。在匆匆一瞥之中所有的东西都呈现到他面前要他去了解。他在仔细考虑一系列事件发生时间的同时，还要将它们和发生地点联系起来。比如，他能同时从时间顺序和疆域范围两方面来考虑大英帝国的扩张。所以可以这样说，他可以画一个地图，一个四维的地图，尽管不能画在纸上，但依然可以描述一系列事件。

让我们用我们所熟悉的二维地图进行类比。我们可以想象有一个描述世界上某个区域的地图，用来制作这个地图的材料可以不受物理限制，随意延展和扭曲以保持它的连续性。不管这种材料如何扭曲变形，它上面表示地点次序的连续直线依然保持连续，沿着这条直线各个地点的排列顺序不变。这种地图无法记录旅行的距离，但是它可以忠实地记录某些特定的事实，比如，伦敦位于英国境内，从伦敦到另一个地方——巴黎，不可能不跨越海洋。但是一般地图所具有的记录任意一小块地方的功能在这种地图中完全丧失了。

在地球表面从一地到另一地的最短路径是沿着大圆的路径；在普通的地图上，一系列的大圆，即经线，是用一系列的直线来表示的。初看起来，我们的可伸缩地图可以被拉伸开，这样，地球表面的所有大圆都可以呈直线。可事实并不是这样。我们可以把经线和另外一组由地球的另外一个直径确定的大圆表示成两组直线，但是这样做之后，所有其他的大圆就只能表示为曲线了。

将上述观点扩展到四维时空，就构成了爱因斯坦的基本概念。我们通常认为在没有万有引力的世界里，自由粒子将永远静止或者做匀速直线运动。我们可以想象一下，在四维的地图中，粒子的历史被记录为一条直线。如果这个粒子是静止的，那么这条直线就平行于时间轴，否则就是倾斜的。现在，如果地图以任意方式被拉伸，那么粒子的路径就不再被表示为直线。如果我们能接受可伸缩地图作为表征事实的方式，就可以把这些弯曲的路径看作是“万有引力场”的证据，但是，这种引力场也可以马上被解释成是由这种特殊的表示方法造成的，因为所有路径都可以变成直线。

但是类似的二维地图告诉我们，没有任何一种变形方式可以使所有记录自由粒子历史的线都同时呈直线；关于这一点，纯数学可以给出它在几何学上的精确证明，还可以给出表达式，用来度量那些不可消除的弯曲。天才的黎曼和克里斯托弗尔给出了我们所需的微积分算法。

Einstein now identifies the presence of curvatures that cannot be smoothed out with the presence of matter. This means that the vanishing of certain mathematical expressions indicates the absence of matter. Thus he writes down the laws of the gravitational field in free space. On the other hand, if the expressions do not vanish, they must be equal to quantities characteristic of matter and its motion. These equalities form the expression of his law of gravitation at points where matter exists.

The reader will ask: What are the quantities which enter into these equations? To this only a very insufficient answer can here be given. If, in the four-dimensional map, two neighbouring points be taken, representing what may be called two neighbouring occurrences, the actual distance between them measured in the ordinary geometrical sense has no physical meaning. If the map be strained, it will be altered, and therefore to the relativist it represents something which is not in the external world of events apart from the observer's caprice of measurement. But Einstein assumes that there is a quantity depending on the relation of the points one to the other which is invariant—that is, independent of the particular map of events. Comparing one map with another, thinking of one being strained into the other, the relative positions of the two events are altered as the strain is altered. It is assumed that the strain at any point may be specified by a number of quantities (commonly denoted g_{rs}), and the invariable quantity is a function of these and of the relative positions of the points.

It is these quantities g_{rs} which characterise the gravitational field and enter into the differential equations which constitute the new law of gravitation.

It is, of course, impossible to convey a precise impression of the mathematical basis of this theory in non-mathematical terms. But the main purpose of this article is to indicate its very general nature. It differs from many theories in that it is not devised to meet newly observed phenomena. It is put together to satisfy a mental craving and an obstinate philosophic questioning. It is essentially pure mathematics. The first impression on the problem being stated is that it is incapable of solution; the second of amazement that it has been carried through; and the third of surprise that it should suggest phenomena capable of experimental investigation. This last aspect and the confirmation of its anticipations will form the subject of the next article.

(**104**, 374-376; 1919)

III. The Crucial Phenomena

In the article last week an attempt was made to indicate the attitude of the complete relativist to the laws which must be obeyed by gravitational matter. The present article deals with particular conclusions.

现在，爱因斯坦认为，无法消除的弯曲表示物质的存在。这意味着，如果某个数学表达式为零则表示没有物质存在。于是他写出了自由空间中万有引力场的定律。另一方面，如果表达式不为零，那么它们一定等于描述物质及其运动的物理量。这些方程就是爱因斯坦在有物质存在的点上构筑的万有引力定律表达式。

读者可能会问：这些方程中都有哪些物理量？在这里我们只能给出一个非常不充分的回答。如果在四维地图中，两个相邻的点被认为代表两个相邻的事件，那么用普通几何方法测量的两点之间的实际距离将没有任何物理意义。如果这个地图发生变形，它就会被改变，所以对于相对论者来说，撇开观察者反复无常的测量结果，它代表了某种不存在于由事件组成的外部世界中的东西。但是爱因斯坦认为有一个物理量依赖于两个点之间的关系，具有不变性，也就是说，它不依赖于某种事件地图。比较一个地图和另一个地图，设想其中一个发生变形而成为另一个，代表两个事件的点的相对位置会随着变形方式的变化而变化。可以假设，任意点的变形可以用一些物理量来表示（一般记做 g_{rs}），而不变量是这些物理量和事件点相对位置的函数。

这个表征万有引力场的物理量 g_{rs} 被引入构成万有引力新定律的微分方程中。

当然，我们不可能用非数学语言将这个理论用数字精确地表达出来。但是这篇文章的主要目的是说明它的普遍特征。这个理论与许多其他理论的不同之处在于，它不是为了解释某个新发现而被构建的。构建它的目的是为了满足精神上的渴望和应对哲学上的质疑。它在本质上是纯数学问题。这个问题给我们的第一印象是，它是不可能被攻破的；第二点出人意料的是，它居然被攻破了；第三个令人惊奇的是，它竟然预言了可以用实验研究的现象。关于最后一方面以及对其预言的证实将是下一篇文章的主题。

III. 关键的现象

上周的文章旨在说明一个完全的相对论者对万有引力物质必须遵循的法则的态度。而本文是要介绍一些特定的结论。

As Minkowski remarked in reference to Einstein's early restricted principle of relativity: "From henceforth, space by itself and time by itself do not exist; there remains only a blend of the two" ("Raum und Zeit", 1908). In this four-dimensional world that portrays all history let (x_1, x_2, x_3, x_4) be a set of coordinates. Any particular set of values attached to these coordinates marks an event. If an observer notes two events at neighbouring places at slightly different times, the corresponding points of the four-dimensional map have coordinates slightly differing one from the other. Let the differences be called (dx_1, dx_2, dx_3, dx_4). Einstein's fundamental hypothesis is this: there exists a set of quantities g_{rs} such that

$$g_{11}dx_1^2 + 2g_{12}dx_1dx_2 + \cdots + g_{44}dx_4^2$$

has the same value, no matter how the four-dimensional map is strained. In any strain g_{rs} is, of course, changed, as are also the differences *dx*.*

If the above expression be denoted by $(ds)^2$, *ds* may conveniently be called the *interval* between two events (not, of course, in the sense of time interval). In the case of a field in which there is no gravitation at all, if dx_4 is taken to be *dt*, it is supposed that ds^2 reduces to the expression $dx_1^2 + dx_2^2 + dx_3^2 - c^2dt^2$, where *c* is the velocity of light. If this is put equal to zero, it simply expresses the condition that the neighbouring events correspond to two events in the history of a point travelling with the velocity of light.

Einstein is now able to write down differential equations connecting the quantities g_{rs} with the coordinates (x_1, x_2, x_3, x_4), which are in complete accord with the requirement of complete relativity. † These equations are assumed to hold at all points of space unoccupied by matter, and they constitute Einstein's law of gravitation.

Planetary Motion

The next step is to find a solution of the equations when there is just one point in space at which matter is supposed to exist, one point which is a singularity of the solution. This can be effected completely ‡ : that is, a unique expression is obtained for the interval between two neighbouring events in the gravitational field of a single mass. This mass is now taken to be the sun.

It is next assumed that in the four-dimensional map (which, by the way, has now a bad twist in it, that cannot be strained out, all along the line of points corresponding to the

* The gravitational field is specified by the set of quantities g_{rs}. When the gravitational field is small, these are all zero, except for g_{44}, which is approximately the ordinary Newtonian gravitational potential.

† These equations take the place of the old Laplace equation $\nabla^2 V = 0$. Just as that equation is the only differential equation of the second order which is entirely independent of any change of ordinary space coordinates, so Einstein equations are uniquely determined by the condition of relativity.

‡ The result is that the invariant interval *ds* is given by $ds^2 = (1 - 2m/r)(dt^2 - dr^2) - r^2(d\theta^2 + \sin^2\theta d\phi^2)$, the four coordinates being now interpreted as time and ordinary spherical polar coordinates.

闵可夫斯基这样评论爱因斯坦早期的狭义相对论："从今以后，单独的空间和单独的时间都将不存在；二者只能作为一个复合体而存在"(《空间和时间》，1908 年)。在这个描述了所有历史事件的四维空间中，(x_1，x_2，x_3，x_4) 被视为一组坐标。把任意一组特定数值代入坐标中，都能表示一个事件。如果一个观察者观察到发生地点和时间都很接近的两个事件，那么四维地图上相应两点的坐标也区别不大。我们把它们之间的差别表示为 (dx_1，dx_2，dx_3，dx_4)。爱因斯坦的基本假设是这样的：存在一组 g_{rs}，无论四维地图发生什么样的变形，

$$g_{11}dx_1^2 + 2g_{12}dx_1dx_2 + \cdots + g_{44}dx_4^2$$

都具有相同的值。在发生变形时，g_{rs} 的值当然会发生变化，差值 dx 也同样会改变。*

如果上面的表达式被记作 $(ds)^2$，为方便起见，我们可以把 ds 称作两个事件的**间隔**（当然不是一般观念中的时间间隔）。在万有引力场不存在的情况下，如果 dx_4 用 dt 来代替，那么 ds^2 就会退化成表达式 $dx_1^2 + dx_2^2 + dx_3^2 - c^2dt^2$，这里 c 是光速。如果这个表达式等于零，则表示一个以光速运动的点所经历的两个相邻的事件。

现在，爱因斯坦就可以写出将物理量 g_{rs} 与坐标 (x_1，x_2，x_3，x_4) 相联系的微分方程了，这与完善相对论的要求完全一致。† 这些方程包含了空间中所有未被物质占据的点，它们构成了爱因斯坦的万有引力定律。

行星的运动

下一步的任务是找到一个空间中只有一个点被物质占据的方程解，而这个点是解的奇点。这完全可以做到‡：也就是说，对于单一质点在引力场中的两个相邻事件的间隔，我们可以得到唯一的表达式。太阳可以当作这样的一个质点。

接下来假设在四维地图（现在这个地图严重扭曲，不能把对应于太阳每个时刻位置的点组成的线拉伸开）中，在太阳引力场中运动的质点的路径将是图上任意两

---

* 万有引力场由一组物理量 g_{rs} 来说明，当万有引力场很小的时候，这些值除了 g_{44} 以外都为零，这就近似成为牛顿的万有引力势场。

† 这些方程代替了旧的拉普拉斯方程 $\nabla^2V=0$。正如拉普拉斯方程是唯一一个完全不受普通空间坐标体系变化影响的二阶微分方程一样，爱因斯坦方程是唯一一个由相对论条件确定的方程。

‡ 结果是：不变的间隔 ds 可以由下式确定，$ds^2 = (1 - 2m/r)(dt^2 - dr^2) - r^2(d\theta^2 + \sin^2\theta d\phi^2)$，这四个坐标可以解释为时间和普通的球面极坐标。

positions of the sun at every instant of time) the path of a particle moving under the gravitation of the sun will be the most direct line between any two points on it, in the sense that the sum of all the intervals corresponding to all the elements of its path is the least possible.* Thus the equations of motion are written down. The result is this:

The motion of a particle differs only from that given by the Newtonian theory by the presence of an additional acceleration towards the sun equal to three times the mass of the sun (in gravitational units) multiplied by the square of the angular velocity of the planet about the sun.

In the case of the planet Mercury, this new acceleration is of the order of 10^{-8} times the Newtonian acceleration. Thus up to this order of accuracy Einstein's theory actually arrives at Newton's laws: surely no dethronement of Newton.

The effect of the additional acceleration can easily be expressed as a perturbation of the Newtonian elliptic orbit of the planet. It leads to the result that the major axis of the orbit must rotate in the plane of the orbit at the rate of 42.9″ per century.

Now it has long been known that the perihelion of Mercury does actually rotate at the rate of about 40″ per century, and Newtonian theory has never succeeded in explaining this, except by *ad hoc* assumptions of disturbing matter not otherwise known.

Thus Einstein's theory almost exactly accounts for the one outstanding failure of Newton's scheme, and, we may note, does not introduce any discrepancy where hitherto there was agreement.

The Deflection of Light by Gravitation

The new theory having justified itself so far, it was thought worth while for British astronomers to devote their main energies at the recent solar eclipse to testing its prediction of an entirely new phenomenon.

As was remarked above, the propagation of light in the ordinary case of freedom from gravitational effect is represented by the equation $ds = 0$.

This Einstein boldly transfer to his generalised theory. After all, it is quite a natural assumption. The propagation of light is a purely objective phenomenon. The emission of a disturbance from one point at one moment, and its arrival at another point at another moment, are events distinct and independent of the existence of an observer. Any law that connects them must be one which is independent of the map the observer uses; ds being an invariant quantity, $ds = 0$ expresses such an invariant law.

* This corresponds to the fact that in a field where there is no acceleration at all the path of a particle is the shortest distance between two points.

点之间最直的线，即与路径上所有组成部分对应的所有间隔之和尽可能最小。* 这样，就可以写出运动方程。结果如下：

一个质点的运动与牛顿理论给出的运动的不同之处在于多出了一个朝向太阳的加速度，这个加速度的值等于三倍的太阳质量（万有引力单位）乘以行星绕太阳运动的角速度的平方。

对于水星，这个新加速度的量级是牛顿加速度的 10^{-8}。这样，低于这个精确度，爱因斯坦理论就还原成了牛顿理论：牛顿理论当然不会失效。

这个多出来的加速度可以被看作是牛顿椭圆行星轨道的一种扰动。这就造成了轨道主轴在轨道平面中以每世纪 42.9 角秒的速度进动。

很久以前我们就知道，水星的近日点的确在以每世纪约 40 角秒的速度进动，牛顿理论从来没有成功地解释过这个现象，除非特意假设有一个在其他情况下未曾出现的干扰物体。

这样，爱因斯坦的理论就彻底解决了牛顿理论框架中的一个重要不足，并且，我们注意到，爱因斯坦理论和牛顿理论没有矛盾，至今它们仍然是统一的。

光在万有引力作用下的偏转

到目前为止，这个新理论的正确性已经得到了证明，英国天文学家们正将他们的主要精力放在近期的日食上，为的是检验此理论所预言的一个全新的现象，大家都认为这是一件值得做的事情。

正如上面所说的那样，在没有万有引力效应的情况下，光的传播可以表示为方程 $ds = 0$。

爱因斯坦大胆地将它移植到了他的广义理论中。毕竟，这是一个很自然的假设。光的传播是一个纯客观的现象。在某一时刻某一点发生的干扰，以及这个干扰在另一时刻到达另一点，这两者是不同的事件，与观察者是否存在无关。任何将它们联系起来的法则都一定不依赖于观察者所使用的地图；ds 是一个不变的量，$ds = 0$ 表示出了这样一个不变的法则。

* 它对应于这样一个事实：在一个没有加速度的场中，粒子运动的路径是两点之间的最短距离。

This leads at once to a law of variation of the velocity of light in the gravitational field of the sun.

$$v = c(1 - 2m/r)$$

Here m, as before, is the mass of the sun in gravitational units, and is equal to 1.47 kilometres, while c is the velocity of light at a great distance from the sun. Thus the path of a ray is the same as that if, on the ordinary view, it were travelling in a medium the refractive index of which was $(1 - 2m/r)^{-1}$. In this medium the refractive index would increase in approaching the sun, so that the rays would be bent round towards the sun in passing through it. The total amount of the deflection for a ray which just grazes the sun's surface works out to be 1.75″, falling off as the inverse of the distance of nearest approach.

The apparent position of a star near to the sun is thus further from the sun's centre than the true position. On the photographic plate in the actual observations made by the Eclipse Expedition the displacement of the star image is of the order of a thousandth of an inch. The measurements show without doubt such a displacement. The stars observed were, of course, not exactly at the edge of the sun's disc; but on reduction, allowing for the variation inversely as the distance, they give for the bend of a ray just grazing the sun the value 1.98″, with a probable error of 6 percent, in the case of the Sobral expedition, and of 1.64″ in the Principe expedition.

The agreement with the theory is close enough, but, of course, alternative possible causes of the shift have to be considered. Naturally, the suggestion of an actual refracting atmosphere surrounding the sun has been made. The existence of this, however, seems to be negatived by the fact that an atmosphere sufficiently dense to produce the refraction in question would extinguish the light altogether, as the rays would have to travel a million miles or so through it. The second suggestion, made by Prof. Anderson in *Nature* of December 4, that the observed displacement might be due to a refraction of the ray in travelling through the earth's atmosphere in consequence of a temperature gradient within the shadow cone of the moon, seems also to be negatived. Prof. Eddington estimates that it would require a change of temperature of about 20℃ per minute at the observing station to produce the observed effect. Certainly no such temperature change as this has ever been noted; and, in fact, in Principe, at which the Cambridge expedition made its observations, there was practically no fall of temperature.

Gravitation and the Solar Spectrum

It was suggested by Einstein that a further consequence of his theory would be an apparent discrepancy of period between the vibrations of an atom in the intense gravitational field of the sun and the vibrations of a similar atom in the much weaker field of the earth. This is arrived at thus. An observer would not be able to infer the intensity of the gravitational field in which he was placed from any observations of atomic vibrations in the same field: that is, an observer on the sun would estimate the period of vibration of

这立刻就引出了在太阳引力场中光速不变的定律。

$$v = c(1-2m/r)$$

这里的 m 和前面一样，是太阳在万有引力单位下的质量，它相当于 1.47 千米，c 是远离太阳处的光速。这样，从通常的观点上看，一束光的行进路线和它在折射率为 $(1-2m/r)^{-1}$的介质中传播一样。在这个介质中，越接近太阳折射率就越大，所以光线在经过太阳的时候就会发生弯曲。一束刚好掠过太阳表面的光的偏转角度是 1.75 角秒，偏转角度会随着光线与太阳之间最小距离的倒数的减小而减小。

靠近太阳的恒星的视位置比它的真实位置离日心更远。在观测日食时所拍的照相底片上，恒星图像位移的数量级为千分之一英尺。测量结果毫无疑问地显示了这样的一个位移。当然，观测的恒星并非恰好在日面的边缘；但是可以利用偏转量和距离成反比的关系进行化规，在索布拉尔的观测队测算出一束刚好掠过太阳表面的光线的偏转角度是 1.98 角秒，允许的误差范围是 6%；在普林西比岛的观测队得到的结果是 1.64 角秒。

实验和理论已经足够吻合，但是，也必须考虑到引起位移的其他可能原因。很自然的，有人怀疑在太阳周围有一个能发生真正的折射效应的大气层。但是，它的存在却可以被以下因素否定，即产生这样的折射作用要求大气层足够厚，而这么厚的大气层会使光线消失，因为光线需要经过 100 万公里左右才能穿过去。第二种可能性是安德森教授在 12 月 4 日的《自然》上提出的。他认为，观察到的位移可能是由光线穿过地球大气层时的折射造成的，因为在月影锥内存在温度梯度，温度梯度可以引发折射现象，这个猜测也不成立。爱丁顿教授估算过，要产生我们观测到的效应，观测站的温度变化应达到每分钟 20℃之多。这样的温度变化当然从未有过；事实上，在剑桥考察队进行观测的普林西比岛，温度并未下降过。

万有引力和太阳光谱

爱因斯坦指出，他的理论还会进一步导出这样的结果：原子在太阳的强引力场中的振动周期和其在弱得多的地球引力场中的振动周期有明显的差别。这个结果是这样得到的：当一个观察者与他所观察的振动原子处在同一个引力场中时，他不可能通过观察原子的振动来判断这个引力场的强度。也就是说，一个在太阳上的观察者测量到的原子振动周期与一个相似原子在地球上的振动周期相同，前提是他本人

an atom there to be the same that he would find for a similar atom in the earth's field if he transported himself thither. But on transferring himself he automatically changes his scale of time; in the new scale of time the solar atom vibrates differently, and, therefore, is not synchronous with the terrestrial atom.

Observations of the solar spectrum so far are adverse to the existence of such an effect. What, then, is to be said? Is the theory wrong at this point? If so, it must be given up, in spite of its extraordinary success in respect of the other two phenomena.

Sir Joseph Larmor, however, is of opinion that Einstein's theory itself does not in reality predict the displacement at all. The present writer shares his opinion. Imagine, in fact, two identical atoms originally at a great distance from both sun and earth. They have the same period. Let an observer A accompany one of these into the gravitational field of the sun, and an observer B accompany the other into the field of the earth. In consequence of A and B having moved into different gravitational fields, they make different changes in their scales of time, so that actually the solar observer A will find a different period for the solar atom from that which B, on the earth, attributes to his atom. It is only when the two observers choose so to measure space and time that they consider themselves to be in identical gravitational fields that they will estimate the periods of the atoms alike. This is exactly what would happen if B transferred himself to the same position as A. Thus, though an important point remains to be cleared up, it cannot be said that it is one which at present weighs against Einstein's theory.

(**104**, 394-395; 1919)

来到地球。但是在他转换观测地点的时候，他会自动调整时间尺度；在新的时间尺度中，太阳引力场中的原子振动周期将发生变化，所以就与地球上的原子不同步了。

迄今为止，太阳光谱的观测结果并不支持这种效应的存在。接下来我们该说些什么呢？这个理论在这一点上是否错了？如果是这样，它必须被放弃，尽管它成功地解释了另外两个现象。

然而，约瑟夫·拉莫尔爵士认为，事实上爱因斯坦的理论本身并没有预言过这种移位效应。本文作者也同意他的观点。事实上，我们可以想象，两个相同的原子最初处于既远离太阳又远离地球的某地。它们具有相同的周期。让观察者 A 伴随着其中一个原子来到太阳引力场中，让观察者 B 伴随着另一个原子来到地球引力场中。由于 A 和 B 进入的引力场不同，他们的时间尺度发生的变化也不同，因此，太阳上的观察者 A 将发现太阳上原子的振动周期与地球上的观察者 B 看到的地球上原子的振动周期不同。只有当两个观察者在同一个引力场中来测量时空时，他们才会判断出两原子的周期相同。如果 B 来到 A 的位置，就会发生以上所说的情况。所以，尽管还有一个要点有待澄清，但我们目前还不能说这是一个与爱因斯坦理论相悖的现象。

（王静 翻译；鲍重光审稿）

A Brief Outline of the Development of the Theory of Relativity*

A. Einstein

Editor's Note

By 1921, Einstein's theory of relativity was widely accepted. Here he describes the theory's historical development, starting from the aim to rid physics of reliance on action at a distance. Maxwell had achieved this for electricity and magnetism, and his mathematical formulation led others to suppose that all space is filled with an ether that carried the electric and magnetic fields. This led to difficulties, which Einstein overcame by abandoning the belief that events may be simultaneous regardless of an observer's motion. But this new understanding didn't encompass gravity. The supposition that gravity and inertia are identical prompted the general theory of relativity. Einstein wonders if gravitational and electrical phenomena might be unified in a theory of all nature's forces—a theory physicists still seek today.

THERE is something attractive in presenting the evolution of a sequence of ideas in as brief a form as possible, and yet with a completeness sufficient to preserve throughout the continuity of development. We shall endeavour to do this for the Theory of Relativity, and to show that the whole ascent is composed of small, almost self-evident steps of thought.

The entire development starts off from, and is dominated by, the idea of Faraday and Maxwell, according to which all physical processes involve a continuity of action (as opposed to action at a distance), or, in the language of mathematics, they are expressed by partial differential equations. Maxwell succeeded in doing this for electro-magnetic processes in bodies at rest by means of the conception of the magnetic effect of the vacuum-displacement-current, together with the postulate of the identity of the nature of electro-dynamic fields produced by induction, and the electro-static field.

The extension of electro-dynamics to the case of moving bodies fell to the lot of Maxwell's successors. H. Hertz attempted to solve the problem by ascribing to empty space (the ether) quite similar physical properties to those possessed by ponderable matter; in particular, like ponderable matter, the ether ought to have at every point a definite velocity. As in bodies at rest, electro-magnetic or magneto-electric induction ought to be determined by the rate of change of the electric or magnetic flow respectively, provided that these velocities of alteration are referred to surface elements moving with the body. But the theory of Hertz was opposed to the fundamental experiment of Fizeau on the

* Translated by Dr. Robert W. Lawson.

相对论发展概述*

爱因斯坦

编者按

爱因斯坦的相对论在1921年得到了大家的广泛认可。在这篇演讲稿中，爱因斯坦从物理学力图摆脱超距作用的影响开始，对这个理论的历史沿革进行了回顾。麦克斯韦的电磁理论完成了这一使命，其他物理学家根据他的数学公式提出了以太假说，即认为所有的空间中都充满了作为电场和磁场媒介的以太。这使相对性原理遇到了困难，而爱因斯坦通过放弃事件的同时性可能与观测者的运动无关的观点摆脱了这个困境。但是这种新的理解没有考虑到重力。爱因斯坦猜测重力和惯性可能具有同一性，这一构想促使他提出了广义相对论。爱因斯坦设想重力和电现象或许也可以用一种适用于所有自然力的理论统一起来，这也是今天的物理学家们正在寻找的理论。

用尽可能简练的语言来阐述一系列观念的演变，但仍充分完整地把这种演变的连续性保留下来，这是一件很吸引人的事。在讲述相对论的发展时，我们将尽力做到这一点，并说明其整个发展过程是由一系列细微而又不言而喻的思维过程构成的。

整个发展历程始于并受制于法拉第和麦克斯韦的观念，按照他们的观念，所有的物理过程都包含连续作用（与超距作用相反），或者用数学语言来表示就是利用偏微分方程来描述物理过程。麦克斯韦利用真空位移电流的磁效应概念以及感生电动力场和静电场在本质上完全相同这一假定，成功地构筑了描述静止介质中电磁过程的偏微分方程。

把电动力学理论推广到运动物体的重任落在了麦克斯韦的后继者的身上。赫兹试图通过赋予虚空（以太）与一般有重物质颇类似的物理性质来解决这个问题。特别是，与有重物质一样，以太在空间的每一点上都应该有确定的速度。正如静止物体那样，如果电流或磁流的变化速度是以随物体一起运动的曲面元作为参考的话，那么电磁感应或者磁电感应应当分别由电流或磁流的变化率决定。但是赫兹的理论与斐索有关光在流动液体中传播的基本实验相矛盾。就是说，麦克斯韦理论对运动

---

* 由罗伯特·劳森博士翻译。

propagation of light in flowing liquids. The most obvious extension of Maxwell's theory to the case of moving bodies was incompatible with the results of experiment.

At this point, H. A. Lorentz came to the rescue. In view of his unqualified adherence to the atomic theory of matter, Lorentz felt unable to regard the latter as the seat of continuous electro-magnetic fields. He thus conceived of these fields as being conditions of the ether, which was regarded as continuous. Lorentz considered the ether to be intrinsically independent of matter, both from a mechanical and a physical point of view. The ether did not take part in the motions of matter, and a reciprocity between ether and matter could be assumed only in so far as the latter was considered to be the carrier of attached electrical charges. The great value of the theory of Lorentz lay in the fact that the entire electro-dynamics of bodies at rest and of bodies in motion was led back to Maxwell's equations of empty space. Not only did this theory surpass that of Hertz from the point of view of method, but with its aid H. A. Lorentz was also pre-eminently successful in explaining the experimental facts.

The theory appeared to be unsatisfactory only in *one* point of fundamental importance. It appeared to give preference to one system of coordinates of a particular state of motion (at rest relative to the ether) as against all other systems of coordinates in motion with respect to this one. In this point the theory seemed to stand in direct opposition to classical mechanics, in which all inertial systems which are in uniform motion with respect to each other are equally justifiable as systems of coordinates (Special Principle of Relativity). In this connection, all experience also in the realm of electro-dynamics (in particular Michelson's experiment) supported the idea of the equivalence of all inertial systems, *i.e.* was in favour of the special principle of relativity.

The Special Theory of Relativity owes its origin to this difficulty, which, because of its fundamental nature, was felt to be intolerable. This theory originated as the answer to the question: Is the special principle of relativity really contradictory to the field equations of Maxwell for empty space? The answer to this question appeared to be in the affirmative. For if those equations are valid with reference to a system of coordinates K, and we introduce a new system of coordinates K′ in conformity with the—to all appearances readily establishable—equations of transformation

$$\left.\begin{aligned} x' &= x - vt \\ y' &= y \\ z' &= z \\ t' &= t \end{aligned}\right\} \text{(Galileo transformation)},$$

then Maxwell's field equations are no longer valid in the new coordinates (x', y', z', t'). But appearances are deceptive. A more searching analysis of the physical significance of space and time rendered it evident that the Galileo transformation is founded on arbitrary assumptions, and in particular on the assumption that the statement of simultaneity has a meaning which is independent of the state of motion of the system of coordinates used. It was shown that the field equations for *vacuo* satisfy the special principle of relativity,

物体的这种最直接的推广与实验结果不符。

正在这时，洛伦兹进行了补救。由于洛伦兹是物质原子理论的忠实支持者，所以他觉得不能把物质看成是连续电磁场的所在地。因此，他设想这些场是连续的以太的某种状态。洛伦兹认为，从力学和物理学两方面的观点来看，以太在本质上与物质无关。以太不参与物质的运动，以太和物质之间的相互关系仅在于，物质被看成是所附电荷的载体。洛伦兹理论的重要价值在于，它使包括静止物体和运动物体在内的整个电动力学回归到了真空中的麦克斯韦方程。该理论不仅在方法论上超越了赫兹的理论，而且洛伦兹还利用它非常成功地解释了许多实验事实。

这个理论似乎仅仅在一个重要的基本点上不能令人满意。这就是，似乎某个具有特殊运动状态的坐标系（它相对于以太是静止的）要比相对于这个坐标系运动的所有其他坐标系更加优越。从这一点上来看，这个理论好像违背了经典力学，因为在经典力学中，所有相互间做匀速运动的惯性系都同样有理由被用来当作坐标系（狭义相对性原理）。在这一点上，包括电动力学领域在内的所有经验（尤其是迈克尔逊实验）都支持所有惯性系均等价这一观点，即都支持狭义相对性原理。

狭义相对论就是为解决这一困难而诞生的，这个困难由于它具有的根本性而无法让人容忍。狭义相对论最初被用于解答下述问题：狭义相对性原理真的与真空中的麦克斯韦场方程矛盾吗？答案似乎是肯定的。因为，如果某些方程对于坐标系 K 是成立的，而且我们引进一个新的坐标系 K′，使它符合于（显然容易做到）如下的变换方程：

$$\left.\begin{aligned} x' &= x - vt \\ y' &= y \\ z' &= z \\ t' &= t \end{aligned}\right\}\text{（伽利略变换）},$$

那么麦克斯韦场方程组在新的坐标系（x'，y'，z'，t'）中不再成立。但表面现象是靠不住的，在更透彻地分析时间和空间的物理意义后发现，伽利略变换是建立在几个相当任意的假设上面的，尤其是假设同时性的陈述与所使用的坐标系的运动状态无关。研究表明，如果我们利用下面的变换方程，则真空中的场方程可以满足狭义

provided we make use of the equations of transformation stated below:

$$\left.\begin{aligned} x' &= \frac{x - vt}{\sqrt{1 - v^2/c^2}} \\ y' &= y \\ z' &= z \\ t' &= \frac{t - vx/c^2}{\sqrt{1 - v^2/c^2}} \end{aligned}\right\} \text{(Lorentz transformation)}$$

In these equations x, y, z represent the coordinates measured with measuring-rods which are at rest with reference to the system of coordinates, and t represents the time measured with suitably adjusted clocks of identical construction, which are in a state of rest.

Now in order that the special principle of relativity may hold, it is necessary that all the equations of physics do not alter their form in the transition from one inertial system to another, when we make use of the Lorentz transformation for the calculation of this change. In the language of mathematics, all systems of equations that express physical laws must be co-variant with respect to the Lorentz transformation. Thus, from the point of view of method, the special principle of relativity is comparable to Carnot's principle of the impossibility of perpetual motion of the second kind, for, like the latter, it supplies us with a general condition which all natural laws must satisfy.

Later, H. Minkowski found a particularly elegant and suggestive expression for this condition of co-variance, one which reveals a formal relationship between Euclidean geometry of three dimensions and the space-time continuum of physics.

Euclidean Geometry of Three Dimensions.	*Special Theory of Relativity.*
Corresponding to two neighbouring points in space, there exists a numerical measure (distance ds) which conforms to the equation $ds^2 = dx_1^2 + dx_2^2 + dx_3^2$	Corresponding to two neighbouring points in space-time (point events), there exists a numerical measure (distance ds) which conforms to the equation $ds^2 = dx_1^2 + dx_2^2 + dx_3^2 + dx_4^2$.
It is independent of the system of coordinates chosen, and can be measured with the unit measuring-rod.	It is independent of the inertial system chosen, and can be measured with the unit measuring-rod and a standard clock. x_1, x_2, x_3 are here rectangular coordinates, whilst $x_4=\sqrt{-1}ct$ is the time multiplied by the imaginary unit and by the velocity of light.
The permissible transformations are of such a character that the expression for ds^2 is invariant, *i.e.* the linear orthogonal transformations are permissible.	The permissible transformations are of such a character that the expression for ds^2 is invariant, *i.e.* those linear orthogonal substitutions are permissible which maintain the semblance of reality of x_1, x_2, x_3, x_4. These substitutions are the Lorentz transformations.
With respect to these transformations, the laws of Euclidean geometry are invariant.	With respect to these transformations, the laws of physics are invariant.

From this it follows that, in respect of its *rôle* in the equations of physics, though not with regard to its physical significance, time is equivalent to the space coordinates (apart from the relations of reality). From this point of view, physics is, as it were, a Euclidean geometry of four dimensions, or, more correctly, a statics in a four-dimensional Euclidean continuum.

相对性原理：

$$\left.\begin{aligned} x' &= \frac{x - vt}{\sqrt{1 - v^2/c^2}} \\ y' &= y \\ z' &= z \\ t' &= \frac{t - vx/c^2}{\sqrt{1 - v^2/c^2}} \end{aligned}\right\}\text{（洛伦兹变换）}$$

在上述方程中 x、y、z 表示位置坐标，用相对于坐标系静止的量尺来测量；t 表示时间，用处于静止状态的经过适当校准并且具有相同构造的时钟来测量。

现在，为了使狭义相对性原理成立，要求所有的物理方程在使用洛伦兹变换来计算它们从一个惯性系到另一个惯性系的转换时其形式保持不变。用数学语言来描述就是，所有描述物理定律的方程相对于洛伦兹变换必须是协变的。因此，从方法论的角度来看，狭义相对性原理可以与第二种永恒运动不能实现的卡诺定理相比拟，因为正如卡诺定理一样，狭义相对性原理为我们提供了所有自然规律必须遵守的一般法则。

随后，闵可夫斯基找到了一个特别简洁而又极具启发性的方式来表述这个协变条件，揭示了三维欧几里德几何学与物理学中时空连续统之间的对应关系。

三维欧几里德几何学	狭义相对论
对应于空间中两个相邻的点，存在一种按如下方程计算的数值度量（距离 ds）： $ds^2 = dx_1^2 + dx_2^2 + dx_3^2$	对应于时空中两个相邻的点（点事件），存在一种按如下方程计算的数值度量（距离 ds）： $ds^2 = dx_1^2 + dx_2^2 + dx_3^2 + dx_4^2$
该距离与所选的坐标系无关，并可以用单位量尺测量。	该距离与所选择的惯性系无关，并可以用单位量尺和标准时钟测量。这里的 x_1，x_2，x_3 是直角坐标系中的坐标，而 $x_4 = \sqrt{-1}ct$ 是时间和虚数单位以及光速的乘积。
保持 ds^2 表达式不变的变换是允许的，即线性正交变换是允许的。	保持 ds^2 表达式不变的变换是允许的，即那些线性正交变换是允许的，它们保持了 x_1，x_2，x_3，x_4 表面上的实数性，这些变换就是洛伦兹变换。
相对于这些变换，欧几里德几何学中的定律保持不变。	相对于这些变换，物理学中的定律保持不变。

由此可以看出，时间坐标在物理方程中的**作用**与空间坐标等价（除了实数性之外），虽然不是就其物理意义而言。按照这种观点，物理学过去和现在都是一种四维

The development of the special theory of relativity consists of two main steps, namely, the adaptation of the space-time "metrics" to Maxwell's electro-dynamics, and an adaptation of the rest of physics to that altered space-time "metrics". The first of these processes yields the relativity of simultaneity, the influence of motion on measuring-rods and clocks, a modification of kinematics, and in particular a new theorem of addition of velocities. The second process supplies us with a modification of Newton's law of motion for large velocities, together with information of fundamental importance on the nature of inertial mass.

It was found that inertia is not a fundamental property of matter, nor, indeed, an irreducible magnitude, but a property of energy. If an amount of energy E be given to a body, the inertial mass of the body increases by an amount E/c^2, where c is the velocity of light *in vacuo*. On the other hand, a body of mass m is to be regarded as a store of energy of magnitude mc^2.

Furthermore, it was soon found impossible to link up the science of gravitation with the special theory of relativity in a natural manner. In this connection I was struck by the fact that the force of gravitation possesses a fundamental property, which distinguishes it from electro-magnetic forces. All bodies fall in a gravitational field with the same acceleration, or—what is only another formulation of the same fact—the gravitational and inertial masses of a body are numerically equal to each other. This numerical equality suggests identity in character. Can gravitation and inertia be identical? This question leads directly to the General Theory of Relativity. Is it not possible for me to regard the earth as free from rotation, if I conceive of the centrifugal force, which acts on all bodies at rest relatively to the earth, as being a "real" field of gravitation, or part of such a field? If this idea can be carried out, then we shall have proved in very truth the identity of gravitation and inertia. For the same property which is regarded as *inertia* from the point of view of a system not taking part in the rotation can be interpreted as *gravitation* when considered with respect to a system that shares the rotation. According to Newton, this interpretation is impossible, because by Newton's law the centrifugal field cannot be regarded as being produced by matter, and because in Newton's theory there is no place for a "real" field of the "Koriolis-field" type. But perhaps Newton's law of field could be replaced by another that fits in with the field which holds with respect to a "rotating" system of coordinates? My conviction of the identity of inertial and gravitational mass aroused within me the feeling of absolute confidence in the correctness of this interpretation. In this connection I gained encouragement from the following idea. We are familiar with the "apparent" fields which are valid relatively to systems of coordinates possessing arbitrary motion with respect to an inertial system. With the aid of these special fields we should be able to study the law which is satisfied in general by gravitational fields. In this connection we shall have to take account of the fact that the ponderable masses will be the determining factor in producing the field, or, according to the fundamental result of the special theory of relativity, the energy density—a magnitude having the transformational character of a tensor.

欧几里德几何学，或者，更确切地说，是四维欧几里德连续统中的一种静力学。

狭义相对论的发展经历了两个主要阶段，这就是使时空"度规"适合于麦克斯韦电动力学，以及使物理学中的其余部分适合于这个新的时空"度规"。第一个阶段的成果有同时性的相对性、运动对量尺及时钟的影响、运动学的修正，特别是还有新的速度相加定理。第二个阶段给我们提供了在高速情况下对牛顿运动定律的修正，以及关于惯性质量本质的具有基本重要性的知识。

研究表明惯性不是物质的基本属性，也不是一个不能分解的基本量，而只是能量的一个属性。如果我们赋予一个物体大小为 E 的能量，则此物体的惯性质量将增加 E/c^2，这里 c 是光在真空中的传播速度。同样，一个质量为 m 的物体将被认为具有 mc^2 的能量。

此外，我们很快发现很难把引力科学同狭义相对论以自然的方式联系起来。这种情况使我意识到引力具有一种不同于电磁力的基本性质。在引力场中，所有物体都以相同的加速度下落，或者说，一个物体的引力质量和惯性质量在数值上是相等的（这只不过是同一个事实的另一种表达方式）。这种数值上的相同暗示着两者本质上的等同。引力和惯性能够等同吗？这个问题直接导致了广义相对论的产生。如果我把作用于所有相对于地球静止的物体上的离心力想象成是一个"真实的"引力场，或者是这种引力场的一部分，那我难道不能认为地球是不转动的吗？如果这个想法能够实现，那么我们已经真正证明了引力和惯性的等同性。在不随地球转动的参考系里看来是**惯性**的这一特性，在随地球一起转动的参考系里可以被解释为**引力**。按照牛顿的观点，这样的解释是说不通的，因为牛顿定律告诉我们，离心力场不能被看作是由物质产生的，而且因为在牛顿的理论中，没有把"科里奥利场"这种类型的场当成是"真实的"场。但是，或许牛顿的有关场的定律可以用场的另外一种定律取代，这种定律既适合于这种场又在"转动"坐标系中成立？我坚信惯性质量与引力质量的等同性，这使我有绝对的信心认为上述解释是正确的。就这一点来说，我从以下观点中受到了鼓舞。我们熟悉"表观的"场，这些场在那些相对于一个惯性系做任意运动的坐标系中是有效的。借助于这些特殊的场，我们应当有可能研究引力场通常所满足的定律。关于这一点，我们不得不考虑这样的事实，即有重物质是产生场的决定性因素。或者可以这样表达，按照狭义相对论的基本结果，能量密度这个具有张量变换特性的物理量是产生场的决定因素。

On the other hand, considerations based on the metrical results of the special theory of relativity led to the result that Euclidean metrics can no longer be valid with respect to accelerated systems of coordinates. Although it retarded the progress of the theory several years, this enormous difficulty was mitigated by our knowledge that Euclidean metrics holds for small domains. As a consequence, the magnitude *ds*, which was physically defined in the special theory of relativity hitherto, retained its significance also in the general theory of relativity. But the coordinates themselves lost their direct significance, and degenerated simply into numbers with no physical meaning, the sole purpose of which was the numbering of the space-time points. Thus in the general theory of relativity the coordinates perform the same function as the Gaussian coordinates in the theory of surfaces. A necessary consequence of the preceding is that in such general coordinates the measurable magnitude *ds* must be capable of representation in the form

$$ds^2 = \sum_{uv} g_{uv}\, dx_u\, dx_v \ ,$$

where the symbols g_{uv} are functions of the space-time coordinates. From the above it also follows that the nature of the space-time variation of the factors g_{uv} determines, on one hand the space-time metrics, and on the other the gravitational field which governs the mechanical behaviour of material points.

The law of the gravitational field is determined mainly by the following conditions: First, it shall be valid for an arbitrary choice of the system of coordinates; secondly, it shall be determined by the energy tensor of matter; and thirdly, it shall contain no higher differential coefficients of the factors g_{uv} than the second, and must be linear in these. In this way a law was obtained which, although fundamentally different from Newton's law, corresponded so exactly to the latter in the deductions derivable from it that only very few criteria were to be found on which the theory could be decisively tested by experiment.

The following are some of the important questions which are awaiting solution at the present time. Are electrical and gravitational fields really so different in character that there is no formal unit to which they can be reduced? Do gravitational fields play a part in the constitution of matter, and is the continuum within the atomic nucleus to be regarded as appreciably non-Euclidean? A final question has reference to the cosmological problem. Is inertia to be traced to mutual action with distant masses? And connected with the latter: Is the spatial extent of the universe finite? It is here that my opinion differs from that of Eddington. With Mach, I feel that an affirmative answer is imperative, but for the time being nothing can be proved. Not until a dynamical investigation of the large systems of fixed stars has been performed from the point of view of the limits of validity of the Newtonian law of gravitation for immense regions of space will it perhaps be possible to obtain eventually an exact basis for the solution of this fascinating question.

(**106**, 782-784; 1921)

另一方面，基于对狭义相对论度规结果的考虑导致了这样的结论，即欧几里德度规不再适用于加速参考系。尽管它使理论的进程延迟了几年，但是这个巨大的困难在我们认识到欧几里德度规对于小的区域依然适用之后就变得比较容易解决了。结果，ds 这个迄今在狭义相对论中定义的物理量，在广义相对论中其物理含义仍保持不变。但是坐标本身失去了直接的意义，而完全退化成没有物理意义的数字，它们的唯一用途是标记时空点，因此广义相对论中的坐标与曲面论中高斯坐标的作用相同。以上的叙述必然给出这样一个结论：可测量的量 ds 必定可以用这种广义坐标表达为以下形式：

$$ds^2 = \sum_{uv} g_{uv}\, dx_u\, dx_v \text{ ,}$$

这里符号 g_{uv} 是时空坐标的函数。以上的分析还表明，因子 g_{uv} 随时空变化的特性，一方面决定了时空度规，另一方面还决定了支配质点力学行为的引力场。

引力场的定律主要取决于以下几个条件：第一，它应该在任意选择的坐标系中都是有效的；第二，它应当由物质的能量张量决定；第三，它所包含的因子 g_{uv} 的微分系数最高不超过二阶，并且它相对于这些微分都是线性的。这样我们就得到了一个定律，虽然它在本质上不同于牛顿定律，但是在由它给出的推论中它与牛顿定律吻合得很好，以至于发现只有很少几个判据能够用来对它进行决定性的实验检验。

下面是一些目前尚待解决的重要问题。电场和引力场在特性上真的有那么不同以至于不能用一个形式上的统一体把它们都包含进去吗？引力场对物质的结构起一定作用吗？原子核内部的连续统在相当程度上可以被看作是非欧氏的吗？最后一个问题涉及宇宙论，即惯性是否源自于远距离物质之间的相互作用？与后者相关的是：宇宙的空间范围是有限的吗？在这一点上我的观点和爱丁顿的观点相悖。与马赫一样，我感到迫切需要一个肯定的答案，但暂时还找不到证据证明。只有从这样的观点（即认为在广袤宇宙空间使用牛顿引力定律具有局限性的观点）出发来对庞大恒星系的动力学进行研究之后，才有可能最终获得解决这一令人困惑的问题的确切依据。

（王锋 翻译；张元仲 审稿）

Atomic Structure

N. Bohr

Editor's Note

Here Danish physicist Niels Bohr ponders the possibility of explaining atomic properties on the basis of the new quantum theory. Physicists had conjectured about how the grouping of elements in the periodic table might reflect specific patterns of electrons arranged around the nucleus, yet without explaining how such patterns arise. As Bohr notes, his model of the hydrogen atom suggested that electrons may fall into distinct shells. Much of the periodic table, says Bohr, can be understood as the successive filling of these shells. The stability of filled shells would explain the chemical inactivity of inert gases such as helium and argon. Bohr's model was supplanted by a more mathematically rigorous quantum theory, yet much of his qualitative picture remains intact today.

IN a letter to *Nature* of November 25 last Dr. Norman Campbell discusses the problem of the possible consistency of the assumptions about the motion and arrangement of electrons in the atom underlying the interpretation of the series spectra of the elements based on the application of the quantum theory to the nuclear theory of atomic structure, and the apparently widely different assumptions which have been introduced in various recent attempts to develop a theory of atomic constitution capable of accounting for other physical and chemical properties of the elements. Dr. Campbell puts forward the interesting suggestion that the apparent inconsistency under consideration may not be real, but rather appear as a consequence of the formal character of the principles of the quantum theory, which might involve that the pictures of atomic constitution used in explanations of different phenomena may have a totally different aspect, and nevertheless refer to the same reality. In this connection he directs attention especially to the so-called "principle of correspondence", by the establishment of which it has been possible—notwithstanding the fundamental difference between the ordinary theory of electromagnetic radiation and the ideas of the quantum theory—to complete certain deductions based on the quantum theory by other deductions based on the classical theory of radiation.

In so far as it must be confessed that we do not possess a complete theory which enables us to describe in detail the mechanism of emission and absorption of radiation by atomic systems, I naturally agree that the principle of correspondence, like all other notions of the quantum theory, is of a somewhat formal character. But, on the other hand, the fact that it has been possible to establish an intimate connection between the spectrum emitted by an atomic system—deduced according to the quantum theory on the assumption of a certain type of motion of the particles of the atom—and the constitution of the radiation, which, according to the ordinary theory of electromagnetism, would result from the same type of motion, appears to me do afford an argument in favour of the reality of the assumptions of the spectral theory of a kind scarcely compatible with Dr. Campbell's suggestion. On

原子结构

玻尔

编者按

在这篇文章中丹麦物理学家尼尔斯·玻尔反复思索是否可以用新的量子理论来解释原子的性质。物理学家们猜测周期表中的元素分组可能在某种程度上反映了核外电子的排布情况，但没有说明这种排布模式是如何形成的。玻尔指出，他的氢原子模型可以说明电子为什么会填入不同的壳层。玻尔说，周期表中大部分元素的核外电子都可以被认为是连续填充这些壳层的。充满电子的壳层结构比较稳定，因而表现出化学反应上的惰性，如氦、氩等惰性气体。虽然玻尔的模型已经被数学上更为严密的量子理论取代，但他对核外电子排布的大部分定性描述至今仍然有效。

诺曼·坎贝尔博士在去年 11 月 25 日写给《自然》的信中，谈到有关原子中电子运动和排布的假说是否可能保持前后一致的问题，这关系到量子理论能否作为原子结构的核心理论对一系列元素的光谱进行解释的问题，为了建立一个能够解释元素其他物理化学性质的原子组成理论，人们在最近的研究中提出了许多差异很大的假设。坎贝尔博士的观点令人振奋，他提出，表面上的矛盾可能不是真的，而是因为量子理论的原理具有表观化的特征，其对原子构成的描述用于解释不同的现象时可能使用完全不同的形式。他把人们对这个问题的注意力转向了所谓的“对应原理”，尽管普通电磁辐射理论与量子理论存在本质上的不同，但在建立了“对应原理”之后，就可能可以用经典辐射理论的结论推导出量子力学理论中的某些推论。

必须承认，我们至今尚未建立起一个完整的理论来详细地描述原子系统发射辐射和吸收辐射的机制，我当然同意“对应原理”像所有其他的量子理论概念一样，在某种意义上也带有形式化的特征。但是，从另一个角度上说，现在已经有可能在一个原子系统——根据基于假设原子中的粒子做某种形式的运动的量子理论推导得到——的发射光谱和由同样类型的运动产生的放射物的构成之间建立一种紧密的关系，根据普通的电磁学理论，我认为它提供了一个论点，这个论点支持一种与坎贝尔博士的建议几乎完全不相符的光谱理论假说。相反，如果我们承认用量子理论解

the contrary, if we admit the soundness of the quantum theory of spectra, the principle of correspondence would seem to afford perhaps the strongest inducement to seek an interpretation of the other physical and chemical properties of the elements on the same lines as the interpretation of their series spectra; and in this letter I should like briefly to indicate how it seems possible by an extended use of this principle to overcome certain fundamental difficulties hitherto involved in the attempts to develop a general theory of atomic constitution based on the application of the quantum theory to the nucleus atom.

The common character of theories of atomic constitution has been the endeavour to find configurations and motions of the electrons which would seem to offer an interpretation of the variations of the chemical properties of the elements with the atomic number as they are so clearly exhibited in the well-known periodic law. A consideration of this law leads directly to the view that the electrons in the atom are arranged in distinctly separate groups, each containing a number of electrons equal to one of the periods in the sequence of the elements, arranged according to increasing atomic number. In the first attempts to obtain a definite picture of the configuration and motion of the electrons in these groups it was assumed that the electrons within each group at any moment were placed at equal angular intervals on a circular orbit with the nucleus at the centre, while in later theories this simple assumption has been replaced by the assumptions that the configurations of electrons within the various groups do not possess such simple axial symmetry, but exhibit a higher degree of symmetry in space, it being assumed, for instance, that the configuration of the electrons at any moment during their motions possesses polyhedral symmetry. All such theories involve, however, the fundamental difficulty that no interpretation is given why these configurations actually appear during the formation of the atom through a process of binding of the electrons by the nucleus, and why the constitution of the atom is essentially stable in the sense that the original configuration is reorganized if it be temporarily disturbed by external agencies. If we reckon with no other forces between the particles except the attraction and repulsion due to their electric charges, such an interpretation claims clearly that there must exist an intimate interaction or "coupling" between the various groups of electrons in the atom which is essentially different from that which might be expected if the electrons in different groups are assumed to move in orbits quite outside each other in such a way that each group may be said to form a "shell" of the atom, the effect of which on the constitution of the outer shells would arise mainly from the compensation of a part of the attraction from the nucleus due to the charge of the electrons.

These considerations are seen to refer to essential features of the nucleus atom, and so far to have no special relation to the character of the quantum theory, which was originally introduced in atomic problems in the hope of obtaining a rational interpretation of the stability of the atom. According to this theory an atomic system possesses a number of distinctive states, the co-called "stationary states", in which the motion can be described by ordinary mechanics, and in which the atom can exist, at any rate for a time, without emission of energy radiation. The characteristic radiation from the atom is emitted only during a transition between two such states, and this process of transition cannot be described by ordinary mechanics, any more than the character of the emitted radiation

释光谱是合理的，那么“对应原理”在解释元素的系列光谱以及元素的其他物理化学性质方面也许能够得出最可信的推论；在这封信中，我将简单地介绍当我们在试图把量子理论应用于原子核系统以构建一个普适的原子组成理论时，是如何广泛地运用这一原则来克服目前遇到的主要难题的。

一系列原子组成理论的共性是都在极力寻找电子的排列和运动规律以解释元素的化学性质为什么会随原子序数的增加而发生周期性变化，正如众所周知的周期率明确指出的那样。周期率使人们猜想，原子中的电子可以被分为不同的组，每一组包括的电子数等于元素序列的一个周期。在早先试图明确描述各组中电子排布和运动的理论中，人们假设每一组中的电子在任一时刻都以相等的角间距排布在以原子核为中心的圆形轨道上，而在后来的理论中，这一简单的假设被新的假设取代，即认为各组中的电子排布不具有这种单一的轴对称，但在空间上则表现出更高程度的对称。例如，假设在电子运动的任一时刻它们的排布呈多面体对称。但是，所有这些理论都难以解释为什么通过原子核束缚电子而生成原子的过程中会出现这样的排布，也不能解释当外部介质暂时侵入时为什么原子的结构在原始排列重组后依然保持基本稳定。如果我们设想，除了由于所带电荷产生的吸引力或排斥力之外，粒子之间再无其他作用力，那么这样的解释就明确要求在原子中不同组的电子之间必须存在一种紧密的相互作用或“耦合”，这完全不同于认为各组电子运动轨道之间的距离也许远到足以使每一组电子单独形成原子的一个“壳层”的假设，耦合效应对外层电子的影响主要是部分地抵消由于电子带负电荷而受到的来自原子核的吸引力。

上述论点被认为涉及核原子的基本特征，与量子理论的特殊性没有关系，人们把量子理论引入原子体系的初衷是希望它能够对原子的稳定性给出合理的解释。根据量子理论，原子体系中存在若干种不同的状态，即所谓的“定态”，电子在定态中的运动不能用一般理论来解释，在某一段时间内，原子可以以任意速率运动，但不会以辐射方式释放能量。只有在两个定态之间发生跃迁时原子才会发射特征辐射，而这种跃迁过程不能用一般理论来描述，更不用说利用普通的电磁理论根据运动规律计算发射的特征辐射了。与普通电磁理论形成鲜明的对照，量子理论假设跃迁通

can be calculated from the motion by the ordinary theory of electro-magnetism, it being, in striking contrast to this theory, assumed that the transition is always followed by an emission of monochromatic radiation the frequency of which is determined simply from the difference of energy in the two states. The application of the quantum theory to atomic problems—which took its starting point from the interpretation of the simple spectrum of hydrogen, for which no *a priori* fixation of the stationary states of the atoms was needed—has in recent years been largely extended by the development of systematic methods for fixing the stationary states corresponding to certain general classes of mechanical motions. While in this way a detailed interpretation of spectroscopic results of a very different kind has been obtained, so far as phenomena which depend essentially on the motion of one electron in the atom were concerned, no definite elucidation has been obtained with regard to the constitution of atoms containing several electrons, due to the circumstance that the methods of fixing stationary states were not able to remove the arbitrariness in the choice of the number and configurations of the electrons in the various groups, or shells, of the atom. In fact, the only immediate consequence to which they lead is that the motion of every electron in the atom will on a first approximation correspond to one of the stationary states of a system consisting of a particle moving in a central field of force, which in their limit are represented by the various circular or elliptical stationary orbits which appear in Sommerfeld's theory of the fine structure of the hydrogen lines. A way to remove the arbitrariness in question is opened, however, by the introduction of the correspondence principle, which gives expression to the tendency in the quantum theory to see not merely a set of formal rules for fixing the stationary states of atomic systems and the frequency of the radiation emitted by the transitions between these states, but rather an attempt to obtain a rational generalization of the electromagnetic theory of radiation which exhibits the discontinuous character necessary to account for the essential stability of atoms.

Without entering here on a detailed formulation of the correspondence principle, it may be sufficient for the present purpose to say that it establishes an intimate connection between the character of the motion in the stationary states of an atomic system and the possibility of a transition between two of these states, and therefore offers a basis for a theoretical examination of the process which may be expected to take place during the formation and reorganisation of an atom. For instance, we are led by this principle directly to the conclusion that we cannot expect in actual atoms configurations of the type in which the electrons within each group are arranged in rings or configurations of polyhedral symmetry, because the formation of such configurations would claim that all the electrons within each group should be originally bound by the atom at the same time. On the contrary, it seems necessary to seek the configurations of the electrons in the atoms among such configurations as may be formed by the successive binding of the electrons one by one, a process the last stages of which we may assume to witness in the emission of the series spectra of the elements. Now on the correspondence principle we are actually led to a picture of such a process which not only affords a detailed insight into the structure of these spectra, but also suggests a definite arrangement of the electrons in the atom of a type which seems suitable to interpret the high-frequency spectra and the chemical properties of the elements. Thus from a consideration of the possible transitions

常伴随着发出单色辐射的过程，频率只取决于两个状态之间的能量差。最初应用量子理论解决原子中的问题是从解释简单的氢原子光谱开始的，因为氢原子不需要先验地确定定态，现在人们根据特定力学运动的类别已经得到了确定定态的系统化方法，因而量子力学理论已经被广泛地用于解决原子中的问题。虽然通过这种方式我们已经获得了许多差异很大的对光谱结果的详细解释，但只局限于研究原子中与单电子运动有关的现象。对于包括几个电子的原子，则由于确定定态的方法不能够排除在原子内不同组（或称壳层）中选择电子数量及其排布时的随意性，而无法得到明确的解释。实际上，这导致的唯一的直接后果是：原子中每个电子的运动情况都可以用一个在中心力场中运动的粒子的一个定态来近似，这种近似只限于用索末菲关于氢线精细结构理论中的圆形或椭圆形轨道来描述定态。但在引入了“对应原理”之后，人们就有了消除随意性的方法，这表明量子理论的发展趋势不仅在于建立一套正式的规则以确定原子体系中的定态和在这些定态之间发生跃迁时发射的辐射频率，而且还要努力总结出合理的辐射电磁理论法则，这一法则对不连续特性的解释必须能够说明原子是稳定的。

在这里不用详述“对应原理”的具体内容，为了解决现在的矛盾，也许可以认为它足以在原子体系不同定态的运动特征与两个定态发生跃迁的可能性之间建立很强的相关性，因而为用理论检验一个原子形成或重组时可能发生的过程提供了依据。例如，根据这个原理我们得出的直接结论是：在实际原子中每一组电子的排布都不可能是环形也都不是多边形对称，因为形成这样的排布要求各组中所有的电子都必须在开始的时候同时被原子束缚住。恰恰相反，我们有必要在原子中寻找使电子有可能是一个一个连续地被原子束缚住的排布方式，我们也许能在元素系列发射光谱中观察到这个过程的最后阶段。现在利用“对应原理”，我们对该过程的描述不仅能够精确地解析这些光谱的结构，还能明确地提出原子中电子的排布方式，既适合解释高频光谱，又能说明元素的化学性质。因此从考虑定态之间可能出现的跃迁入手，根据束缚每一个电子的不同步骤，我们首先假设只有最开始的 2 个电子在可以被称为 1–量子的轨道上运动，1–量子轨道近似于一个中心体系的定态，即一个电子绕一个原子核旋转的体系的基准状态。在最开始的 2 个电子之后被束缚的电子将不能通

between stationary states, corresponding to the various steps of the binding of each of the electrons, we are led in the first place to assume that only the two first electrons move in what may be called one-quantum orbits, which are analogous to that stationary state of a central system which corresponds to the normal state of a system consisting of one electron rotating round a nucleus. The electrons bound after the first two will not be able by a transition between two stationary states to procure a position in the atom equivalent to that of these two electrons, but will move in what may be called multiple-quanta orbits, which correspond to other stationary states of a central system.

The assumption of the presence in the normal state of the atom of such multiple-quanta orbits has already been introduced in various recent theories, as, for instance, in Sommerfeld's work on the high-frequency spectra and in that of Landé on atomic dimensions and crystal structure; but the application of the correspondence principle seems to offer for the first time a rational theoretical basis for these conclusions and for the discussion of the arrangement of the orbits of the electrons bound after the first two. Thus by means of a closer examination of the progress of the binding process this principle offers a simple argument for concluding that these electrons are arranged in groups in a way which reflects the periods exhibited by the chemical properties of the elements within the sequence of increasing atomic numbers. In fact, if we consider the binding of a large number of electrons by a nucleus of high positive charge, this argument suggests that after the first two electrons are bound in one-quantum orbits, the next eight electrons will be bound in two-quanta orbits, the next eighteen in three-quanta orbits, and the next thirty-two in four-quanta orbits.

Although the arrangements of the orbits of the electrons within these groups will exhibit a remarkable degree of spatial symmetry, the groups cannot be said to form simple shells in the sense in which this expression is generally used as regards atomic constitution. In the first place, the argument involves that the electrons within each group do not all play equivalent parts, but are divided into sub-groups corresponding to the different types of multiple-quanta orbits of the same total number of quanta, which represents the various stationary states of an electron moving in a central field. Thus, corresponding to the fact that in such a system there exist two types of two-quanta orbits, three types of three-quanta orbits, and so on, we are led to the view that the above-mentioned group of eight electrons consists of two sub-groups of four electrons each, the group of eighteen electrons of three sub-groups of six electrons each, and the group of thirty-two electrons of four sub-groups of eight electrons each.

Another essential feature of the constitution described lies in the configuration of the orbits of the electrons in the different groups relative to each other. Thus for each group the electrons within certain sub-groups will penetrate during their revolution into regions which are closer to the nucleus than the mean distances of the electrons belonging to groups of fewer-quanta orbits. This circumstance, which is intimately connected with the essential features of the processes of successive binding, gives just that expression for the "coupling" between the different groups which is a necessary condition for the

过两个定态之间的跃迁进入原子中与最开始的 2 个电子等同的位置，但将在可以被称作多量子的轨道上运动，相当于一个中心体系中的其他定态。

最近的一些理论已经假定过这样的多量子轨道存在于正常状态的原子中，如在索末菲关于高频光谱的论文中，以及在朗代关于原子大小和晶体结构的著作中；但正是对应原理第一次为这些推论提供了合理的理论依据，也为在最开始的 2 个电子之后被束缚的电子轨道如何分配这一问题提供了解答。因此通过对束缚过程的进一步研究，对应原理提出了一个简单的规则，认为这些电子分组排列的方式也是元素化学性质随原子序数递增而表现出周期性的反映。实际上，如果我们考虑的是大量电子被一个带较多正电荷的原子核束缚，该规则指出：最开始的 2 个电子位于 1–量子轨道，后 8 个电子被束缚于 2–量子轨道，然后 18 个电子在 3–量子轨道，再后面 32 个电子位于 4–量子轨道。

尽管各组中电子轨道的排列呈现出惊人的空间对称性，但不能因此而认为这些组在原子构造上显示出大家普遍接受的简单壳层结构。首先，该规则认为，同一壳层中的电子并不都扮演同样的角色，总量子数相等的多量子轨道有不同的类型，对应于这些不同的类型，电子又被归入不同的子壳层，子壳层也是描绘电子在中心力场中运动的定态。因此，根据一个体系中存在 2 类 2–量子轨道，3 类 3–量子轨道，以此类推，我们可以得到这样的结论——上面提到的 8 电子壳层由 2 个 4 电子子壳层构成，18 电子壳层由 3 个 6 电子子壳层构成，32 电子壳层由 4 个 8 电子子壳层构成。

原子组成的另一个基本特征是：不同壳层中电子轨道的构型相互关联。因而对每一个壳层来说，其中某些子壳层中的电子在绕核旋转过程中会进入离核距离比量子数较低轨道（即更内层的轨道）中电子离核的平均距离更近的区域。该现象与连续成键过程的基本特征联系紧密，说明不同壳层之间存在“耦合”现象，这种耦合正是使原子构型保持稳定的必要条件。事实上，这样的耦合是整个理论的主要特征，

stability of atomic configurations. In fact, this coupling is the predominant feature of the whole picture, and is to be taken as a guide for the interpretation of all details as regards the formation of the different groups and their various sub-groups. Further, the stability of the whole configuration is of such a character that if any one of the electrons is removed from the atom by external agencies not only may the previous configuration be reorganised by a successive displacement of the electrons within the sequence in which they were originally bound by the atom, but also the place of the removed electron may be taken by any one of the electrons belonging to more loosely bound groups or sub-groups through a process of direct transition between two stationary states, accompanied by an emission of a monochromatic radiation. This circumstance—which offers a basis for a detailed interpretation of the characteristic structure of the high-frequency spectra of the elements—is intimately connected with the fact that the electrons in the various sub-groups, although they may be said to play equivalent parts in the harmony of the inter-atomic motions, are not at every moment arranged in configurations of simple axial or polyhedral symmetry as in Sommerfeld's or Landé's work, but that their motions are, on the contrary, linked to each other in such a way that it is possible to remove any one of the electrons from the group by a process whereby the orbits of the remaining electrons are altered in a continuous manner.

These general remarks apply to the constitution and stability of all the groups of electrons in the atom. On the other hand, the simple variations indicated above of the number of electrons in the groups and sub-groups of successive shells hold only for that region in the atom where the attraction from the nucleus compared with the repulsion from the electrons possesses a preponderant influence on the motion of each electron. As regards the arrangements of the electrons bound by the atom at a moment when the charges of the previously bound electrons begin to compensate the greater part of the positive charge of the nucleus, we meet with new features, and a consideration of the conditions for the binding process forces us to assume that new, added electrons are bound in orbits of a number of quanta equal to, or fewer than, that of the electrons in groups previously bound, although during the greater part of their revolution they will move outside the electrons in these groups. Such a stop in the increase, or even decrease, in the number of quanta characterising the orbits corresponding to the motion of the electrons in successive shells takes place, in general, when somewhat more than half the total number of electrons is bound. During the progress of the binding process the electrons will at first still be arranged in groups of the indicated constitution, so that groups of three-quanta orbits will again contain eighteen electrons and those of two-quanta orbits eight electrons. In the neutral atom, however, the electrons bound last and most loosely will, in general, not be able to arrange themselves in such a regular way. In fact, on the surface of the atom we meet with groups of the described constitution only in the elements which belong to the family of inactive gases, the members of which from many points of view have also been acknowledged to be a sort of landmark within the natural system of the elements. For the atoms of these elements we must expect the constitutions indicated by the following symbols:

Helium	(2_1),	Krypton	$(2_1 8_2 18_3 8_2)$,
Neon	$(2_1 8_2)$	Xenon	$(2_1 8_2 18_3 18_3 8_2)$,
Argon	$(2_1 8_2 8_2)$,	Niton*	$(2_1 8_2 18_3 32_4 18_3 8_2)$,

* "Niton" was the provisional name in 1921 for the radioactive gas now called "radon".

也是理解不同壳层及其子壳层结构所有细节的基础。另外，整个原子结构具有保持稳定的特点——如果原子中的任何一个电子由于外界因素被移走，不仅被原子束缚的电子会在以前原子构型的基础上通过连续位移而进行重新组合，而且那些受束缚较弱的壳层或子壳层中的电子可以通过两个定态之间的直接跃迁而占据被移走的电子的位置，同时发出单色辐射。这个能为详细解释元素高频光谱特征结构提供依据的现象与以下的事实密切相关：尽管我们认为不同子壳层中的电子在原子内的简谐运动中起着同样的作用，但它们并非每时每刻都在外形上保持索末菲和朗代的著作中所说的简单轴对称或者多面体对称，相反，这些电子的行为是通过这样的方式相互联系的——当外力从壳层中移走任何一个电子后，其余电子的轨道将发生连续的变化。

这些观点可用于解释原子中所有壳层电子的组成方式和稳定性。另外，在连续的壳层和子壳层中，上述电子数目的简单变化只会发生在原子中的某个区域，在这个区域内，原子核的吸引力对每个电子运动的影响远远大于电子之间的排斥力。考虑到之前被原子核束缚的电子的电荷开始抵消掉原子核的大部分正电荷时的电子排布情况，我们得到了一些新的观点，成键过程要求的条件使我们不得不假设新增电子所在轨道的量子数等于或小于以前该壳层被束缚的电子，尽管它们在绕原子核旋转的大部分时间内都在该壳层中其他电子的外侧运动。一般说来，表征连续壳层中电子运动状态的轨道量子数或许在一半以上的电子被束缚后就不再增减了。在成键过程中，电子仍会首先排布在确定结构的壳层中，因此 3–量子轨道壳层还会有 18 个电子，而 2–量子轨道壳层还会有 8 个电子。然而，在中性原子中，最后成键且受束缚最弱的电子通常不会按照这样的规则排布。事实上，在原子表层，我们仅在惰性气体元素中观察到上面描述的壳层结构，因此，从各方面看，惰性气体家族都可以称得上是人们认识天然元素体系的里程碑。我们料想这些元素的原子组成可以用符号表示如下：

氦 (2_1),	氪 $(2_1 8_2 18_3 8_2)$,
氖 $(2_1 8_2)$,	氙 $(2_1 8_2 18_3 18_3 8_2)$,
氩 $(2_1 8_2 8_2)$,	氡 *$(2_1 8_2 18_3 32_4 18_3 8_2)$,

* “Niton”是现在被称为“radon”的放射性气体在 1921 年的临时名称。

where the large figures denote the number of electrons in the groups starting from the innermost one, and the small figures the total number of quanta characterising the orbits of electrons within each group.

These configurations are distinguished by an inherent stability in the sense that it is especially difficult to remove any of the electrons from such atoms so as to form positive ions, and that there will be no tendency for an electron to attach itself to the atom and to form a negative ion. The first effect is due to the large number of electrons in the outermost group; hence the attraction from the nucleus is not compensated to the same extent as in configurations where the outer group consists only of a few electrons, as is the case in those families of elements which in the periodic table follow immediately after the elements of the family of the inactive gases, and, as is well known, possess a distinct electro-positive character. The second effect is due to the regular constitution of the outermost group, which prevents a new electron from entering as a further member of this group. In the elements belonging to the families which in the periodic table precede the family of the inactive gases we meet in the neutral atom with configurations of the outermost group of electrons which, on the other hand, exhibit a great tendency to complete themselves by the binding of further electrons, resulting in the formation of negative ions.

The general lines of the latter considerations are known from various recent theories of atomic constitution, such as those of A. Kossel and G. Lewis, based on a systematic discussion of chemical evidence. In these theories the electro-positive and electro-negative characters of these families in the periodic table are interpreted by the assumption that the outer electrons in the atoms of the inactive gases are arranged in especially regular and stable configurations, without, however, any attempt to give a detailed picture of the constitution and formation of these groups. In this connection it may be of interest to direct attention to the fundamental difference between the picture of atomic constitution indicated in this letter and that developed by Langmuir on the basis of the assumption of stationary or oscillating electrons in the atom, referred to in Dr. Campbell's letter. Quite apart from the fact that in Langmuir's theory the stability of the configuration of the electrons is considered rather as a postulated property of the atom, for which no detailed *a priori* interpretation is offered, this difference discloses itself clearly by the fact that in Langmuir's theory a constitution of the atoms of the inactive gases is assumed in which the number of electrons is always largest in the outermost shell. Thus the sequence of the number of electrons within the groups of a niton atom is, instead of that indicated above, assumed to be 2, 8, 18, 18, 32, such as the appearance of the periods in the sequence of the elements might seem to claim at first sight.

The assumption of the presence of the larger groups in the interior of the atom, which is an immediate consequence of the argument underlying the present theory, appears, however, to offer not merely a more suitable basis for the interpretation of the general properties of the elements, but especially an immediate interpretation of the appearance of such families of elements within the periodic table, where the chemical properties of

其中较大的数字代表从最内层开始每一壳层上的电子数，较小的数字代表每一壳层中电子轨道的总量子数。

惰性气体的原子结构具有很强的内在稳定性，从某种意义上说，很难从这样的原子中移走任何一个电子使其变成正离子；同样，一个电子与这类原子结合而使其变成负离子也是不可能的。前者是由于最外层有大量的电子，因此对原子核吸引力的抵消程度大于那些外层只有较少电子的原子，如元素周期表中紧跟在惰性气体之后的那些元素，正如大家所知道的，这种结构具有明显的正电特性。后者是由于最外层电子的规则排列阻止了一个外来电子加入该壳层成为新成员。另一方面，我们发现在元素周期表中排在惰性气体之前的那些族的元素，它们的中性原子的最外层电子结构非常倾向于吸引外来电子以填满该壳层从而形成负离子。

后面提到的种种看法都是根据最近关于原子组成的各种理论得到的，如科塞尔和刘易斯建立在化学实验系统分析基础上的理论。这些理论认为，如果假定惰性气体原子的外层电子结构非常规则和稳定，就可以理解为什么元素周期表中某些族的元素带有正电特性而另一些族的元素带有负电特性，但没有人试图对原子壳层的结构和成因给予详细解释。关于这一点也许更值得关注的是：本文中所阐述的原子结构理论与坎贝尔博士在来信中提到的朗缪尔的理论有本质的不同，朗缪尔假设电子在原子中处于定态或振荡态。本文提到的理论与朗缪尔的理论的不同之处在于：电子排布的稳定性被视为是原子的必要属性，但关于这一点没有给出详细的**先验的**解释，这种不同本身很明确地表明，朗缪尔的理论认为惰性气体原子的最外层电子数总是最大的。因此，氡原子每一壳层上的电子数不再是上文提到的，而是 2、8、18、18、32，乍一看似乎在表面上仍然保留了元素的周期性。

然而，作为当前理论的一个直接推论，原子内部存在较大壳层的假设不仅更适合于解释各种元素的一般性质，而且能马上解释元素周期表中为什么存在相邻元素的化学性质相差非常小的族。事实上，这些族的存在是因为随着原子序数的递增，增加的电子填充到了原子内部的可以容纳大量电子的壳层中。如此说来，我们也许

successive elements differ only very slightly from each other. The existence of such families appears, in fact, as a direct consequence of the formation of groups containing a larger number of electrons in the interior of the atom when proceeding through the sequence of the elements. Thus in the family of the rare earths we may be assumed to be witnessing the successive formation of an inner group of thirty-two electrons at that place in the atom where formerly the corresponding group possessed only eighteen electrons. In a similar way we may suppose the appearance of the iron, palladium, and platinum families to be witnessing stages of the formation of groups of eighteen electrons. Compared with the appearance of the family of the rare earths, however, the conditions are here somewhat more complicated, because we have to do with the formation of a group which lies closer to the surface of the atom, and where, therefore, the rapid increase in the compensation of the nuclear charge during the progress of the binding process plays a greater part. In fact, we have to do in the cases in question, not, as in the rare earths, with a transformation which in its effects keeps inside one and the same group, and where, therefore, the increase in the number in this group is simply reflected in the number of the elements within the family under consideration, but we are witnesses of a transformation which is accompanied by a confluence of several outer groups of electrons.

In a fuller account which will be published soon the questions here discussed will be treated in greater detail. In this letter it is my intention only to direct attention to the possibilities which the elaboration of the principles underlying the spectral applications of the quantum theory seems to open for the interpretation of other properties of the elements. In this connection I should also like to mention that it seems possible, from the examination of the change of the spectra of the elements in the presence of magnetic fields, to develop an argument which promises to throw light on the difficulties which have hitherto been involved in the explanation of the characteristic magnetic properties of the elements, and have been discussed in various recent letters in *Nature*.

(**107**, 104-107; 1921)

N. Bohr: Copenhagen, February 14.

可以认为在稀土族元素中，电子在最多可以容纳 32 个电子而原来仅含有 18 个电子的内壳层中连续地填充。我们同样可以猜测到铁、钯、铂族元素是在逐步填充最多可容纳 18 个电子的壳层。然而，与稀土族相比，铁、钯、铂族元素的情况更复杂一些，因为在靠近原子表面的壳层填充电子时我们不得不面临的难题是，在成键过程中对核电荷的补偿的增加速度非常快。事实上，我们还必须用一种在同一壳层内部发生的转变来解决这些在稀土元素中不存在的难题，因此该壳层中电子数目的增加只是反映了被考虑的族中原子序数的增加，但是我们发现了伴随着几个外层电子汇合的转化过程。

我在即将发表的报告中将更详细地说明这里提到的问题。我写这封信的目的只是让大家注意，应用量子理论解释光谱的那些基本原理的详细阐述，看起来也有可能同时解释了元素的其他性质。就此而言，我还想提醒大家，通过研究磁场中元素光谱的变化，也许可以提出一种有望解决迄今为止人们在解释元素磁特性时遇到的困难的观点，《自然》近期的几篇快报已经对此问题进行了讨论。

（王锋 翻译；李淼 审稿）

The Dimensions of Atoms and Molecules

W. L. Bragg and H. Bell

Editor's Note

In the early 1920s, physicists were struggling to understand the physical structure of atoms using a model introduced by Bohr in 1913, in which electrons filled up shells around the nucleus. Bohr's arguments predicted that elements within any particular period (row) of the periodic table should have roughly the same atomic size, but that atomic size should jump markedly when moving from one period to the next. Here Lawrence Bragg and H. Bell report data on the dimensions of atoms that support Bohr's ideas. Estimating these dimensions from crystal densities and liquid viscosity, they find that elements near the end of rows in the table have almost identical dimensions, while a definite increase happens from one period to the next.

CERTAIN relations which are to be traced between the distances separating atoms in a crystal make it possible to estimate the distance between their centres when linked together in chemical combination. On the Lewis-Langmuir theory of atomic constitution, two electro-negative elements when combined hold one or more pairs of electrons in common, so that the outer electron shell of one atom may be regarded as coincident with that of the other at the point where the atoms are linked together. From this point of view, estimates may be made (W. L. Bragg, *Phil. Mag.*, vol. XI, August, 1920) from crystal data of the diameters of these outer shells. The outer shell of neon, for example, was estimated from the apparent diameters of the carbon, nitrogen, oxygen, and fluorine atoms, which show a gradual approximation to a minimum value of 1.30×10^{-8} cm. The diameters of the inert gases as found in this way are given in the second column of the following table:

Gas	Diameter 2σ (Crystals)	Diameter $2\sigma'$ (Viscosity)	Difference $2\sigma' - 2\sigma$
Helium	—	1.89	—
Neon	1.30	2.35	1.05
Argon	2.05	2.87	0.82
Krypton	2.35	3.19	0.84
Xenon	2.70	3.51	0.81

In the third column are given Rankine's values (A. O. Rankine, *Proc. Roy. Soc.*, A, vol. XCVIII, 693, pp. 360–374, February, 1921) for the diameters of the inert gases calculated from their viscosities by Chapman's formula (S. Chapman, *Phil. Trans. Roy. Soc.*, A, vol. CCXVI, pp. 279–348, December, 1915). These are considerably greater than the diameters calculated from crystals, but this is not surprising in view of our ignorance both of the field of force surrounding the outer electron shells and of the nature of the

原子和分子的尺度

布拉格，贝尔

编者按

20 世纪 20 年代初期的物理学家总想应用 1913 年玻尔提出的模型来努力理解原子的物理结构，在玻尔的模型中，电子填充原子核外的壳层。玻尔的理论预言，在周期表中任意一个周期（一行）的元素应该具有大致相同的原子尺寸，但是从一个周期变化到下一个周期时，原子的大小将发生显著的变化。在这篇文章中，劳伦斯·布拉格和贝尔列举了各种原子的尺寸数据，这些数据支持玻尔的理论。根据这些由晶体密度和液体黏度估算得到的数据，他们发现周期表中每行靠近行尾的元素具有几乎相同的尺寸，而当元素从一个周期过渡到下一个周期时，原子尺寸出现了一定的增加。

通过探索晶体中原子间距离之间的特定关系，可以估算出它们在化学结合中彼此连接时两者中心之间的距离。根据原子结构的刘易斯–朗缪尔理论，两种负电性的元素在结合时共用一对或若干对电子，这可以看作是一个原子的外层电子与另一个原子的外层电子在两个原子连接处重合。从这种观点出发，我们可以利用某些外壳直径的晶体数据作出估计（布拉格，《哲学杂志》，第 11 卷，1920 年 8 月）。例如，氖的外壳直径就是根据碳、氮、氧和氟原子的表观直径估计得到的——估算逐渐逼近于一个最小值 1.30×10^{-8} cm。下表中的第二列给出了以上述方式得到的惰性气体的直径：

气体	直径2σ（晶体）	直径$2\sigma'$（黏度）	差值 $2\sigma'-2\sigma$
氦	—	1.89	—
氖	1.30	2.35	1.05
氩	2.05	2.87	0.82
氪	2.35	3.19	0.84
氙	2.70	3.51	0.81

第三列中给出了兰金的值（兰金，《皇家学会学报》，A 辑，第 98 卷，第 693 期，第 360～374 页，1921 年 2 月），这是根据查普曼公式由黏度计算出来的惰性气体的直径（查普曼，《皇家学会自然科学会报》，A 辑，第 216 卷，第 279~348 页，1915 年 12 月），它们明显大于利用晶体数据计算出的直径。但是，考虑到我们忽略了外部电子层周围的力场以及将原子连接在一起的共用电子的性质，这就不足为奇了，

electron-sharing which links the atoms together, for it is quite possible that their structures might coalesce to a considerable extent. The constancy of the differences between the two estimates given in the fourth column shows that the *increase* in the size of the atom as each successive electron shell is added is nearly the same (except in the case of neon), whether measured by viscosity or by the crystal data. Further, Rankine has shown that the molecule Cl_2 behaves as regards its viscosity like two argon atoms with a distance between their centres very closely equal to that calculated from crystals, and that the same is true for the pairs Br_2 and krypton, I_2 and xenon.

We see, therefore, that the evidence both of crystals and viscosity measurements indicates that (*a*) the elements at the end of any one period in the periodic table are very nearly identical as regards the diameters of their outer electron shells, and (*b*) in passing from one period to the next there is a definite increase in the dimensions of the outer electron shell, the absolute amount of this increase estimated by viscosity agreeing closely with that determined from crystal measurements.

A further check on these measurements is afforded by the infra-red absorption spectra of HF, HCl, and HBr. The wave-number difference δv between successive absorption lines determines the moment of inertia I of the molecule in each case, the formula being

$$\delta v = \frac{h}{4\pi^2 c\mathrm{I}},$$

where h is Planck's constant and c the velocity of light.

It is therefore possible to calculate the distances between the centres of the nuclei in each molecule, for

$$s^2 = \frac{m + m'}{mm'} \cdot \frac{h}{4\pi^2 c m_{\mathrm{H}} \delta v},$$

where m and m' are the atomic weights relative to hydrogen and m_{H} the mass of the hydrogen atom. The following table gives these distances (E. S. Imes, *Astroph. Journal*, vol. 1, p. 251, 1919). It will be seen that there are again increases in passing from F to Cl and Cl to Br, which agree closely with the increases in the radii σ of the electron shells given by the crystal and viscosity data.

	$s \times 10^8$				$\sigma \times 10^8$ (Crystals)		$\sigma' \times 10^8$ (Viscosity)	
HF	0.93		Neon	(=F)	0.65		1.17	
		0.35				*0.37*		*0.26*
HCl	1.28		Argon	(=Cl)	1.02		1.43	
		0.15				*0.15*		*0.15*
HBr	1.43		Krypton	(= Br)	1.17		1.58	
						0.18		*0.17*
HI	—		Xenon	(=I)	1.35		1.75	

The increase from fluorine to chlorine of 0.35×10^{-8} cm confirms the estimate given by crystals of 0.37×10^{-8} cm, as against the estimate 0.26×10^{-8} cm given by viscosity data.

因为很有可能它们的结构会有一定程度的重叠。第四列显示了两种估计值之间差值的恒定性，这表明，不论结果是通过黏度还是通过晶体数据测得的，在相继加入各个电子层时，原子尺寸的**增加**几乎是一样的（氖的情况例外）。兰金还进一步指出，Cl_2 在黏度性质方面的表现如同两个氩原子，其中心间距与根据晶体数据计算出来的几乎相等，对于 Br_2 与氪以及 I_2 与氙来说也是如此。

由此我们看到，晶体和黏度的测量结果都指出：(*a*) 就其外部电子层的直径来说，周期表中任一周期末尾的元素几乎是一样的；(*b*) 从一个周期过渡到下一个周期时，外部电子层尺寸会有一个确定的增加量，通过黏度估计出的这一增量的绝对数值与利用晶体数据确定的结果非常接近。

HF、HCl 和 HBr 的红外吸收光谱可以进一步证实上述测量结果。相继的吸收谱线的波数差 δv 决定了各种情况下分子的转动惯量 I，公式为

$$\delta v = \frac{h}{4\pi^2 c\mathrm{I}},$$

其中，h 为普朗克常数，c 为光速。

由此就有可能计算出每个分子中核心间的距离，因为

$$s^2 = \frac{m + m'}{mm'} \cdot \frac{h}{4\pi^2 c m_{\mathrm{H}} \delta v},$$

其中，m 和 m' 为相对于氢的原子量，m_{H} 为氢原子的质量。下面的表格给出了这些距离（艾姆斯，《天体物理学杂志》，第 1 卷，第 251 页，1919 年）。我们将看到，从 F 过渡到 Cl 以及从 Cl 过渡到 Br，该距离都有所增加，并与通过晶体数据和黏度给出的电子壳层半径 σ 的增量非常接近。

$s \times 10^8$				$\sigma \times 10^8$（晶体）		$\sigma' \times 10^8$（黏度）	
HF	0.93		氖 (=F)	0.65		1.17	
		0.35			*0.37*		*0.26*
HCl	1.28		氩 (=Cl)	1.02		1.43	
		0.15			*0.15*		*0.15*
HBr	1.43		氪 (=Br)	1.17		1.58	
					0.18		*0.17*
HI	—		氙 (=I)	1.35		1.75	

从氟到氯的增量 0.35×10^{-8}cm 肯定了晶体测量给出的估计值 0.37×10^{-8} cm，但与黏度数据给出的估计值 0.26×10^{-8} cm 不一致。根据上述结果可知，要得到氢原子

It follows from the above that the distance between the hydrogen nucleus and the centre of an electro-negative atom to which it is attached is obtained by adding 0.26×10^{-8} cm to the radius of the electro-negative atom as given by crystal structures. The radius of the inner electron orbit, according to Bohr's theory, is 0.53×10^{-8} cm, double this value. The crystal data, therefore, predict the value $\delta v = 13.0\ \mathrm{cm}^{-1}$ for the HI molecule, corresponding to a distance 1.61×10^{-8} cm between their atomic centres.

This evidence is interesting as indicating that the forces binding the atoms together are localised at that part of the electron shell where linking takes place.

(**107**, 107; 1921)

W. L. Bragg, H. Bell: Manchester University, March 16.

核与附着其上的负电性原子中心之间的距离，只需将该原子根据晶体结构得到的半径加上 0.26×10^{-8} cm 即可。根据玻尔理论，内部电子轨道半径为 0.53×10^{-8} cm，是这一数值的两倍。因此，晶体数据预言 HI 分子的 $\delta v = 13.0\ \text{cm}^{-1}$，与其原子中心间的距离 1.61×10^{-8} cm 相符。

这个证据是引人关注的，因为它表明，将原子束缚在一起的力就位于电子层中发生连接的地方。

（王耀杨 翻译；李芝芬 审稿）

Waves and Quanta

L. de Broglie

Editor's Note

By 1923, physicists were facing up to the implications of Planck's and Einstein's discoveries about the quantized nature of light. Although a wave phenomenon, light also seemed particulate. Bohr's model of the atom had exploited the quantization principle for electrons. Here Louis de Broglie suggests that particles with mass, such as electrons, may also have associated waves, and that this idea could put Bohr's view on firmer ground. De Broglie says that Bohr's results can be obtained by demanding that an integral number of such electron waves must fit into its orbit around the nucleus. De Broglie's suggestion was confirmed in dramatic fashion by the discovery in 1927 of electron diffraction by crystals.

THE quantum relation, energy = $h \times$ frequency, leads one to associate a periodical phenomenon with any isolated portion of matter or energy. An observer bound to the portion of matter will associate with it a frequency determined by its internal energy, namely, by its "mass at rest." An observer for whom a portion of matter is in steady motion with velocity βc, will see this frequency lower in consequence of the Lorentz–Einstein time transformation. I have been able to show (*Comptes rendus*, September 10 and 24, of the Paris Academy of Sciences) that the fixed observer will constantly see the internal periodical phenomenon in phase with a wave the frequency of which $\nu = \frac{m_0 c^2}{h\sqrt{1-\beta^2}}$ is determined by the quantum relation using the whole energy of the moving body—provided it is assumed that the wave spreads with the velocity c/β. This wave, the velocity of which is greater than c, cannot carry energy.

A radiation of frequency ν has to be considered as divided into atoms of light of very small internal mass ($<10^{-50}$ gm) which move with a velocity very nearly equal to c given by $\frac{m_0 c^2}{\sqrt{1-\beta^2}} = h\nu$. The atom of light slides slowly upon the non-material wave the frequency of which is ν and velocity c/β, very little higher than c.

The "phase wave" has a very great importance in determining the motion of any moving body, and I have been able to show that the stability conditions of the trajectories in Bohr's atom express that the wave is tuned with the length of the closed path.

The path of a luminous atom is no longer straight when this atom crosses a narrow opening; that is, diffraction. It is then *necessary* to give up the inertia principle, and we must suppose that any moving body follows always the ray of its "phase wave"; its path will then bend by passing through a sufficiently small aperture. Dynamics must undergo

波与量子

德布罗意

编者按

物理学家们在 1923 年的主要工作是探讨普朗克和爱因斯坦关于光的量子性质的发现。虽然光是一种波，但它似乎也具有粒子的特性。玻尔的原子模型已经用到了电子的量子化原则。路易斯·德布罗意在本文中提出像电子这样有质量的粒子可能也有相对应的波，这一观点为玻尔的理论奠定了更坚实的基础。德布罗意认为，玻尔理论中的结果可以由电子波波长取整数以适合核外电子轨道得到。1927 年，德布罗意的假设就被电子在晶体中的衍射实验成功证实了。

量子关系式，即能量 = 普朗克常数 h × 频率，使人们可以把一个周期现象与任一孤立的物质或能量联系起来。与物体一起运动的观察者观测到的周期现象的频率将由物体的内部能量，即“静止质量”决定。当物体相对于观察者以 βc 的速度匀速运动时，由洛伦兹–爱因斯坦时间变换公式，观察者观测到的频率变低。我已经指出（巴黎科学院的《法国科学院院刊》，9 月 10 日和 24 日），这位固定不动的观察者总能通过波的相位看到频率为 $\nu = \frac{m_0 c^2}{h\sqrt{1-\beta^2}}$ 的内禀周期性现象，该公式是依据量子关系式推导出来的，利用了运动物体的总能量，并假设波以 c/β 的速度传播。这个速度大于 c 的波不能携带能量。

频率为 ν 的辐射必须被看成由内部质量极小（小于 10^{-50} 克）的光原子组成，其运动速度（由公式 $\frac{m_0 c^2}{\sqrt{1-\beta^2}} = h\nu$ 决定）很接近于 c。这些光原子沿着频率为 ν，速度为 c/β（仅比 c 略高一点）的非实物波缓慢行进。

“相位波”对于确定任何物体的运动都是至关重要的。我已经指出，玻尔原子轨道的稳定条件表明相位波的波长应与闭合轨道的长度相匹配。

当光原子穿过一个小孔时，它的路径就不再是一条直线，即产生了衍射现象。因此**必须**摒弃惯性原理，我们必须假定，任何运动物体总沿着它的“相位波”的放

the same evolution that optics has undergone when undulations took the place of purely geometrical optics. Hypotheses based upon those of the wave theory allowed us to explain interferences and diffraction fringes. By means of these new ideas, it will probably be possible to reconcile also diffusion and dispersion with the discontinuity of light, and to solve almost all the problems brought up by quanta.

(**112**, 540; 1923)

Louis de Broglie: Paris, September 12.

射路径行进。因此，当它经过一个足够小的孔时，其轨迹会发生弯曲。如同几何光学被波动光学取代一样，实物粒子的动力学也应经历相应的变革。我们可以利用基于波动理论的假设解释干涉和衍射产生的条纹。借助这些新思想，还有可能把漫射和散射现象与光的不连续性联系起来，解决由量子引出的几乎所有问题。

（王锋 翻译；刘纯 审稿）

Australopithecus africanus: the Man-Ape of South Africa

R. A. Dart

Editor's Note

Raymond Dart's discovery of the face and brain cast of a juvenile ape-like creature in South Africa, reported here, can be marked as the beginning of the modern era of the study of fossil man. Until that date, all members of the fossil human family were either definitely apes, such as Dryopithecus, or clearly close to humans, such as Neanderthal Man or Pithecanthropus (nowadays *Homo erectus*). Having something so clearly transitional raised challenging questions about the course of human evolution. One was the very human-like teeth associated with a small, ape-like brain, at complete variance with the then-current dogma that human ancestors evolved bigger brains before human-like teeth—amplified by Piltdown Man, now known to have been a hoax.

TOWARDS the close of 1924, Miss Josephine Salmons, student demonstrator of anatomy in the University of the Witwatersrand, brought to me the fossilised skull of a cercopithecid monkey which, through her instrumentality, was very generously loaned to the Department for description by its owner, Mr. E. G. Izod, of the Rand Mines Limited. I learned that this valuable fossil had been blasted out of the limestone cliff formation—at a vertical depth of 50 feet and a horizontal depth of 200 feet—at Taungs, which lies 80 miles north of Kimberley on the main line to Rhodesia, in Bechuanaland, by operatives of the Northern Lime Company. Important stratigraphical evidence has been forthcoming recently from this district concerning the succession of stone ages in South Africa (Neville Jones, *Jour. Roy. Anthrop. Inst.*, 1920), and the feeling was entertained that this lime deposit, like that of Broken Hill in Rhodesia, might contain fossil remains of primitive man.

I immediately consulted Dr. R. B. Young, professor of geology in the University of the Witwatersrand, about the discovery, and he, by a fortunate coincidence, was called down to Taungs almost synchronously to investigate geologically the lime deposits of an adjacent farm. During his visit to Taungs, Prof. Young was enabled, through the courtesy of Mr. A. F. Campbell, general manager of the Northern Lime Company, to inspect the site of the discovery and to select further samples of fossil material for me from the same formation. These included a natural cercopithecid endocranial cast, a second and larger cast, and some rock fragments disclosing portions of bone. Finally, Dr. Cordon D. Laing, senior lecturer in anatomy, obtained news, through his friend Mr. Ridley Hendry, of another primate skull from the same cliff. This cercopithecid skull, the possession of Mr. De Wet, of the Langlaagte Deep Mine, has also been liberally entrusted by him to the Department for scientific investigation.

南方古猿非洲种：南非的人猿

达特

编者按

这篇文章报道的是雷蒙德·达特在南非发现了一件幼年类人猿的头骨，这被认为是现代人类化石研究的开端。在此之前，所有的人类化石要么确定无疑地属于猿类，比如森林古猿，要么非常接近人类，比如尼安德特人和爪哇猿人（现在称为直立人）。这块化石具有非常明确的过渡性特征，这对人类的进化历程提出了质疑。这块化石中的牙齿与人类的牙齿非常相似，脑比较小，类似于古猿，这与当时的主流观点是完全相悖的。由于皮尔当人的发现，当时人们广泛接受的主流观点认为，人类祖先脑量的增大先于类似现代人牙齿的出现，不过现在看来皮尔当人不过是场骗人的闹剧。

在 1924 年岁末年终之际，威特沃特斯兰德大学解剖学专业的学生助教约瑟芬·萨蒙斯小姐给我带来了一件狒的头骨化石。因为她的关系，这件化石的主人兰德矿业有限公司的伊佐德先生才非常慷慨地把头骨化石借给学院用于描述研究。我得知这块珍贵的化石是从垂直高 50 英尺，水平宽 200 英尺的一个石灰岩悬崖中炸出来的，地点是在汤恩。汤恩位于金伯利北部 80 英里，在通往贝专纳兰的罗得西亚市的主干线上，为北方石灰公司所有。最近出现的重要的地层学证据表明，这一地区与南非岩层年代的连续性有关（内维尔·琼斯，《皇家人类学研究院院刊》，1920 年），我想这个石灰岩沉积层可能像罗得西亚的布罗肯希尔山一样，可能包含原始人类的化石遗迹。

我立即与威特沃特斯兰德大学的地质学教授扬博士讨论了相关发现。非常巧合的是，几乎同时，他被派遣到汤恩附近的一个农场去调查石灰岩沉积层的地质情况。在汤恩，经北方石灰公司总经理坎贝尔先生的首肯，扬教授获准探查化石发现地，并从同一形成层中给我挑选了更多的化石标本。这些标本包括一个天然的狒颅内模，另一个更大的颅内模，和一些漏出部分骨头的岩石碎块。后来，一位年长的解剖学教师科登·莱恩博士通过他的朋友里德利·亨德里先生获知，同一悬崖中又发现了一件灵长类头骨。这块头骨来自兰拉格特深矿，已经由他的拥有者德威特先生委托给学院作科研之用。

Fig. 1. Norma facialis of *Australopithecus africanus* aligned on the Frankfort horizontal

The cercopithecid remains placed at our disposal certainly represent more than one species of catarrhine ape. The discovery of Cercopithecidae in this area is not novel, for I have been informed that Mr. S. Haughton has in the press a paper discussing at least one species of baboon from this same spot (Royal Society of South Africa). It is of importance that, outside of the famous Fayüm area, primate deposits have been found on the African mainland at Oldaway (Hans Reck, *Silsungsbericht der Gesellsch. Naturforsch. Freunde*, 1914), on the shores of Victoria Nyanza (C. W. Andrews, *Ann. Mag. Nat. Hist.*, 1916), and in Bechuanaland, for these discoveries lend promise to the expectation that a tolerably complete story of higher primate evolution in Africa will yet be wrested from our rocks.

In manipulating the pieces of rock brought back by Prof. Young, I found that the larger natural endocranial cast articulated exactly by its fractured frontal extremity with another piece of rock in which the broken lower and posterior margin of the left side of a mandible was visible. After cleaning the rock mass, the outline of the hinder and lower part of the facial skeleton came into view. Careful development of the solid limestone in which it was embedded finally revealed the almost entire face depicted in the accompanying photographs.

It was apparent when the larger endocranial cast was first observed that it was specially important, for its size and sulcal pattern revealed sufficient similarity with those of the chimpanzee and gorilla to demonstrate that one was handling in this instance an anthropoid and not a cercopithecid ape. Fossil anthropoids have not hitherto been recorded south of the Fayüm in Egypt, and living anthropoids have not been discovered in recent times south of Lake Kivu region in Belgian Congo, nearly 2,000 miles to the north, as the crow flies.

All fossil anthropoids found hitherto have been known only from mandibular or maxillary fragments, so far as crania are concerned, and so the general appearance of the types they

图 1. 南方古猿非洲种的前面观（已在眼耳平面上对齐）

我们手里有的这些猴化石绝不仅仅只代表狭鼻猴的一个种。在这一地区发现猴类物种并不是什么新闻，就我所知，霍顿先生的一篇已投稿的文章中讨论了至少一种来自这一地区的狒狒（南非皇家学会）。更重要的是，在著名的法尤姆地区之外，非洲大陆的奥德威（汉斯 · 雷克，《研究者协会会刊》，1914 年），维多利亚–尼亚萨湖岸（安德鲁斯，《自然史年鉴》，1916 年），以及贝专纳兰都有灵长类化石发现，这些发现使我有希望从我们这些化石研究中获得一个关于非洲高等灵长类动物进化的相对完整的故事，当然这需要费一番辛苦。

在处理扬教授带回的这些化石岩块时，我发现较大的那块颅内模在额骨前端破裂了，但它正好与另一块相连，其中可以清楚地看到下颌骨左侧的后下缘。清理完这些岩块，面部骨骼的后下部分轮廓就呈现出来了。然后，经过小心处理包埋着头骨的坚硬的石灰石，最终一张几乎完整的面部出现了（如图所示）。

很明显，这个大的颅内模的首次发现相当重要，因为它的尺寸和沟回形状与黑猩猩和大猩猩非常相似，这表明我们手中的并非猴类化石，而是一个类人猿。到目前为止在埃及法尤姆以南至今也没有发现类人猿化石，比属刚果的基伍湖（按照直线距离计算距北部将近 2,000 英里）以南的地区也没有关于现存类人猿的记录。

到目前为止，所有关于类人猿化石的知识，就头骨而言，仅仅来源于上颌骨或者下颌骨，这些化石所代表的类型也都不清楚。因此事实上，长满牙齿的完整面部

represented has been unknown; consequently, a condition of affairs where virtually the whole face and lower jaw, replete with teeth, together with the major portion of the brain pattern, have been preserved, constitutes a specimen of unusual value in fossil anthropoid discovery. Here, as in *Homo rhodesiensis*, Southern Africa has provided documents of higher primate evolution that are amongst the most complete extant.

Apart from this evidential completeness, the specimen is of importance because it exhibits an extinct race of apes *intermediate between living anthropoids and man*.

In the first place, the whole cranium displays *humanoid* rather than anthropoid lineaments. It is markedly dolichocephalic and leptoprosopic, and manifests in a striking degree the *harmonious relation* of calvaria to face emphasised by Pruner-Bey. As Topinard says, "A cranium elongated from before backwards, and at the same time elevated, is already in harmony by itself; but if the face, on the other hand, is elongated from above downwards, and narrows, the harmony is complete." I have assessed roughly the difference in the relationship of the glabella-gnathion facial length to the glabella-inion calvarial length in recent African anthropoids of an age comparable with that of this specimen (depicted in Duckworth's "Anthropology and Morphology", second edition, vol. I), and find that, if the glabella-inion length be regarded in all three as 100, then the glabella-gnathion length in the young chimpanzee is approximately 88, in the young gorilla 80, and in this fossil 70, which proportion suitably demonstrates the enhanced relationship of cerebral length to facial length in the fossil (Fig. 2).

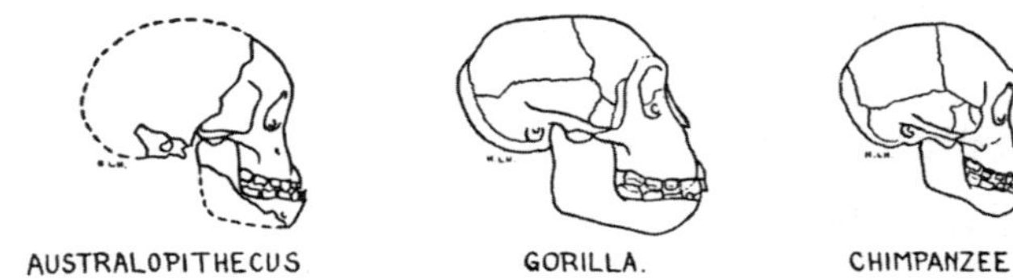

Fig. 2. Cranial form in living anthropoids of similar age (after Duckworth) and in the new fossil. For this comparison, the fossil is regarded as having the same calvarial length as the gorilla.

The glabella is tolerably pronounced, but any traces of the salient supra-orbital ridges, which are present even in immature living anthropoids, are here entirely absent. Thus the relatively increased glabella-inion measurement is due to brain and not to bone. Allowing 4 mm for the bone thickness in the inion region, that measurement in the fossil is 127 mm; *i.e.* 4 mm less than the same measurement in an adult chimpanzee in the Anatomy Museum at the University of the Witwatersrand. The orbits are not in any sense detached from the forehead, which rises steadily from their margins in a fashion amazingly human. The interorbital width is very small (13 mm) and the ethmoids are not blown out laterally as in modern African anthropoids. This lack of ethmoidal expansion causes the lacrimal fossae to face posteriorly and to lie relatively far back in the orbits, as in man. The orbits, instead of being subquadrate as in anthropoids, are almost circular, furnishing an orbital index of 100, which is well within the range of human variation (Topinard,

和下颌，连同脑结构的主要部分，被一起保留了下来，这构成了类人猿发现史上一件不寻常的宝贵的标本。在这里，如同罗德西亚人，南非提供了有关高等灵长类动物进化的现存最完整的资料。

抛开证据的完整性不谈，这个标本的重要性在于它揭示了一种已经灭绝的猿类，**介于现存类人猿与人类之间**的中间类型。

首先，整个头骨轮廓显示的是**人类的**特征，而非类人猿的。其显著特征：颅长，面窄，这些都非常符合普瑞纳贝（人类学家）强调的颅顶与脸的**和谐关系**。如托皮纳尔所说："头盖骨从前向后延伸，同时抬高，本身已经是和谐的；但另一方面如果面部 也从下向上延伸，并且变窄，就和谐完整了。"在粗略估计了这个标本和与之年龄相近的近代非洲类人猿的面长（眉间至颔下点）与颅长（眉间至枕骨隆突）比例的差异后（详见迪克沃斯的《人类学与形态学》，第 2 版，第 1 卷），我发现，如果三个物种的颅长都为 100，那么年轻黑猩猩的面长为 88，年轻大猩猩的面长为 80，这个化石标本的面长则为 70。这一比例很好地证明了化石中颅长与面长之比不断增加的关系（图 2）。

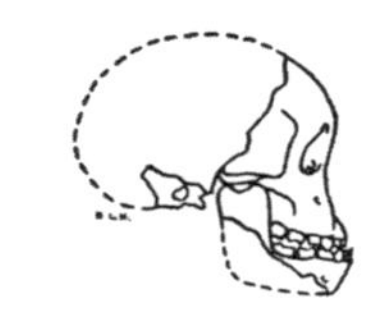

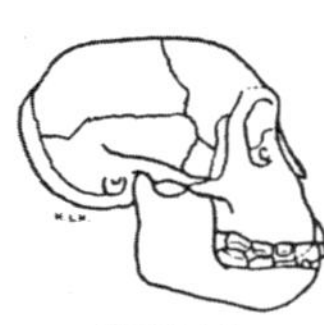

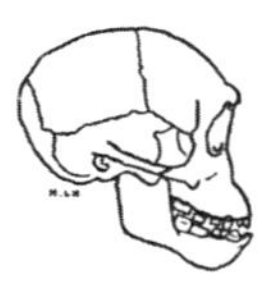

图 2. 新发现化石及年岁接近的现存类人猿（根据迪克沃斯的著作）的颅骨形状。通过这个比较可以看到，新发现化石的颅盖骨与大猩猩的一样长。

在这个化石标本中，眉间还算突出，然而突出的眶上脊却完全缺失，这一特征即使在现存的幼年类人猿中也是存在的。这就是说相对增长的眉间至枕外隆突之间的距离是由于脑量增长而并非骨骼生长所致。留出 4 毫米作为枕外隆突区域的骨骼厚度，这个化石的颅长为 127 毫米，也只比威特沃特斯兰德大学解剖学博物馆中成年黑猩猩的颅长少 4 毫米。其眼眶没有任何从前额分离的迹象，而是从边缘平稳突出，与人类的眼眶模式极其相似。眼间宽度很小（13 毫米），而筛骨也没有像现代非洲类人猿一样从侧面鼓起。由于筛骨没有膨大，使得其泪腺沟朝后，像人一样位于眼眶相对较后的位置。眼眶几乎是圆形的，不像类人猿的方形。设定眼眶指数为 100，则它完全位于人类变化范围内（托皮纳尔，《人类学》）。颧骨，颧弓，上颌骨和下颌骨，所有这些显示的是精巧的人类的特征。面部突颌度相对轻微，弗劳尔颌指数

"Anthropology"). The malars, zygomatic arches, maxillae, and mandible all betray a delicate and humanoid character. The facial prognathism is relatively slight, the gnathic index of Flower giving a value of 109, which is scarcely greater than that of certain Bushmen (Strandloopers) examined by Shrubsall. The nasal bones are not prolonged below the level of the lower orbital margins, as in anthropoids, but end above these, as in man, and are incompletely fused together in their lower half. Their maximum length (17 mm) is not so great as that of the nasals in *Eoanthropus dawsoni*. They are depressed in the median line, as in the chimpanzee, in their lower half, but it seems probable that this depression has occurred post-mortem, for the upper half of each bone is arched forwards (Fig. 1). The nasal aperture is small and is just wider than it is high (17 mm × 16 mm). There is no nasal spine, the floor of the nasal cavity being continuous with the anterior aspect of the alveolar portions of the maxillae, after the fashion of the chimpanzee and of certain New Caledonians and negroes (Topinard, *loc. cit.*).

In the second place, the dentition is *humanoid* rather than anthropoid. The specimen is juvenile, for the first permanent molar tooth only has erupted in both jaws on both sides of the face; *i.e.* it corresponds anatomically with a human child of six years of age. Observations upon the milk dentition of living primates are few, and only one molar tooth of the deciduous dentition in one fossil anthropoid is known (Gregory, "The Origin and Evolution of the Human Dentition", 1920). Hence the data for the necessary comparisons are meagre, but certain striking features of the milk dentition of this creature may be mentioned. The tips of the canine teeth transgress very slightly (0.5–0.75 mm) the general margin of the teeth in each jaw, *i.e.* very little more than does the human milk canine. There is no diastema whatever between the premolars and canines on either side of the lower jaw, such as is present in the deciduous dentition of living anthropoids; but the canines in this jaw come, as in the human jaw, into alignment with the incisors (Gregory, *loc. cit.*). There is a diastema (2 mm on the right side, and 3 mm on the left side) between the canines and lateral incisors of the upper jaw; but seeing, first, that the incisors are narrow, and, secondly, that diastemata (1 mm–1.5 mm) occur between the central incisors of the upper jaw and between the medial and lateral incisors of both sides in the lower jaw, and, thirdly, that some separation of the milk teeth takes place even in mankind (Tomes, "Dental Anatomy", seventh edition) during the establishment of the permanent dentition, it is evident that the diastemata which occur in the upper jaw are small. The lower canines, nevertheless, show wearing facets both for the upper canines and for the upper lateral incisors.

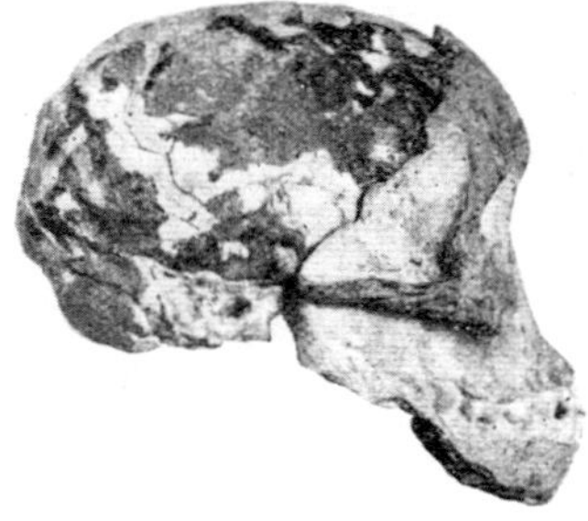

Fig. 3. Norma lateralis of *Australopithecus africanus* aligned on the Frankfort horizontal.

为 109，这一指数几乎不大于舒本萨尔检测过的布希曼人。鼻骨向下延伸不低于眶下沿水平，这点像类人猿，而止于眶下沿之上，这点又像人类，并且下半部分不完全地融合在一起。鼻骨的最大长度为 17 毫米，不像道森曙人的那么大。鼻骨下半部分从中线部位下陷，这点像黑猩猩，然而这种下陷也可能是发生在死后，因为其上半部分的每块骨头都向前拱起（图 1）。鼻孔小，其宽度刚好大于其高度（17 毫米 ×16 毫米）。没有鼻棘，鼻腔底面与上颌齿槽部的前面相连，这与黑猩猩和一些新苏格兰人及黑人的样式相仿（如前面托皮纳尔所述）。

其次，其齿系是**人类的**而非类人猿的。从两侧第一恒臼齿刚刚萌出可以判断这一标本还是幼年，从解剖学上说大约相当于人类的 6 岁儿童。对现存灵长类乳齿系的研究还很少，而化石类人猿中也只有一个乳臼齿的记录（格雷戈里，《人类齿系的起源与进化》，1920 年）。因此对乳齿作必要比较的数据太少，但是这个物种乳齿系显著的特征还是值得一提的。在两颌中，犬齿尖端略微超出整个齿列（0.5 ~ 0.75 毫米），例如，只比人类的犬齿突出一点点。下颌两侧的前臼齿和犬齿之间没有齿隙，现存类人猿的乳齿也是如此；然而其犬齿与门齿排列在一起，这点与人类一样（格雷戈里，见上述引文）。上颌犬齿与侧门齿之间有齿隙（右侧为 2 毫米，左侧为 3 毫米），但应注意到：（1）门齿狭窄；（2）上颌中门齿之间以及下颌两侧中门齿与侧门齿之间都有间隙裂（1~1.5 毫米）；（3）在恒齿形成过程中，即使在人类中，乳齿分开的现象也有发生（托姆斯，《牙齿解剖学》，第 7 版）。显然上颌的牙间隙比较窄。然而，下犬齿有相对于上犬齿以及上侧门齿的磨损面。

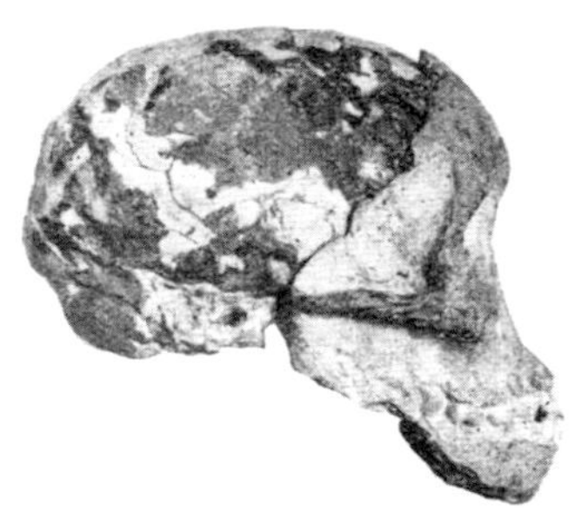

图 3. 南方古猿非洲种的侧面观（已在眼耳平面上对齐）

The incisors as a group are irregular in size, tend to overlap one another, and are almost vertical, as in man; they are not symmetrical and well spaced, and do not project forwards markedly, as in anthropoids. The upper lateral incisors do project forwards to some extent and perhaps also do the upper central incisors very slightly, but the lateral lower incisors betray no evidence of forward projection, and the central lower incisors are not even vertical as in most races of mankind, but are directed slightly backwards, as *sometimes* occurs in man. Owing to these remarkably human characters displayed by the deciduous dentition, when contour tracings of the upper jaw are made, it is found that the jaw and the teeth, as a whole, take up a parabolic arrangement comparable only with that presented by mankind amongst the higher primates. These facts, together with the more minute anatomy of the teeth, will be illustrated and discussed in the memoir which is in the process of elaboration concerning the fossil remains.

In the third place, the mandible itself is *humanoid* rather than anthropoid. Its ramus is, on the whole, short and slender as compared with that of anthropoids, but the bone itself is more massive than that of a human being of the same age. Its symphyseal region is virtually complete and reveals anteriorly a more vertical outline than is found in anthropoids or even in the jaw of Piltdown man. The anterior symphyseal surface is scarcely less vertical than that of Heidelberg man. The posterior symphyseal surface in living anthropoids differs from that of modern man in possessing a pronounced posterior prolongation of the lower border, which joins together the two halves of the mandible, and so forms the well-known *simian shelf* and above it a deep genial impression for the attachment of the tongue musculature. In this character, *Eoanthropus dawsoni* scarcely differs from the anthropoids, especially the chimpanzee; but this new fossil betrays no evidence of such a shelf, the lower border of the mandible having been massive and rounded after the fashion of the mandible of *Homo heidelbergensis*.

Fig. 4. Norma basalis of *Australopithecus africanus* aligned on the Frankfort horizontal.

其门齿大小不规则，倾向于彼此重叠，几乎垂直，这点像人类；它们呈不对称分布，空间分布也不甚合理，没有显著向前突出，这点像类人猿。上侧门齿的确有一定程度的向前突出，上中门齿似乎也略为有点向前突出；然而下侧门齿丝毫没有向前突出的迹象，并且下中门齿不是像大多数人种中那样整齐地垂直向上，而是像人类中**有时**发生的那样略为向后倾斜。基于乳齿系所显示的这些显著的人类特征，当绘出上颌的轮廓线之后，发现颌与牙齿在整体上呈抛物线的排列方式，仅能与高等灵长类动物中的人类相比拟。这些事实，以及更多的牙齿微细解剖特征将在即将发表的有关化石的论文中阐述和讨论。

第三，下颌骨本身是**人类的**而非类人猿的。从整体上看，下颌支短而纤细，与类人猿相仿，但骨头本身比同龄的人类的大。其联合区近乎完整，并且从前面显示比类人猿甚至比皮尔当人更为垂直的轮廓。其前端联合面几乎与海德堡人的一样垂直。现存类人猿下颌骨后联合面与现代人类的区别在于其下沿明显地向后延伸，将下颌骨后部连接在一起形成著名的**猿板**结构，这一结构上的深印迹在于舌头肌肉组织的附着处。在这一特征上，道森曙人与类人猿，尤其是黑猩猩几乎没有差别，但在这一新化石标本中没有找到这一结构，其下颌骨下沿粗壮而圆隆，类似于海德堡人的下颌骨。

图 4. 南方古猿非洲种的底面观（已在眼耳平面上对齐）

That hominid characters were not restricted to the face in this extinct primate group is borne out by the relatively forward situation of the foramen magnum. The position of the basion can be assessed within a few millimetres of error, because a portion of the right exoccipital is present alongside the cast of the basal aspect of the cerebellum. Its position is such that the basi-prosthion measurement is 89 mm, while the basi-inion measurement is at least 54 mm. This relationship may be expressed in the form of a "head-balancing" index of 60.7. The same index in a baboon provides a value of 41.3, in an adult chimpanzee 50.7, in Rhodesian man 83.7, in a dolichocephalic European 90.9, and in a brachycephalic European 105.8. It is significant that this index, which indicates in a measure the poise of the skull upon the vertebral column, points to the assumption by this fossil group of an attitude appreciably more erect than that of modern anthropoids. The improved poise of the head, and the better posture of the whole body framework which accompanied this alteration in the angle at which its dominant member was supported, is of great significance. It means that a greater reliance was being placed by this group upon the feet as organs of progression, and that the hands were being freed from their more primitive function of accessory organs of locomotion. Bipedal animals, their hands were assuming a higher evolutionary role not only as delicate tactual, examining organs which were adding copiously to the animal's knowledge of its physical environment, but also as instruments of the growing intelligence in carrying out more elaborate, purposeful, and skilled movements, and as organs of offence and defence. The latter is rendered the more probable, in view, first, of their failure to develop massive canines and hideous features, and, secondly, of the fact that even living baboons and anthropoid apes can and do use sticks and stones as implements and as weapons of offence ("Descent of Man", p. 81 *et seq.*).

Lastly, there remains a consideration of the endocranial cast which was responsible for the discovery of the face. The cast comprises the right cerebral and cerebellar hemispheres (both of which fortunately meet the median line throughout their entire dorsal length) and the anterior portion of the left cerebral hemisphere. The remainder of the cranial cavity seems to have been empty, for the left face of the cast is clothed with a picturesque lime crystal deposit; the vacuity in the left half of the cranial cavity was probably responsible for the fragmentation of the specimen during the blasting. The cranial capacity of the specimen may best be appreciated by the statement that the length of the cavity could not have been less than 114 mm, which is 3 mm greater than that of an adult chimpanzee in the Museum of the Anatomy Department in the University of the Witwatersrand, and only 14 mm less than the greatest length of the cast of the endocranium of a gorilla chosen for casting on account of its great size. Few data are available concerning the expansion of brain matter which takes place in the living anthropoid brain between the time of eruption of the first permanent molars and the time of their becoming adult. So far as man is concerned, Owen ("Anatomy of Vertebrates", vol. III) tells us that "The brain has advanced to near its term of size at about ten years, but it does not usually obtain its full development till between twenty and thirty years of age." R. Boyd (1860) discovered an increase in weight of nearly 250 grams in the brains of male human beings after they had reached the age of seven years. It is therefore reasonable to believe that the adult forms typified by our present specimen possessed brains which were larger than

在这一灭绝的灵长类中这样的人类特征不仅仅局限于面部，其枕骨大孔处于相对朝前的位置也是一个很好的例证。由于沿小脑基部轮廓的外侧处的一部分右枕骨保存了下来，对颅底点的测量可以控制在几毫米的误差之内。它的位置可以参考两个位置点：即颅底点距上颌齿槽前缘点 89 毫米，颅底点至枕外隆突至少 54 毫米。这可以用头“平衡指数”的方式表达为 60.7。狒狒这一指数为 41.3，成年黑猩猩为 50.7，罗得西亚人为 83.7，长头欧洲人为 90.9，短头欧洲人为 105.8。很重要的一点是，这一指数反映的是头骨在脊柱上的姿态，从这一指数可以推论，化石标本所代表的类群表现出比现代类人猿更直立的一种姿态。头部姿势的提升，以及伴随这一角度改变而来的整个身体框架的姿态的改进，对于生物体本身来说太重要了。这意味着这一类群的生物更多地依赖于脚作为身体行进的器官，而手则被解放出来，其功能不再只是原始的移动器官的附属品。两足动物的手被认为是一种高等的进化，因为手已经不仅仅只是一个精妙的触觉上的感知器官，使动物获得更丰富的物理环境知识，更重要的是手已经成为提高智力的一种工具，能够承担更精细的，更有目的性的，更有技巧性的运动，并作为防御以及进攻的器官。而后者的可能性居多，这些体现在，首先，没有形成巨大的犬齿和丑陋的面貌特征，其次，一个不容忽略的事实是，即使现存的狒狒和类人猿也能够并确实在使用树枝和石头作为工具以及攻击的武器（《人类的由来》，第 81 页）。

最后，除了考虑面部特征外，对于颅内模型还有一些新的考虑。这个标本包含右大脑半球和小脑半球（很幸运两者都贯穿整个背侧长度达到中线位置）以及左大脑半球的前面部分。化石标本的左面部包裹了一层别致的石灰石晶体沉积物，由此推断颅腔的其余部分可能是空的；而颅腔左半部分的中空可能是因为爆破中标本碎裂了。标本的颅容量可以通过腔体长度计算，这一标本的颅腔长度不小于 114 毫米，这比威特沃特斯兰德大学解剖学博物馆中成年黑猩猩的颅腔长度长 3 毫米，而比一只由于其体形巨大被挑选出来用于制作颅内模型的大猩猩的颅腔最大长度仅仅少 14 毫米。目前，在长出第一恒臼齿到成为成体这段时间，有关现存类人猿脑量扩张的数据还基本没有。就人类而言，欧文（《脊椎动物解剖学》，第 3 卷）指出，“在大约 10 岁时脑能够发育到接近成年的体积大小，而通常要到 20~30 岁才能发育完全。”博伊德（1860 年）发现 7 岁以后人类男性个体的脑重量将增长近 250 克。因此有理由相信，目前我们手头的这个化石标本代表的成年脑量，应该比这个幼年标本的大，如果不超过的话，也应该等于一个发育完全的成年大猩猩的脑量。

that of this juvenile specimen, and equalled, if they did not actually supersede, that of the gorilla in absolute size.

Whatever the total dimensions of the adult brain may have been, there are not lacking evidences that the brain in this group of fossil forms was distinctive in type and was an instrument of greater intelligence that that of living anthropoids. The face of the endocranial cast is scarred unfortunately in several places (cross-hatched in the dioptographic tracing—see Fig. 5). It is evident that the relative proportion of cerebral to cerebellar matter in this brain was greater than in the gorilla's. The brain does not show that general pre- and post-Rolandic flattening characteristic of the living anthropoids, but presents a rounded and well-filled-out contour, which points to a symmetrical and balanced development of the faculties of associative memory and intelligent activity. The pithecoid type of parallel sulcus is preserved, but the sulcus lunatus has been thrust backwards towards the occipital pole by a pronounced general bulging of the parieto-temporo-occipital association areas.

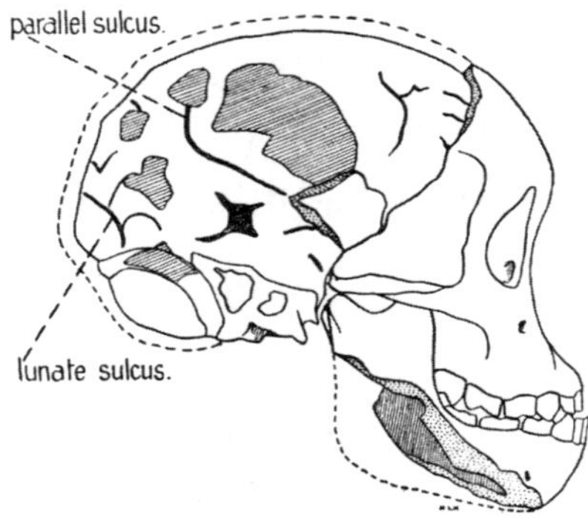

Fig. 5. Dioptographic tracing of *Australopithecus africanus* (right side), $\times\frac{1}{3}$.

To emphasise this matter, I have reproduced (Fig. 6) superimposed coronal contour tracings taken at the widest part of the parietal region in the gorilla endocranial cast and in this fossil. Nothing could illustrate better the mental gap that exists between living anthropoid apes and the group of creatures which the fossil represents than the flattened atrophic appearance of the parietal region of the brain (which lies between the visual field on one hand, and the tactile and auditory fields on the other) in the former and its surgent vertical and dorso-lateral expansion in the latter. The expansion in this area of the brain is the more significant in that it explains the posterior *humanoid* situation of the sulcus lunatus. It indicates (together with the narrow interorbital interval and human characters of the orbit) the fact that this group of beings, having acquired the faculty of stereoscopic vision, had profited beyond living anthropoids by setting aside a relatively much larger area of the cerebral cortex to serve as a storehouse of information concerning their objective environment as its details were simultaneously revealed to the senses of vision and touch, and also of hearing. They possessed to a degree unappreciated by living anthropoids the use of their hands and ears and the consequent faculty of associating with the colour,

无论成体脑的大小如何，我们都不难发现形成这种化石的种群的脑的类型，与现存类人猿相比不仅是明显不同的，而且是更加高级的。不幸的是颅腔标本表面有多处破损（图 5，用交叉平行线画出的阴影部分）。显然，这个颅腔中大脑与小脑的相对比例高于大猩猩。这个脑没有表现出一般现存类人猿的罗蓝氏区前后扁平的模式特征，而是显示出一个变圆的、更充盈的轮廓，表明与记忆和智力活动相关的官能得到了对称的、平衡的发展。类人猿类型的平行沟得以保存下来，然而由于顶骨–颞骨–枕骨联合区域显著的整体突出使得月状沟被向后推向枕侧。

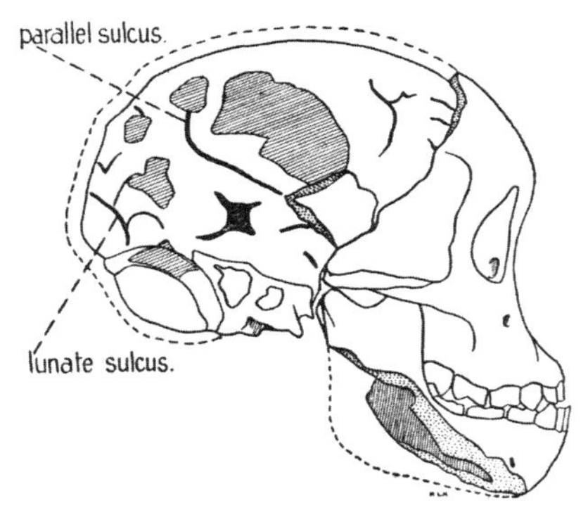

图 5. 南方古猿非洲种透视素描图（右侧），× $\frac{1}{3}$

为了强调这一点，我绘制了大猩猩和这个化石标本的颅内模顶区最宽处的轮廓叠加线（图 6）。大猩猩脑顶区（处于视觉区与触觉区及听觉区之间）扁平、萎缩，而化石标本相应区域陡立并向背部膨大，最好地说明了现存类人猿与化石标本代表的生物类群之间的智力差异。更重要的是这一脑区的膨大揭示了月状沟靠后这种**人类化**情况的原因。这一特征，加上窄的眶间隔，具有人类特征的眼眶，表明这一类群的生物已经获得了立体视觉的能力，这一超越现存类人猿的能力得益于将一块相对较大的大脑皮层区域设置为同时向视觉、触觉以及听觉传递有关客观环境细节的信息储藏库。一定程度上它们拥有类人猿不能相比的支配它们手、耳的使用的能力，并随之而来获知有关颜色、形状和物体的总体面貌、重量、质地、弹性和柔韧性，以及物体发出的声音所代表的意义的能力。换言之，与近代的猿类相应的器官相比，更有意识性和目的性地用它们的眼睛看，用它们的耳朵听，用它们的手进行操作。

form, and general appearance of objects, their weight, texture, resilience, and flexibility, as well as the significance of sounds emitted by them. In other words, their eyes saw, their ears heard, and their hands handled objects with greater meaning and to fuller purpose than the corresponding organs in recent apes. They had laid down the foundations of that discriminative knowledge of the appearance, feeling, and sound of things that was a necessary milestone in the acquisition of articulate speech.

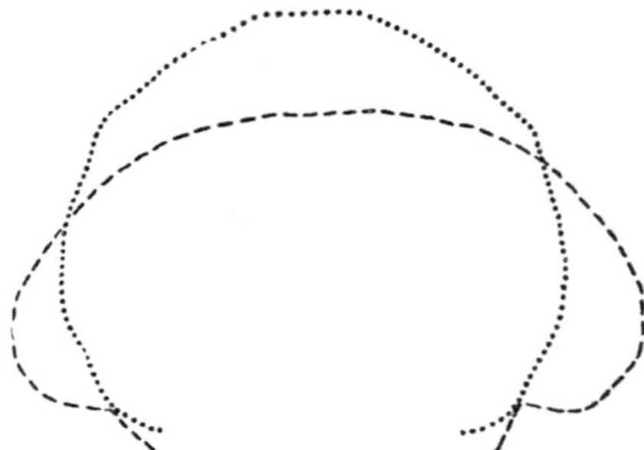

Fig. 6. Contour tracings of coronal sections through the widest part of the parietal region of the endocranial casts in Australopithecus ... and in a gorilla....

There is, therefore, an ultra-simian quality of the brain depicted in this immature endocranial cast which harmonises with the ultra-simian features revealed by the entire cranial topography and corroborates the various inferences drawn therefrom. The two thousand miles of territory which separate this creature from its nearest living anthropoid cousins is indirect testimony to its increased intelligence and mastery of its environment. It is manifest that we are in the presence here of a pre-human stock, neither chimpanzee nor gorilla, which possesses a series of differential characters not encountered hitherto in any anthropoid stock. This complex of characters exhibited is such that it cannot be interpreted as belonging to a form ancestral to any living anthropoid. For this reason, we may be equally confident that there can be no question here of a primitive anthropoid stock such as has been recovered from the Egyptian Fayüm. Fossil anthropoids, varieties of Dryopithecus, have been retrieved in many parts of Europe, Northern Africa, and Northern India, but the present specimen, despite its youth, cannot be confused with anthropoids having the dryopithecid dentition. Other fossil anthropoids from the Siwalik hills in India (Miocene and Pliocene) are known which, according to certain observers, may be ancestral to modern anthropoids and even to man.

Whether our present fossil is to be correlated with the discoveries made in India is not yet apparent; that question can only be solved by a careful comparison of the permanent molar teeth from both localities. It is obvious, meanwhile, that it represents a fossil group distinctly advanced beyond living anthropoids in those two dominantly human characters of facial and dental recession on one hand, and improved quality of the brain on the other. Unlike Pithecanthropus, it does not represent an ape-like man, a caricature of precocious hominid failure, but a creature well advanced beyond modern anthropoids in just those characters, facial and cerebral, which are to be anticipated in an extinct

它们已经具有了辨别事物外貌、触感以及声音的基础，这对于语言能力（能够清晰发音）的获得是一个重要的里程碑。

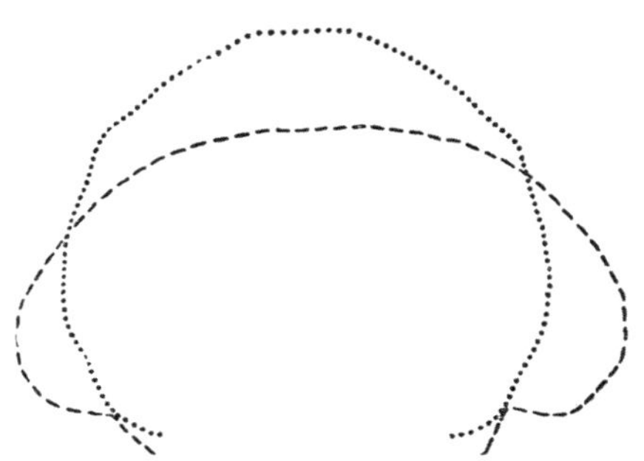

图 6. 南方古猿非洲种和大猩猩颅内模顶区最宽处的冠状面轮廓图

因此，这个未成熟的颅内模描绘的是一个具有超类人猿性质的脑，这与整个头骨解剖形态显示的超类人猿特征是一致的，并进一步证实了以此而来的各种推论。这个生物和它最近的现存类人猿兄弟们相隔了 2,000 英里的区域，这也间接证实了其智力的提高以及对环境的掌握。显然，摆在我们面前的是一个前人类类群，既不是黑猩猩也不是大猩猩，它已经拥有了一系列至今为止任何类人猿都不具有的特征。这些特征的复杂性表明，它不能被认为是任何现存类人猿的祖先。因此，我们同样可以确信，如同在埃及法尤姆所发现的，这里毫无疑问有一个原始类人猿类群的存在。在欧洲、南非、北印度的多个地区曾发现多种森林古猿的化石类人猿，然而我们的标本，尽管年轻，也不可能与具有森林古猿齿系的类人猿相混淆。根据其他人的观察，在印度（中新世和上新世）西瓦利克山脉发现的一些其他化石类人猿可能是现代类人猿甚至人类的祖先。

我们目前的化石是否与印度的发现相互关联还不清楚，这一问题只有通过仔细比较来自两地的恒臼齿才能解决。然而，很显然的是它所代表的化石类群毫无疑问超越了现存类人猿，主要表现在，一方面已经具有显著的人类特征的面部和齿系，另一方面脑质的提高。它不像猿人属代表的是与猿相像的人类，一种早熟的人类的失败类型，而是代表一种从面部及脑部特征上已经远远超过当代类人猿的生物，这些面部及脑部特征被认为可能是人类及其猿类祖先的已经灭绝的连接类型中应有的

link between man and his simian ancestor. At the same time, it is equally evident that a creature with anthropoid brain capacity, and lacking the distinctive, localised temporal expansions which appear to be concomitant with and necessary to articulate man, is no true man. It is therefore logically regarded as a man-like ape. I propose tentatively, then, that a new family of *Homo-simiadae* be created for the reception of the group of individuals which it represents, and that the first known species of the group be designated *Australopithecus africanus*, in commemoration, first, of the extreme southern and unexpected horizon of its discovery, and secondly, of the continent in which so many new and important discoveries connected with the early history of man have recently been made, thus vindicating the Darwinian claim that Africa would prove to be the cradle of mankind.

It will appear to many a remarkable fact that an ultra-simian and pre-human stock should be discovered, in the first place, at this extreme southern point in Africa, and, secondly, in Bechuanaland, for one does not associate with the present climatic conditions obtaining on the eastern fringe of the Kalahari desert an environment favourable to higher primate life. It is generally believed by geologists (*vide* A. W. Rogers, "Post-Cretaceous Climates of South Africa", *South African Journal of Science*, vol. XIX, 1922) that the climate has fluctuated within exceedingly narrow limits in this country since Cretaceous times. We must therefore conclude that it was only the enhanced cerebral powers possessed by this group which made their existence possible in this untoward environment.

In anticipating the discovery of the true links between the apes and man in tropical countries, there has been a tendency to overlook the fact that, in the luxuriant forests of the tropical belts, Nature was supplying with profligate and lavish hand an easy and sluggish solution, by adaptive specialisation, of the problem of existence in creatures so well equipped mentally as living anthropoids are. For the production of man a different apprenticeship was needed to sharpen the wits and quicken the higher manifestations of intellect—a more open veldt country where competition was keener between swiftness and stealth, and where adroitness of thinking and movement played a preponderating role in the preservation of the species. Darwin has said, "no country in the world abounds in a greater degree with dangerous beasts than Southern Africa", and, in my opinion, Southern Africa, by providing a vast open country with occasional wooded belts and a relative scarcity of water, together with a fierce and bitter mammalian competition, furnished a laboratory such as was essential to this penultimate phase of human evolution.

In Southern Africa, where climatic conditions appear to have fluctuated little since Cretaceous times, and where ample dolomitic formations have provided innumerable refuges during life, and burial-places after death, for our troglodytic forefathers, we may confidently anticipate many complementary discoveries concerning this period in our evolution.

In conclusion, I desire to place on record my indebtedness to Miss Salmons, Prof. Young, and Mr. Campbell, without whose aid the discovery would not have been made; to

特征。同时，同样可以确信，一个生物只具有与类人猿同样大小的脑量并且缺少局部的颞区扩张（而这些是成为具有语言能力的人所必需的），它还不是真正意义上的人。因此，从逻辑上讲它应当被称为像人的猿。这样我试探性地建议，创立一个名为人猿科的新科来接纳标本代表的生物类群，并且将这一类群第一个已知的种命名为南方古猿非洲种，以纪念：第一，其极南的发现地和出人意料的地层，第二，其所在的大陆。在这一大陆上近来有非常多的重要发现与人类的早期历史相关，这将为达尔文所主张的非洲是人类的摇篮这一提议提供依据。

鉴于卡拉哈里沙漠东缘现在的气候条件无法与适宜高等灵长类生存的环境条件相联系，将会发生的一个毋庸置疑的事实是，超级猿类和前人类类群的发现应当在，第一，非洲大陆的极南端，第二，贝专纳兰。地质学家普遍承认（参见罗杰斯的文章《后白垩纪南非的气候》，《南非科学杂志》，第19卷，1922年），从白垩纪以来这片大陆的气候只是在一个非常小的范围内波动。因此我们必然能推出，只有这个大脑能力已经提高的类群才能在这种不利的环境下存在。

在预期位于热带地区的国家将会发现连接猿类与人类真正的“接环”时，存在一种忽略如下事实的倾向，那就是在热带繁茂的森林中，大自然用其多产而慷慨的手，通过适应性的特化，为具有类人猿智力程度的生物们的生存问题提供了一个简单但却迟缓的解决方案。为了人类的产生，需要一个不同的学徒期以磨砺它们的智力，提高它们的理解力。然而正是一片更为广阔的草原地带为之提供了必要的条件，在这里迅捷与隐秘之间的竞争更加尖锐，在这里敏捷的思维和运动在物种生存中扮演了极为重要的角色。达尔文说过，“世界上没有任何国家比南非拥有更多的危险野兽”，以我的观点，南非，由于拥有广阔的空旷地带和稀少的林带，水源相对匮乏，加上哺乳动物之间的残忍严酷的竞争，提供了对于人类进化史上这倒数第二个阶段来说至关重要的实验室。

南非，这里的气候从白垩纪以来波动就很小，而丰富的白云岩层则为我们的穴居人祖先提供了无数的生活居所和死亡墓地，我们可以自信地预期，还将会继续发现有关这一时期的更多的人类进化的补充证据。

最后，我要感谢萨蒙斯小姐、扬教授和坎贝尔先生，没有他们的帮助就没有这些发现；感谢莱恩·理查森先生提供的照片；感谢莱恩博士和我实验室的同事们的热

Mr. Len Richardson for providing the photographs; to Dr. Laing and my laboratory staff for their willing assistance; and particularly to Mr. H. Le Helloco, student demonstrator in the Anatomy Department, who has prepared the illustrations for this preliminary statement.

(**115**, 195-199; 1925)

Prof. Raymond A. Dart: University of the Witwatersrand, Johannesburg, South Africa.

情帮助；特别要感谢解剖学学院的学生管理员勒埃洛克先生为这个初步报告准备了插图。

（刘晓辉 翻译；赵凌霞 审稿）

The Fossil Anthropoid Ape from Taungs

Editor's Note

Raymond Dart's announcement of *Australopithecus africanus* in *Nature* the previous week, and his assertion that it was intermediate between apes and humans, was greeted with a chorus of very faint praise in this quartet of letters from the anthropological establishment. All welcomed the discovery, but preferred to consider Australopithecus as very definitely an ape whose human-like features could be attributed to the fact that the fossil was of a child, whose adult form could not yet be discerned. The tone of all letters was courteous—except for Smith Woodward's criticism of the "barbarous" merger of Latin and Greek to create "Australopithecus".

THE discovery of fossil remains of a "man ape" in South Africa raises many points of great interest for those who are studying the evolution of man and of man-like apes. No doubt when Prof. Dart publishes his full monograph of his discovery, he will settle many points which are now left open, but from the facts he has given us, and particularly from the accurate drawing of the endocranial cast and skull in profile, it is even now possible for an onlooker to assess the importance of his discovery. I found it easy to enlarge the profile drawing just mentioned to natural size and to compare it with corresponding drawings of the skulls of children and of young apes. When this is done, the peculiarities of Australopithecus become very manifest.

Prof. Dart regrets he has not access to literature which gives the data for gauging the age of young anthropoids. In the specimen he has discovered and described, the first permanent molar teeth are coming into use. Data which I collected 25 years ago show that these teeth reach this stage near the end of the 4th year, two years earlier than is the rule in man and two years later than is the rule in the higher monkeys. In evolution towards a human form there is a tendency to prolong the periods of growth. Man and the gorilla have approximately the same size of brain at birth; the rapid growth of man's brain continues to the end of the 4th year; in the gorilla rapid growth ceases soon after birth.

Prof. Dart recognises the many points of similarity which link Australopithecus to the great anthropoid apes—particularly to the chimpanzee and gorilla. Those who are familiar with the facial characters of the immature gorilla and of the chimpanzee will recognise a blend of the two in the face of Australopithecus, and yet in certain points it differs from both, particularly in the small size of its jaws.

In size of brain this new form is not human but anthropoid. In the 4th year a child has reached 81 percent of the total size of its brain; at the same period a young gorilla has obtained 85 percent of its full size, a chimpanzee 87 percent. From Prof. Dart's accurate diagrams one estimates the brain length to have been 118 mm—a dimension common in

汤恩发现的类人猿化石

编者按

一周前，雷蒙德·达特在《自然》上宣布了南方古猿非洲种的发现，他断言这是介于古猿与人类之间的中间类型。在下述 4 封信中，人类学研究领域的重要人物一致对此给出了明褒实贬的评价。所有人都表示欢迎这个发现，但同时也提出更倾向于认为南方古猿非常明显就是古猿，其与人类相像的特征可能是因为这块化石属于幼年的个体，其成年的形态还不能被辨认出来。史密斯·伍德沃德批评道，造出“南方古猿”这个名字属于“野蛮”拼合拉丁语和希腊语，除此之外，这几封信的语气都还算客气。

南非发现的“人猿”化石激起了那些正在研究人类和类人猿进化的人们的巨大兴趣。当达特教授发表关于其发现的详尽专著时，他无疑将解决现在尚待进一步讨论的许多问题，但是从他向我们展示的事实来看，尤其是从他对颅腔模型和头骨剖面精确的绘图来看，即使是现在，旁观者也可以掂量出其发现的重要性。我发现很容易将刚刚提到的剖面图放大到真实大小，也很容易将其与相应的人类的孩子和幼年猿类的头骨绘图进行比较。当这样做的时候，南方古猿的特性就会变得十分明显了。

达特教授遗憾的是自己没能使用那些提供鉴定幼年猿类年龄资料的文献。在由他发现并进行描述的这个标本中，第一恒臼齿正开始被使用。我在 25 年前搜集的数据表明这些牙齿达到这一阶段应该是接近 4 岁末的时候，比人类早了两年，比高等猴子则晚了两年。在向人类形式进化的过程中，存在发育期延长的趋势。人类和大猩猩在出生时具有大约相同大小的脑；人脑的快速生长一直持续到 4 岁末；而大猩猩的脑的快速生长在出生后不久就停止了。

达特教授识别出了将南方古猿和大型类人猿联系起来的许多相似点——尤其是将南方古猿与黑猩猩和大猩猩联系起来的相似之处。对未成年大猩猩和黑猩猩的面部特征熟悉的人会在南方古猿的脸上看到二者的混合体，然而有些方面南方古猿与这两者都不同，尤其是其颌骨较小。

在脑尺寸方面，这个新类型与人类不同，是类人猿式的。儿童的脑在 4 岁时达到了其脑总尺寸的 81%；而同样年龄的幼年大猩猩已经达到了全部尺寸的 85%，黑猩猩则达到了 87%。根据达特教授精确的图表可以估计出脑的长度已经达到了

the brains of adult and also juvenile gorillas. The height of the brain above the ear-holes also corresponds in both Australopithecus and the gorilla—about 70 mm. But in width, as Prof. Dart has noted, the gorilla greatly exceeds the new anthropoid; in the gorilla the width of brain is usually about 100 mm; in Australopithecus the width is estimated at 84 mm. The average volume of the interior of gorillas' skulls (males and females) is 470 c.c., but occasional individuals run up to 620 c.c. One may safely infer that the volume of the brain in the juvenile Australopithecus described by Prof. Dart must be less than 450 c.c., and if we allow a 15 percent increase for the remaining stages of growth, the size of the adult brain will not exceed 520 c.c. At the utmost the volume of brain in this new anthropoid falls short of the gorilla maximum. Even if it be admitted, however, that Australopithecus is an anthropoid ape, it is a very remarkable one. It is a true long-headed or dolichocephalic anthropoid—the first so far known. In all living anthropoids the width of the brain is 82 percent or more of its length; they are round-brained or brachycephalic; but in Australopithecus the width is only 71 percent of the length. Here, then, we find amongst anthropoid apes, as among human races, a tendency to roundness of brain in some and to length in others. On this remarkable quality of Australopithecus Prof. Dart has laid due emphasis.

This side-to-side compression of the head taken in conjunction with the small size of jaws throw a side light on the essential features of Australopithecus. The jaws are considerably smaller than those of a chimpanzee of a corresponding age, and much smaller than those of a young gorilla. There is a tendency to preserve infantile characters, a tendency which has had much to do with the shaping of man from an anthropoid stage. The relatively high vault of the skull of Australopithecus and its narrow base may also be interpreted as infantile characters. It is not clearly enough recognised that the anthropoid and human skulls undergo remarkable growth changes leading to a great widening of the base and a lowering or flattening of the roof of the skull. In Australopithecus there is a tendency to preserve the foetal form.

When Prof. Dart produces his evidence in full he may convert those who, like myself, doubt the advisability of creating a new family for the reception of this new form. It may be that Australopithecus does turn out to be "intermediate between living anthropoids and man", but on the evidence now produced one is inclined to place Australopithecus in the same group or sub-family as the chimpanzee and gorilla. It is an allied genus. It seems to be near akin to both, differing from them in shape of head and brain and in a tendency to the retention of infantile characters. The geological evidence will help to settle its relationships. One must suppose we are dealing with fossil remains which have become embedded in the stalagmite of a filled-up cave or fissure of the limestone cliff.

May I, in conclusion, thank Prof. Dart for his full and clear description, and particularly for his accurate drawings. One wishes that discoverers of such precious relics would follow his example, and, in place of reproducing crude tracings and photographs, give the same kind of drawings as an engineer or an architect prepares when describing a new engine or a new building.

Arthur Keith

118 毫米——这个尺寸在成年和幼年大猩猩的脑中也是很常见的。耳孔之上的脑高度在南方古猿和大猩猩中是相当的——大约 70 毫米。但是宽度方面，正如达特教授注明的那样，大猩猩远远超过了新发现的类人猿；大猩猩中，脑的宽度通常是 100 毫米左右；而南方古猿的脑的宽度估计在 84 毫米左右。大猩猩头骨（雄性和雌性）的平均内容量是 470 毫升，但是个别个体达到了 620 毫升。我们可以有把握地推测达特教授描述的幼年南方古猿的脑量肯定不足 450 毫升，如果我们容许其在剩余的生长阶段还有 15% 的增长空间的话，那么成年脑将不会超过 520 毫升。这种新型类人猿的最大脑量比大猩猩的最大脑量小。然而，即使承认南方古猿是一种类人猿，它们也是一种非常特别的类型。它们是一种真正的长头型或长颅型的类人猿——这是目前为止知道的第一种。现存的所有类人猿的脑宽度都是其长度的 82% 以上；它们的脑都是圆形的或短颅型的；而南方古猿的脑宽度只有其长度的 71%。因此我们发现在类人猿中，就像各人种之间那样，有些具有圆头型趋势，而另一些则具有长头型趋势。达特教授突出强调了南方古猿的这一显著特征。

这种与颌骨较小有关的对头颅两侧的挤压从侧面为阐明南方古猿的本质特征提供了线索。其颌骨比相应年龄的黑猩猩小得多，比幼年大猩猩的小得更多。在进化过程中有着保留婴儿特征的趋势，这种趋势对于从类人猿阶段逐渐形成人类具有紧密关联。南方古猿相对较高的头骨顶盖以及狭窄的颅底也可以理解成是婴儿时期的特征。现在还不能十分清楚地认识到类人猿和人类头骨经历了导致颅底显著变宽以及颅顶变低或变平的显著的生长变化。南方古猿具有保留胎儿形式的趋势。

当达特教授将其证据悉数列出时，他可能转变了那些像我一样怀疑过创建一个新科来接纳这种新类型是否明智的人的观点。可能结果南方古猿确实是“介于现存的类人猿和人类之间的一种中间类型”，但是依据现在所列出的证据，人们倾向于将南方古猿放到与黑猩猩和大猩猩同样的群体或亚科中。南方古猿是一个同源的属，它似乎与黑猩猩和大猩猩都具有很近的亲缘关系，但是在头部和脑形状以及保留婴儿特征的趋势等方面有所不同。地质学证据将有助于解决其关系问题。必须想到我们正在研究的是那些埋藏在被填满了的山洞或者石灰岩悬崖裂缝的石笋中的化石。

最后，请允许我感谢达特教授给出的详实而清晰的描述，尤其是他那精确的绘图。人们希望这样一类珍贵化石的发现者会遵循他的榜样，像工程师或者建筑师在描述一种新发动机或新大楼时绘制的图画那样提供同样水平的绘图，而非只是复制粗糙的描图和照片。

阿瑟·基思

* * *

It is a great tribute to Prof. Dart's energy and insight to have discovered the only fossilised anthropoid ape so far obtained from Africa, excepting only the jaw of the diminutive Oligocene Propliopithecus from the Egyptian Fayum. Whether or not the interpretation of the wider significance he has claimed for the fossil should be corroborated in the light of further information and investigation, the fact remains that his discovery is of peculiar interest and importance.

The simian infant discovered by him is an unmistakable anthropoid ape that seems to be much on the same grade of development as the gorilla and the chimpanzee without being identical with either. So far Prof. Dart does not seem to have "developed" the specimen far enough to expose the crowns of the teeth and so obtain the kind of evidence which in the past has provided most of our information for the identification of the extinct anthropoids. Until this has been done and critical comparisons have been made with the remains of Dryopithecus and Sivapithecus, the two extinct anthropoids, that approach nearest to the line of man's ancestry, it would be rash to push the claim in support of the South African anthropoid's nearer kinship with man. Prof. Dart is probably justified in creating a new species and even a new genus for his interesting fossil: for if such wide divergences between the newly discovered anthropoid and the living African anthropoids are recognisable in an infant, probably not more than four years of age, the differences in the adults would surely be of a magnitude to warrant the institution of a generic distinction.

Many of the features cited by Prof. Dart as evidence of human affinities, especially the features of the jaw and teeth mentioned by him, are not unknown in the young of the giant anthropoids and even in the adult gibbon.

The most interesting, and perhaps significant, distinctive features are presented by the natural endocranial cast. They may possibly justify the claim that Australopithecus has really advanced a stage further in the direction of the human status than any other ape. But until Prof. Dart provides us with fuller information and full-size photographs revealing the details of the object, one is not justified in drawing any final conclusions as to the significance of the evidence.

The size of the brain affords very definite evidence that the fossil is an anthropoid on much the same plane as the gorilla and the chimpanzee. But while its brain is not so large as the big gorilla-cast used for comparison by Prof. Dart, it is obvious that it is bigger than a chimpanzee's brain and probably well above the average for the gorilla. But the fossil is an imperfectly developed child, whose brain would probably have increased in volume to the extent of a fifth had it attained the adult status. Hence it is probable the brain would have exceeded in bulk the biggest recorded cranial capacity for an anthropoid ape, about 650 c.c. As the most ancient and primitive human brain case, that of Pithecanthropus, is at least 900 c.c. in capacity, one might regard even a small advance on 650 c.c. as a definite approach to the human status. The most suggestive feature (in Prof. Dart's Fig. 5,

* * *

除了在埃及法尤姆发现的小型渐新世原上猿的唯一颌骨之外，达特教授的发现是迄今为止在非洲得到的唯一一个类人猿化石，他的精力和洞察力令人万分敬佩。无论他声明的这具化石的重要意义是否应该根据更多的信息与研究加以确认，他的发现具有特殊意义和重要性这一事实都不会改变。

达特教授发现的猿孩肯定是一只类人猿，该类人猿似乎处于与大猩猩和黑猩猩同样的发育阶段，但是与两者又都不相同。迄今为止，达特教授似乎并没有对该标本进行足够的“开发”以揭示其牙冠状况，所以得到的都是过去为我们鉴定已灭绝类人猿提供了绝大部分信息的那一类证据。除非完成了揭示牙冠状况的工作，并且对另两种与人类祖先世系最接近的已灭绝类人猿（森林古猿和西瓦古猿）的化石进行比较研究之后，我们才能有理由支持南非类人猿与人类的亲缘关系更近，否则这种说法就太轻率了。达特教授在为其感兴趣的化石创建一个新物种甚至一个新属方面可能是有道理的：因为，如果新发现的类人猿与现存的非洲类人猿在可能不超过4岁的婴儿期阶段就存在如此大的差异的话，那么它们的成年个体之间肯定差异更大，能确保形成属一级的区别。

达特教授引用了许多特征作为与人类具有亲缘关系的证据，尤其是他提到的颌骨和牙齿的特征，我们并不是不知道这些特征在巨型类人猿的幼年期是什么样子，甚至成年长臂猿的也很清楚。

最有趣的，可能也是最重要的独特特征是由天然的颅腔模型呈现出来的。这些颅腔模型可能可以证明如下说法的合理性，即南方古猿在向人类状态进化的方向上确实比其他猿类都迈进了更大一步。但是除非达特教授能够提供更充分的信息以及揭示我们研究对象形态细节的真实尺寸的照片，否则我们不能信服于对证据重要性所作的任何定论。

脑的大小提供了非常明确的证据，证明了该化石是与大猩猩和黑猩猩处于大致同一水平的一种类人猿。但是它的脑没有达特教授用来进行比较的大型大猩猩的模型大，不过它肯定比黑猩猩的脑大，并且可能远远超过了大猩猩的平均尺寸。但是该化石是一个尚未发育完全的孩子的，在其达到成年状态之前其脑量可能还有1/5的增长幅度。因此它的脑可能远远超过了现有大量记录的类人猿的约650毫升最大颅腔容量。作为最古老、最原始的人类脑壳，爪哇猿人的颅容量至少有900毫升，有人可能认为甚至还需要从650毫升经过小幅增长才能达到确定的接近于人类的状

p. 197) is the position of the sulcus lunatus and the extent of the parietal expansion that has pushed asunder the lunate and parallel sulci—a very characteristic human feature.

When fuller information regarding the brain is forthcoming—and no one is more competent than Prof. Dart to observe the evidence and interpret it—I for one shall be quite prepared to admit that an ape has been found the brain of which points the way to the emergence of the distinctive brain and mind of mankind. Africa will then have purveyed one more surprise—but only a real surprise to those who do not know their Charles Darwin. But what above all we want Prof. Dart to tell us is the geological evidence of age, the exact conditions under which the fossil was found, and the exact form of the teeth.

G. Elliot Smith

* * *

The new fossil from Taungs is of special interest as being the first-discovered skull of an extinct anthropoid ape, and Prof. Dart is to be congratulated on his lucid and suggestive preliminary description of the specimen. As usual, however, there are serious defects in the material for discussion, and before the published first impressions can be confirmed, more examples of the same skull are needed.

First, as Prof. Dart remarks, the fossil belongs to an immature individual with the milk-dentition, and, so far as can be judged from the photograph, I see nothing in the orbits, nasal bones, and canine teeth definitely nearer to the human condition than the corresponding parts of the skull of a modern young chimpanzee. The face seems to be relatively short, but the lower jaw of the Miocene Dryopithecus has already shown that this must have been one of the characters of the ancestral apes. The symphysis of the lower jaw may owe its shape and the absence of the "simian shelf" merely to immaturity; but it may be noted that a nearly similar symphysis has been described in an adult Dryopithecus, of which it may also be said that "the anterior symphyseal surface is scarcely less vertical than that of Heidelberg man" (see diagrams in *Quart. Journ. Geol. Soc.*, vol. 70, 1914, pp. 317, 319).

Secondly, the Taungs skull lacks the bones of the brain-case, so that the amount and direction of distortion of the specimen cannot be determined. I should therefore hesitate to attach much importance to rounding or flattening of any part of the brain-cast, and would even doubt whether the relative dimensions of the cast of the cerebellum can be relied on. Confirmatory evidence is needed of the reality of appearances in such a fossil.

In the absence of knowledge of the skulls of the fossil anthropoid apes represented by teeth and fragmentary jaws in the Tertiary formations of India, it is premature to express any opinion as to whether the direct ancestors of man are to be sought in Asia or in Africa. The new fossil from South Africa certainly has little bearing on the question.

态。最具提示性的特征（达特教授的图 5，第 197 页）是月状沟的位置和顶骨的扩展范围，后者将月状沟和平行沟推离开了——这是一种人类特有的特征。

当将来出现更全面的关于脑的信息时——没有人会比达特教授更应该在这个化石上观察证据并且给出解释——就我个人而言，将充分做好准备：承认已经发现一种猿类，它的脑指向人类特有的脑和智能出现的道路。那时非洲将提供又一个惊喜——但是只是对于那些不知道他们的查尔斯·达尔文为何许人的人来说，才是一个真正的惊喜。但是我们最希望达特教授告诉我们的是关于年代的地质学证据、发现化石的地点的准确状况以及牙齿的精确形式。

埃利奥特·史密斯

* * *

在汤恩发现的新化石特别有意思，因为这是第一次发现的一种已灭绝类人猿的头骨，在此祝贺达特教授对标本进行了清晰而具有提示性的初步描述。然而，与通常一样，在讨论部分还是存在严重的缺陷，所以在发表的第一印象能被确认之前，还需要更多同样头骨的例子。

首先，正如达特教授论述的，该化石属于一只具有乳牙齿系的未成年个体，就能从照片判断出来的信息而言，我看不到其具有比现代幼年黑猩猩头骨的相应部分明显更接近于人类状况的眼眶、鼻骨和犬齿。它的面部似乎相对较短，但是中新世森林古猿的下颌骨已经显示出这肯定是猿类祖先的特征之一。下颌联合部位的形状及没有“猿板”仅是因为该个体尚未成年；但是大家也许注意到一只成年森林古猿曾经被描述为具有一个很相似的下颌联合的情况，这也可以说是“下颌联合的前表面简直与海德堡人的一样垂直”（见《地质学会季刊》中的图，第 70 卷，1914 年，第 317、319 页）。

其次，汤恩头骨缺少脑壳骨骼，所以标本扭曲变形的程度和方向都无从确定。因此我对于认为脑模型任何部分的变圆和变得扁平有重要意义的观点都深表怀疑，甚至怀疑小脑模型的相对尺寸是否可信。对于这样一个化石，要想确定其真实的外观，还需要进一步证据来确认。

由于缺乏对在印度的第三纪地层发现的牙齿和颌骨断片所代表的类人猿头骨化石的了解，所以想要表达应该在亚洲还是非洲寻找人类直接祖先的任何观点还为时尚早。南非发现的这个新化石对于这一问题的解答毫无帮助。

Palaeontologists will await with interest Prof. Dart's detailed account of the new anthropoid, but cannot fail to regret that he has chosen for it so barbarous (Latin-Greek) a name as Australopithecus.

Arthur Smith Woodward

* * *

Prof. Dart's description of the fossil skull found at Taungs in Bechuanaland shows that this specimen possesses exceptional interest and importance. Should the claims made on its behalf prove good, then its discovery will in effect be comparable to those of the Pithecanthropus remains, of the Mauer mandible and the Piltdown fragments. In the following paragraphs I venture to make some comments based upon perusal of the article published in *Nature* of February 7.

First of all, the fact that the fragments came immediately under notice of so competent an anatomist as Prof. Dart establishes confidence in the thoroughness of the scrutiny to which they have been subjected. That the history of the specimen should be known precisely from the time of its release from the limestone matrix, provides another cause for satisfaction.

The specimen itself at once raises a number of questions, and, as Prof. Dart evidently realises, these fall into at least two categories. The first question arising out of the discovery is the status of the individual represented by these remains. But the answer to that question, and the presence of such a creature in South Africa, affect other problems. The latter include inquiry into the probable locality of origin of the simian and human types, and the search for evidence of dispersion from a centre, or along a line of successive migrations.

In dealing with the first problem, Prof. Dart has surveyed a considerable number of structural details, and he concludes that the specimen represents an extinct race of apes intermediate between living anthropoid apes and mankind. The specimen comprises the greater part of a skull with the lower jaw still in place (or nearly so). The number and characters of the teeth testify to the immaturity of the individual. The evidence on the last-mentioned point is quite definite, and interest thus comes to be centred in the status assigned to the specimen; namely, that of a form intermediate between the living anthropoid apes and man himself.

Prof. Dart places the specimen on the side of the living anthropoid apes in relation to the interval separating these from man. At the same time, it is claimed that this new form of ape is more man-like than any of the existing varieties of anthropoid apes; and so it comes about that the decision turns on the claims made for the superiority of the new ape to these other forms.

古生物学家会继续怀着兴趣等待达特教授对该新型类人猿作出更详细的说明，但是对于他为其选择了南方古猿这样一个如此野蛮的(拉丁–希腊语)名字不得不感到遗憾。

阿瑟·史密斯·伍德沃德

* * *

达特教授对发现于贝专纳兰的汤恩的头骨化石的描述表明该标本具有特别的影响和重要性。如果就这件化石发表的这个主张被证明是对的，那么实际上这件头骨的发现就堪比爪哇猿人化石、毛尔下颌骨化石和皮尔当颅骨破片化石的发现了。在接下来的段落中，我冒昧地根据自己对《自然》2 月 7 日发表的文章的精读进行评论。

首先，以上那些骨骼破片出现后很快就引起了像达特教授这样杰出的解剖学家的注意，这个事实使我们相信对这件标本的研究会是彻底的。应该由这件标本从石灰岩基质挖掘出来的时间开始精确地了解它的历史，这将提供另一个使人们对此发现感到满意的理由。

标本本身马上就引出了许多问题，正如达特教授明显意识到的，问题至少可以分为两类。从这个发现引出的第一个问题是这些化石所代表的个体的身份。但是这个问题的答案以及这样一种生物存在于南非影响了其他问题。后者包括对猿猴和人类诸多类型起源的可能地点的探究，以及对从中心扩散开来或沿着相继迁徙的路线向外扩散的证据的寻找。

在研究第一个问题时，达特教授调查了大量结构细节，他得出的结论是，标本代表的是一种已灭绝的猿类，介于现存的类人猿和人类之间的中间类型。该标本由头骨的大部分组成，该头骨上还连有仍然处于原位（或接近原位）的下颌骨。牙齿的数目和特征说明该个体是未成年的。最后提到的这个论点的证据是非常确定的，因此兴趣就集中在了这件化石应该属于什么身份；也就是，一种介于现存类人猿和人类自身的中间类型。

达特教授将该标本置于把现存的类人猿与人类分开的间隔当中的类人猿这一边。同时，他主张这种新类型的猿比任何现存种类的类人猿都更像人类；所以就决定转而主张这种新型猿比其他类型的猿更加优越。

The report shows that (as noted above) many structural details have been scrutinised, and that all accessible parts of the specimen have been examined. The observations relate not only to the external parts of the skull and lower jaw, but also to the endocranial parts exposed to view by the partial shattering of the brain-case. The claims advanced on behalf of the higher status of the specimen are based, therefore, upon a number and variety of such details. Should Prof. Dart succeed in justifying these claims, the status he proposes for the new ape-form should be conceded. Much will depend on the interpretation of the features exhibited by the surface of the brain, as also upon that of all the characters connected therewith; and since Prof. Dart is so well equipped for that aspect of the inquiry, his conclusions must needs carry special weight there. In regard to the brain and its characters, I find the tracing of the contour of an endocranial cast in a gorilla-skull shown in Fig. 6 rather surprisingly flattened, and almost suggestive of the influence of age.

Among the anatomical characters enumerated in the article, some appear to me to possess a higher value in evidence than others. As good points in favour of the claims, there may be cited, in addition to the cerebral features to which reference has just been made, the level of the lower border of the nasal bones in relation to the lower orbital margins, the (small) length of the nasal bones, the lack of brow-ridges (even though the first permanent tooth has appeared fully), the steeply-rising forehead, and the relatively short canine teeth.

On the other hand, I feel fairly certain that some of the other characters mentioned are related preponderantly to the youthfulness of the specimen. Fully to appreciate the latter, demands not only the handling of it, but also thorough survey of a collection of immature (anthropoid ape) crania. The development of the "shelf" at the back of the symphysis of the lower jaw may almost certainly be delayed in some individuals (gorillas). Even the level of the lower border of the nasal bones is subject to some variation, and in young gorillas before the first permanent tooth has emerged fully, that level may be (as in man) above the level of the orbital margin. Generally, the elimination and detachment of features influenced largely by the factor of age demand special attention.

If, however, the good points can be justified, then these characters of youth will not gravely affect the final decision.

However these discussions may end, the record remains of the occurrence of an anthropoid ape some two thousand miles to the south of the nearest region providing a record of their presence. So far as the illustrations allow one to judge, the new form resembles the gorilla rather than the chimpanzee, that is, an African, not an Asiatic form of anthropoid ape. In this respect the new ape does not introduce an obviously disturbing factor. Disturbance, and the recasting of disturbed views, might nevertheless be caused in two other directions. Thus, the determination of the geological antiquity of the embedding of the fossil remains might have such an effect, were the estimate such as to carry that event very far back in time. Again, a comparison of the new ape with the fossil forms from India (Siwaliks) remains to be made, and it may be productive of results bearing on the relation of the African and the Asiatic groups. In any case, opinion must needs conform to the situation created by this discovery.

该报告表明（正如上文所述），许多结构性细节都已经被仔细观察过了，标本所有可及的部分也都已经被查看过了。这些观察不仅涉及头骨和下颌骨的外部，还涉及脑壳一部分毁损后暴露出来的颅腔部分。因此为了表示该标本具有比较高级的身份而提出的主张是建立在许多种这类细节的基础上的。如果达特教授成功地证明这些说法是合理的，他建议的新型猿类的身份应当得到承认。更多信息将依赖于对脑表面所展示出的特征的解释，还依赖于对与此相关的所有特征的解释；而且由于达特教授对此方面的调查进行了非常充分的准备，所以他的结论肯定具有格外重要的分量。至于脑及其特征，我觉得图 6 所展示的大猩猩头骨的颅腔模型轮廓线的扁平程度堪称惊人，也暗示了年龄的影响。

在文章中列举的解剖学特征中，对于我来说，就证据价值而言，有些特征的价值比其他特征的更高。除了刚刚已经提到的大脑特征以外，可能被引用的作为支持这些说法的很好的证据还有与眼眶下缘相关的鼻骨下缘位置的水平、鼻骨的长度(短)、缺乏眉脊(尽管第一恒牙已经完全出现)、陡峭上升的前额以及相对短小的犬牙。

另一方面，我非常肯定地认为，提到的其他特征中有些肯定也是与标本的年轻性有关。为了充分理解后者，不仅需要对标本进行处理，还需要对收藏的未成年（类人猿）头盖骨进行彻底的调查。几乎可以肯定的是，下颌联合部位背面的“猿板”的发育在某些个体（大猩猩）中被延迟了。甚至鼻骨下缘的位置水平趋向于发生一定的变化，幼年大猩猩在第一颗恒牙完全出现之前，其位置水平可能（与人类一样）位于眶缘水平之上。通常来说，需要特别注意那些很大程度上受年龄因素影响而造成的特征的消失和分离。

然而，如果这些有利的证据可以被证实的话，那么这些幼年个体的特征将不会严重影响最终的决定。

无论这些讨论将会如何结束，这块化石记录了一只类人猿的存在，它距离现存最近的类人猿发现区域还要偏南 2,000 英里。就允许我们进行判断的现有说明而言，这种新类型与大猩猩的相像程度大于与黑猩猩的，也就是说，这是一种非洲类型的类人猿，而不是亚洲类型的。在这方面，这种新型猿并没有带来明显的干扰因素。干扰和对受到干扰的观点的重新认识可能来自另外两个方向。确定化石的埋藏发生在哪个地质学时期可能具有这样一种作用，这样的估计可能会将这一事件带回到非常远古的时期。再者，现在还没有将这种新型猿类与印度西瓦利克发现的化石类型进行比较，这种比较可能会得到关于非洲和亚洲群体之间关系的丰富结果。无论如何，意见必须与这一发现所产生的局面相符。

If in these notes there have been passed over those observations and reflections wherewith Prof. Dart has illustrated and supported his views, such omissions are not due to want of appreciation, but to lack of capacity and space for their adequate treatment.

W. L. H. Duckworth

(**115**, 234-236; 1925)

如果在这些说明中出现了忽略达特教授描述过并用来支持自己观点的观察和思考，那么这种忽略不是由于想索要赞赏，而是缺乏对它们进行充分处理的能力和空间。

迪克沃斯

（刘皓芳 翻译；吴新智 审稿）

Some Notes on the Taungs Skull

R. Broom

Editor's Note

While anthropologists in distant London debated the significance of Raymond Dart's new "Man-Ape", *Australopithecus africanus,* from South Africa, local palaeontologist Robert Broom went to see the actual specimen. This note looks at the geological setting, suggesting that the skull was of Pleistocene to Recent date; and also the cranial and dental anatomy, asserting the intermediate status of the creature. However, Broom presciently notes that if specimens of adults were to be discovered, "the light thrown on human evolution would be very great". A decade was to elapse before such finds came to light: and the discoverer would be Broom himself.

A few days ago I visited Johannesburg to have a look at the remarkable new skull discovered by Prof. Dart, and named by him *Australopithecus africanus*. Prof. Dart not only allowed me every facility for examining the skull, but also gave me with almost unexampled generosity full permission to publish any observations I made on it, and suggested further that I might send to *Nature* any notes that might amplify the account he had already given. As the skull is one of extreme importance, a full account with measurements and very detailed figures will in due course be published by Prof. Dart, but the world already realises the unique character of the discovery and is anxious for more immediate information.

From the cablegrams received in South Africa, it is manifest that the first demand is for further light on the geological age of the being, and unfortunately complete information on this point cannot now be given, and will possibly never be available. Though I have not myself visited the Taungs locality, I am fairly familiar with many similar deposits farther south along the Kaap escarpment. This escarpment runs for more than 150 miles along the west side of the Harts River and lower Vaal River valleys from a little south of Vryburg to 20 miles south of Douglas. The escarpment is formed for the most part of huge cliffs of dolomitic limestone of the Campbell Rand series, in most places some hundreds of feet thick. The wide valley has an interesting geological history. Originally it was carved out in Upper Carboniferous or Lower Permian times by the Dwyka glaciers. For millions of years it was steadily refilled by Dwyka, Ecca, and Beaufort beds until the whole valley was perhaps buried by more than 2,000 feet of Permian and Triassic shales. Then conditions changed and the valley was re-excavated, by denudation, until today we find it not unlike what it must have been when originally carved out by the Dwyka glaciers.

The dolomite escarpment forms the most striking feature of the landscape in this part of the world. All along the west of the Harts-Vaal valley lies the high dead-level Kaap

汤恩头骨的几点说明

布鲁姆

编者按

当伦敦的人类学家们正在为雷蒙德·达特在南非发现的新“人猿”（即南方古猿非洲种）的意义争论不休的时候，身在南非的古生物学家罗伯特·布鲁姆去看了真实的化石标本。这篇文章着眼于地质学背景，提出该头骨介于更新世到现代之间，同时也考虑了头盖骨和牙齿的解剖学特征，主张该生物处于中间的位置。不过，布鲁姆很有预见性地指出，如果能发现这种生物的成年个体，那么“其对于人类进化带来的启示将是非常重要的”。10 年之后迎来了这样的发现，而发现者正是布鲁姆本人。

几天前我访问了约翰内斯堡，参观了达特教授发现并将其命名为南方古猿非洲种的著名新头骨。达特教授不仅允许我使用查看该头骨的所有设备，而且完全许可我发表自己对该头骨进行的任何观察，这种慷慨几乎是史无前例的，他还建议我可以向《自然》投递可以扩充他发表过的报告内容的任何稿件。因为该头骨是极具重要性的标本之一，所以一份兼具测量尺寸和详细图像的完整报告将会由达特教授在适当的时间发表，但是世人已经意识到了这一发现的独特特征，并且急于得到更多的即时信息。

从南非收到的海底电报来看，很显然第一个要求就是希望对这种生物的地质学年代作更进一步的说明，不幸的是，对于这一点，目前还不能给出完整的信息，而且可能永远都无法得到。尽管我并没有亲自参观过汤恩遗址，但是我非常熟悉远在开普断崖沿线南部的许多相似的堆积物。这一断崖从弗雷堡稍南部沿着哈茨河和瓦尔河下游河谷西侧绵延 150 多英里直至道格拉斯南部 20 英里处。该断崖大部分由坎贝尔–兰德系列的巨大白云灰岩山崖组成，很多地方都有几百英尺厚。这里宽阔的河谷有一段很有趣的地质学史。最初它是在上石炭纪或早二叠纪时期由德怀卡冰川开拓出来的。数百万年间它不断地被德维卡、埃卡和博福特河床回填，直到整条河谷多半被 2,000 多英尺的二叠纪和三叠纪页岩掩埋掉为止。后来环境发生了改变，河谷受到剥蚀作用而被重新挖掘出来了，直到现在，我们发现这条河谷与最初被德维卡冰川开拓出来时的样子几乎一样。

白云石断崖在世界的这个部分形成了最具独特特征的景观。沿着哈茨–瓦尔河谷西部的所有部分都位于高而平坦的开普高原上，从 20 英里外眺望时，整个断崖看

plateau, and when viewed from 20 miles away the escarpment looks like a high black wall bounding the lower plain of the valley. Every five or ten miles along the black wall are to be seen large light-coloured patches which on examination prove to be great masses of calc-sinter formed by calcareous springs. These, of course, must have been formed after the dolomite cliffs had been denuded of their covering Dwyka shales, and may in some cases be of considerable age—perhaps even dating from moderately early Tertiary times. Other masses of this secondary limestone may be of comparatively recent date. In places the great masses of calc-sinter have been excavated by underground water and moderately large caves are formed.

At Taungs the mass of secondary limestone is some hundreds of feet thick and about 70 feet high where it is being worked. Already 250 feet have been quarried away. On the face about 50 feet below the top of the mass, an old cave is cut across which is filled up with sand partly cemented together with lime, and it is in this old cave that the skull of Australopithecus has been found. The only other bones that I have seen or heard of are skulls and bones of a baboon, a jaw of a hyrax, and remains of a tortoise. I have not seen the hyrax jaw, so cannot say if it belongs to one of the living species. The baboon has been examined by Dr. Haughton, who regards it as an extinct species and has named it *Papio capensis*. I have seen a number of imperfect skulls of this baboon, and while they belong to a different species from the living local *Papio porcarius,* the difference between them is not so very striking.

I think it can be safely asserted that the Taungs skull is thus not likely to be geologically of great antiquity—probably not older than Pleistocene, and perhaps even as recent as the *Homo rhodesiensis* skull. When later or other associated mammalian bones are discovered, it may be possible to give the age with greater definiteness. At present all we can say is that the skull is not likely to be older than what we regard as the human period. But the age of the specimen in no way interferes with its being a true "missing link", and the most important hitherto discovered.

Prof. Dart in his photographs has given the general features of the skull and the brain, but there are a number of important characters in the skull and dentition to which I should like to direct attention.

Though the parietals and occipital are almost completely lost from the brain cast, most of the sutures can be clearly made out, and are as I indicate in Fig. 1. The sutures in the temporal region can also be clearly seen. The suture between the temporal bone and the parietal is fairly horizontal as in the anthropoid apes, but in the upward development of the squamous portion we have a character which is human and not met with in the gorilla, the chimpanzee, the orang, or the gibbon.

起来就像一堵以河谷的下流平原为界的高大黑墙。沿着黑墙，每 5~10 英里就可以看到巨大的浅颜色的点缀，经检验证实它们是由石灰泉形成的大片钙华。当然，这些肯定是在白云石山崖被剥蚀掉了覆盖着的德维卡页岩之后才形成的，而且在有些情况下，它们可能有着相当长的年代——甚至可能一直追溯到第三纪的早期。这种次生石灰石的其他块体的年代可能比较晚近。有些地方，地下水挖掘了大片钙华，形成了中等的大型山洞。

在汤恩，次生石灰石有几百英尺厚，70 英尺高。采石场已经挖掉了 250 英尺。在大块石灰石顶部之下约 50 英尺处的表面上，挖掘中穿过了一个旧山洞，洞中填满了与石灰一起形成水泥的沙子，就在这个老山洞里发现了南方古猿头骨。我见过或听说过的骨骼只有狒狒的头骨和骨骼、岩狸的颌骨和龟的残骸。我没有见到岩狸的颌骨，所以不能说它是否属于现存物种之一。狒狒的遗骸已经由霍顿博士查看过了，他认为它是一种已灭绝的物种，并将其命名为狒狒开普种。我见过这只狒狒的许多不完整的头骨，尽管它们属于一种与现存的本地浅灰狒狒不同的物种，但二者之间的差异不是十分明显。

我认为可以有把握地说，从地质学角度而言，汤恩头骨不可能是非常古老的——可能不早于更新世，甚至可能与罗德西亚人头骨一样晚。在发现比较晚的或其他与之共生的哺乳动物骨骼时，就有可能给出更确定的年代了。现在我们能说的就是这个头骨不可能早于我们所认为的人类时期。但是该标本的年代绝不妨碍它作为一个真正的“缺失环节”和迄今为止发现的最重要的标本。

达特教授在他发表的照片中给出了头骨和脑的一般特征，但是我想关注的事情是，头骨和齿系还有许多重要特征。

尽管在脑的模型上几乎找不到顶骨和枕骨的影子，但是可以清楚地辨认出大部分骨缝，这些骨缝我在图 1 中都指明了。颞区的骨缝也可以清楚看到。颞骨和顶骨之间的骨缝与类人猿中的一样非常水平，但是它的鳞部向上发育，这是一个属于人类的特征，在大猩猩、黑猩猩、猩猩和长臂猿中都没有遇见过。

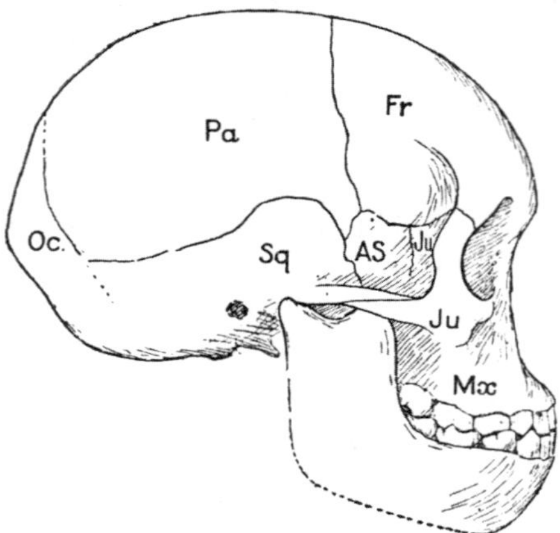

Fig. 1. Side view of skull of *Australopithecus africanus*, Dart. About $\frac{1}{3}$ natural size.

The arrangement of the sutures in the temporal region is also remarkably interesting. The upper part of the sphenoid articulates with both the parietal and the frontal. In the gorilla and chimpanzee in all the drawings I can find, the temporal bone meets the frontal and prevents the meeting of the sphenoid and the parietal. In the orang the condition varies, and I have in my possession a skull which has on the right side a spheno-parietal suture and on the left a fronto-temporal. In the baboon there is a large fronto-temporal suture, and in Cercopithecus a spheno-parietal suture. In the gibbon there is also a spheno-parietal suture. While the arrangement of the sutures in this region may not be of very great fundamental importance, it is interesting to note that Australopithecus agrees with man, the gibbon, and Cercopithecus, but differs from the gorilla, the chimpanzee, and the baboon.

The jugal or malar arch is interesting in that there is a long articulation between the jugal and squamosal. In this Australopithecus agrees rather with the anthropoids than with man.

On the face there are one or two striking characters, and of these perhaps the most important is the fusion of the premaxilla with the maxilla. On the palate the suture between these bones is seen almost as in the human child, the suture running out about two-thirds of the way towards the diastema between the second incisor and the canine. On the face there is no trace of any suture in the dental region, but on the left side of the nasal opening there is what is probably the upper part of the original premaxilla-maxillary suture. On the right side there is a faint indication of a suture just inside the nostril. In the chimpanzee the suture becomes obliterated in the dental region early, as apparently is the case in Australopithecus. In the orang and gorilla the suture remains distinct until a much later stage. In man, as is well known, all trace of the suture is obliterated from the face long before birth.

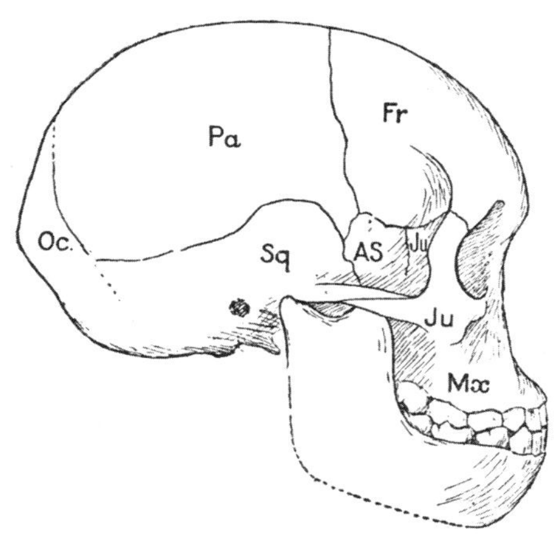

图 1. 南方古猿非洲种头骨的侧面观，达特。约相当于真实尺寸的 $\frac{1}{3}$。

颞区骨缝的分布也非常有意思。蝶骨的上部与顶骨和额骨都连接着。在我能找到的所有大猩猩和黑猩猩的图画中，颞骨都与额骨连接，而且颞骨阻止了蝶骨和顶骨相遇。猩猩则不是这种情况，我拥有的一个头骨的右侧有一条蝶顶缝，左侧有一条额颞缝。狒狒有一条大型的额颞缝，长尾猴有一条蝶顶缝。长臂猿也有一条蝶顶缝。尽管这一区域的骨缝的分布不具有重大的关键意义，但有趣的是，我们注意到南方古猿与人类、长臂猿和长尾猴都是一致的，而与大猩猩、黑猩猩和狒狒则是不同的。

颧弓的有趣之处在于，颧骨和颞鳞之间有一条长的关节。在这点上南方古猿与类人猿的一致性要多过与人类的一致性。

面部有一两点显著的特征，这些特征中最重要的可能是前颌骨和上颌骨之间的融合。可以看出硬腭的这些骨骼间的骨缝几乎与人类孩子的一样，骨缝向第二门齿和犬齿间的齿隙延伸约 2/3 的距离后消失。在面部，牙齿区域没有任何骨缝的痕迹，但是在鼻腔开口的左侧有一条骨缝可能是原先的前颌骨–上颌骨骨缝的上半部分。右侧在鼻孔里有一条微弱的骨缝迹象。黑猩猩这条骨缝早早地就在牙齿区域消失了，南方古猿中表面上也是这样。猩猩和大猩猩中这条骨缝直到很晚的阶段依然明显。众所周知，人类的所有骨缝的痕迹在出生之前就已经从脸上消失了。

Australopithecus agrees with man and the chimpanzee in having a single foramen for the superior maxillary nerve. In the orang, gibbon, and other apes there are usually two or more foramina. In the gorilla sometimes there is one foramen; sometimes two.

In the shortness of the nasal bones and the high position of the nasal opening the Taungs skull agrees more with the chimpanzee than with the gorilla.

The dentition is beautifully preserved, and the teeth have been cleared of matrix by Prof. Dart with the greatest care. Though, owing to the lower jaw being in position, a full view of the crowns of the teeth could only be obtained by detaching the lower jaw, a sufficiently satisfactory view can be obtained to give us practically all we require of the structure.

The whole deciduous denture is present in practically perfect condition. The incisors, which are small, have been much worn down by use, and most of the crowns of the median ones have been worn off. Prof. Dart has directed attention to the vertical position of the teeth, which is a human character and differs considerably from the conditions found in the chimpanzee and gorilla. The small size of the incisors is also a human character.

The relatively small size of the canine is a character in which Australopithecus agrees with both the chimpanzee and man, and lies practically between the two.

The deciduous molars agree more closely with those of man than with those of any of the apes.

The first permanent molars of both upper and lower jaws are perfectly preserved and singularly interesting.

The first molar of the upper jaw (Fig. 2) has four large cusps arranged as in man and the anthropoid apes.

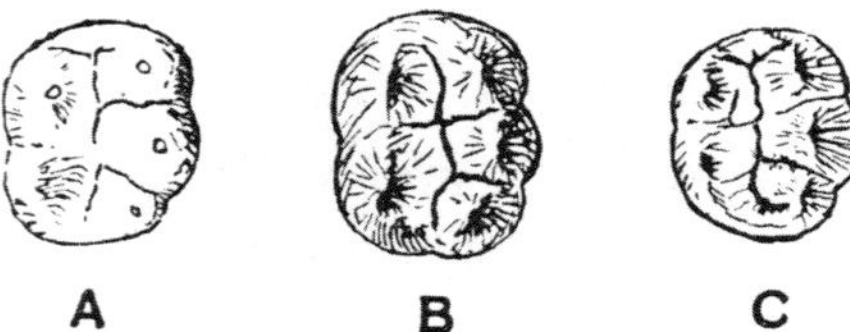

Fig. 2. First right upper molars: A, orang (after Röse); B, *Australopithecus africanus*, Dart, unworn; C, Bushman child, unworn. All natural size.

在上颌神经具有单一小孔这方面，南方古猿与人类和黑猩猩一致。而猩猩、长臂猿和其他猿类通常有两个或更多个小孔。大猩猩有时有一个小孔，有时有两个。

在鼻骨短小和鼻腔开口位置偏高这方面，汤恩头骨与黑猩猩的一致性比与大猩猩的高。

齿系的保存状况很好，而且达特教授已经十分慎重地清除掉了附着在牙齿上的基质。由于下颌骨仍保留在原位，所以要想对牙冠进行全面观察，只有将下颌骨分离下来才可以，这样就能得到一幅足以满足我们对全部结构的需要的视图。

全部乳牙的保存状况都相当完美。门齿小，由于使用而有了很大程度的磨损，大部分内侧门齿的牙冠都被磨损掉了。达特教授注意到牙齿的方位是垂直的，这是一种人类特征，与黑猩猩和大猩猩中观察到的情况差异很大。门齿尺寸较小也是一种人类特征。

相对小尺寸的犬齿是一种南方古猿与黑猩猩和人类都一致的特征，实际上南方古猿的犬齿尺寸介于这二者之间。

与所有猿类的乳臼齿相比，南方古猿的乳臼齿与人类的更加相符。

上颌骨和下颌骨的第一恒臼齿的保存状况都很好，它们都非常有意思。

上颌骨的第一臼齿（图 2）有 4 个大的牙尖，其排列情况与人类和类人猿的一样。

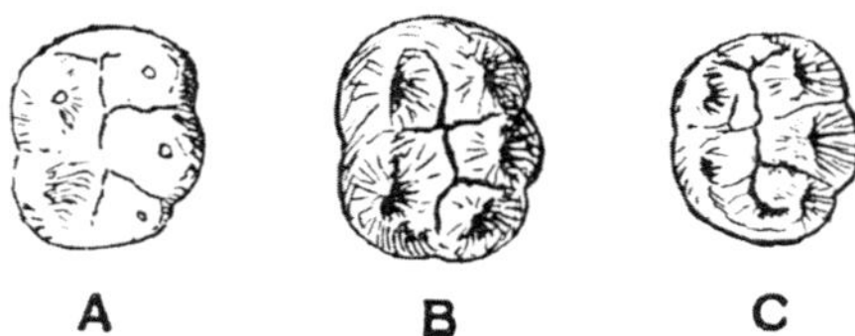

图 2. 第一右上臼齿：A，猩猩（依照罗斯）；B，南方古猿非洲种，达特，未磨损；C，儿童布希曼人，未磨损。全部都是真实尺寸。

The first lower molars (Fig. 3) has three well-developed sub-equal cusps on the outer side and two on the inner. Though in its great length and in the large development of the third outer cusp or hypoconulid the tooth differs considerably from the typical first lower molar of man, teeth of this pattern not infrequently occur in man. In general structure, however, the tooth more closely resembles that of the chimpanzee. It is interesting to compare this tooth with the corresponding tooth in Eoanthropus.

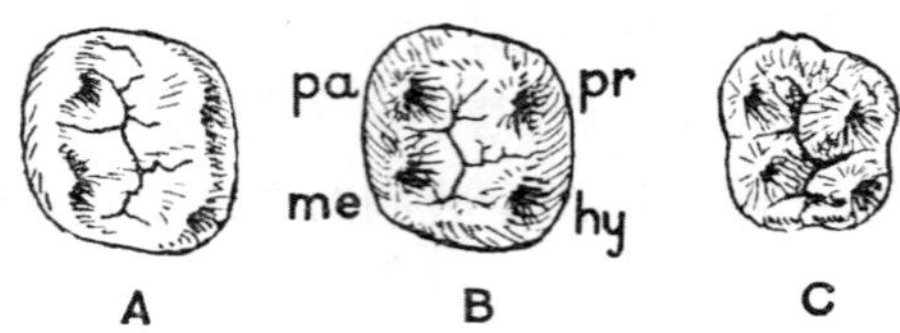

Fig. 3. First right lower molars: A, old chimpanzee, worn (after Miller); B, *Australopithecus africanus*, Dart; C, Bushman child. All natural size.

The arrangement of the furrows on the crown of the molar of Australopithecus is almost exactly similar to that in both the orang and the Bushman. In the chimpanzee and gorilla, there is usually a well-marked ridge passing from the protocone to the metacone, of which there is an indication in the Bushman tooth.

It will be seen that in *Australopithecus africanus* we have a large anthropoid ape resembling the chimpanzee in many characters, but approaching man in others. We can assert with considerable confidence that it could not have been a forest-living animal, and that almost certainly it lived among the rocks and on the plains, as does the baboon of today. Prof. Dart has shown that it must have walked more upright than the chimpanzee or gorilla, and it must thus have approached man more nearly than any other anthropoid hitherto discovered.

Eoanthropus has a human brain with still the chimpanzee jaw. In Australopithecus we have a being also with a chimpanzee-like jaw, but with a sub-human brain. We seem justified in concluding that in this new form discovered by Prof. Dart we have a connecting link between the higher apes and one of the lowest human types.

The accompanying table (Fig. 4) shows what I believe to be the relationships of Australopithecus. If an attempt be made to reconstruct the adult skull (Fig. 5), it is surprising how near it appears to come to *Pithecanthropus erectus*—differing only in the somewhat smaller brain, and less erect attitude.

While nearer to the anthropoid apes than man, it seems to be the forerunner of such a type as Eoanthropus, which may be regarded as the earliest human variety, the other probably branching off in different directions.

第一下臼齿（图 3）外侧有 3 个发育完好的几乎相等的牙尖，内侧有 2 个。尽管在牙尖的巨大长度以及第三外侧牙尖或下次小尖的发达程度上，这颗牙与典型的人类第一下臼齿非常不同，但是人类中这种形式的牙齿并不罕见。然而，就总体结构而言，这颗牙与黑猩猩的更像。将这颗牙与曙人相应的牙齿进行比较会得到很有趣的发现。

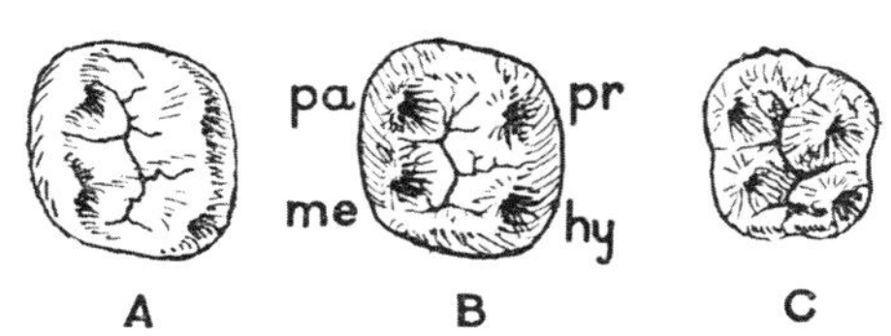

图 3. 第一右下臼齿：A，年老黑猩猩，有磨损（依照米勒）；B，南方古猿非洲种，达特；C，儿童布希曼人。全部都是真实尺寸。

南方古猿臼齿牙冠上沟的排列几乎与猩猩和布希曼人的完全一样。黑猩猩和大猩猩中，通常有一条从上原尖到后尖的明显的脊，布希曼人牙齿中也有这种迹象。

我们将会看到，在南方古猿非洲种中出现了一只许多特征与黑猩猩相像、但其他方面又与人类相像的大型类人猿。我们可以充满信心地断言它不是一只生活在森林中的动物，并且几乎可以肯定它是生活在岩石间和平原上的动物，就像现在的狒狒一样。达特教授已经说明了，它行走的姿势肯定比黑猩猩和大猩猩的更加直立，因此它肯定比迄今发现的其他类人猿更接近于人类。

曙人具有人类的脑、黑猩猩的颌骨。南方古猿中我们也有一只黑猩猩式的颌骨和次人的脑。我们似乎有理由作出如下结论：这种由达特教授发现的新类型使得我们拥有了一个连接高等猿类和一种人类的最低级类型的环节。

附表（图 4）展示了我相信的南方古猿的亲缘关系图。如果试图对成年头骨进行复原（图 5），那么就会惊奇地发现它与爪哇猿人是非常接近的——只是在脑稍小、姿势欠直立方面有所不同。

尽管与人类相比它更接近于类人猿，但是似乎仍旧可以将其认为是像曙人这样的类型的先祖，并将其当作是最早期的人类物种，即可能是另一个与人类发生了分歧而向不同方向进化的物种。

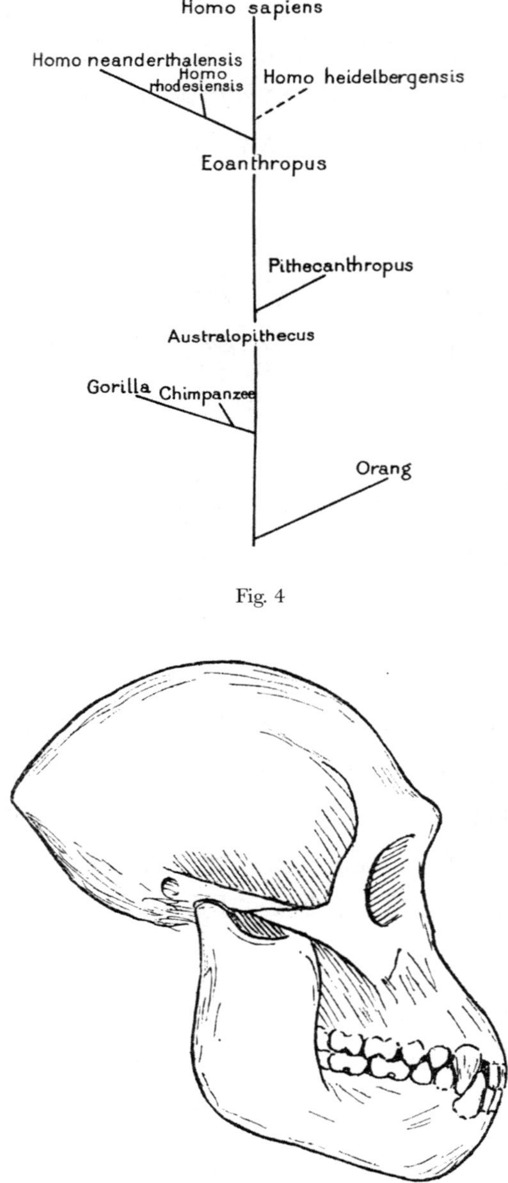

Fig. 4

Fig. 5. Attempted reconstruction of adult skull of *Australopithecus africanus*, Dart. About $\frac{1}{3}$ natural size.

There seems considerable probability that adult specimens will yet be secured, and if the skeleton as well as the skull is preserved, the light thrown on human evolution will be very great.

(**115**, 569-571; 1925)

R. Broom: Douglas, South Africa.

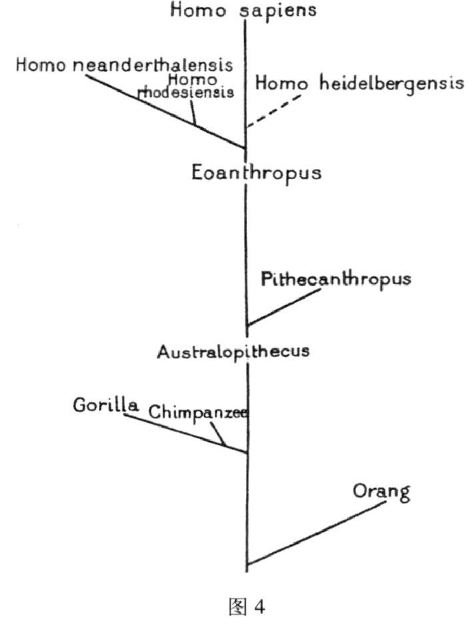

图 4

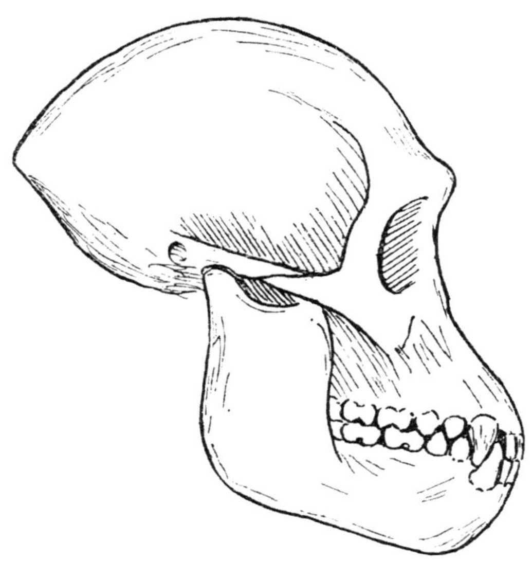

图 5. 成年南方古猿非洲种尝试性的复原头骨，达特。约相当于真实尺寸的 $\frac{1}{3}$。

似乎有足够的可能性认为成年标本会使我们更安心，如果骨架和头骨都被保存下来的话，那么其对于人类进化带来的启示将是非常重要的。

（刘皓芳 翻译；吴新智 审稿）

The Taungs Skull

A. Keith

Editor's Note

Months after the discovery of *Australopithecus africanus,* London anthropologists remained frustrated that they could not see the specimen—not even a cast. This may explain the rage of this letter from the eminent Scottish anatomist and anthropologist Arthur Keith, that the closest he had got to the Taungs skull was to go to an exhibition and "peer at [the casts] in a glass case" along with ordinary members of the public. Keith's conviction that Australopithecus was a fossil ape rather than a transitional form between apes and humans hardened.

THE account which Prof. Dart published of the Taungs skull (*Nature*, Feb. 7, p. 195) left many of us in doubt as to the true status of the animal of which it had formed part, and we preferred, before coming to a decision, to await an examination of the fossil remains, or failing such an opportunity, to study exact casts of them. For some reason, which has not been made clear, students of fossil man have not been given an opportunity of purchasing these casts; if they wish to study them they must visit Wembley and peer at them in a glass case which has been given a place in the South African pavilion.

The chief point which awaited decision relates to the position which must be assigned in the animal kingdom to this newly discovered form of primate. Prof. Dart, in writing of it, has used the name of anthropoid ape; he has described it as representing "an extinct race of apes intermediate between living anthropoids and man"—which is tantamount to saying that at Taungs there has been discovered the form of being usually spoken of as the "missing link". That this is his real decision is evident from the fact that he speaks of it as "ultrasimian and prehuman" and proposes the creation of a new family for its reception.

An examination of the casts exhibited at Wembley will satisfy zoologists that this claim is preposterous. The skull is that of a young anthropoid ape—one which was in the fourth year of growth—a child—and showing so many points of affinity with the two living African anthropoids—the gorilla and chimpanzee—that there cannot be a moment's hesitation in placing the fossil form in this living group. At the most it represents a genus in the Gorilla-Chimpanzee group. It is true that it shows in the development of its jaws and face a refinement which is not met with in young gorillas and chimpanzees at a corresponding age. In these respects it does show human-like traits. It is true that it is markedly narrow-headed while the other African anthropoids are broad-headed—but we find the same kind of difference in human beings of closely allied races. Prof. Dart claimed that the brain showed certain definite human traits. This depends upon whether or not he had correctly identified the position of a certain fissure of the brain—the parallel fissure. In the show-case at Wembley a drawing is placed side by side with the "brain cast"; but when we examine the brain cast at the site where the fissure is shown on the drawing, we find only a broken surface where identification becomes a matter of guess-work.

汤恩头骨

基思

编者按

在达特发现南方古猿非洲种几个月之后，伦敦的人类学家们依旧很沮丧，因为他们看不到那个标本，甚至连模型也看不到。这大概可以解释苏格兰杰出解剖学家、人类学家阿瑟·基思在这封来信中表达的愤怒，他说自己距离那件汤恩头骨最近的时候就是随着普通观众参观头骨展览时“盯着看玻璃柜里的[模型]”。基思坚信，南方古猿更像是古猿的化石，而不是古猿与人类之间的过渡类型。

达特教授就汤恩头骨发表的报道（《自然》，2月7日，第195页）中关于这种动物（这件化石就是它的一个部分）的真实身份给我们留下了许多可疑之处，在得到结论之前，我们更倾向于等待对化石进行查看，如果没有这样的机会，研究其精确的模型也行。没有人清楚是出于什么原因，研究人类化石的学者们还没有得到购买这些模型的机会；如果他们希望研究这些模型的话，他们就必须去温布利参观南非展览馆，然后盯着看玻璃柜里的模型。

有待确定的最主要的一点与将这种新发现的灵长类类型归属到动物界的哪个位置有关。达特教授写这部分的时候，使用了类人猿的名称，他将其描述为代表着“一种已经灭绝的猿类，介于现存的类人猿和人类之间的中间类型”——这种说法相当于是说汤恩发现了通常被称作是“缺失环节”的生物类型。显而易见这是他的真实判断，从他将其称为“超猿和前人类”的说法以及提议创建一个新科来接纳这种动物的事实就可以看出来。

对温布利陈列的模型进行观察将会使动物学家相信这种主张是荒谬可笑的。这具头骨是一只幼年类人猿的——该个体处于生长阶段的第4年——即一个孩子——它显示出了许多与两种现存的非洲类人猿——大猩猩和黑猩猩——的亲缘关系，于是将该化石类型列入这种现存群体中就变得不容置疑了。它最多代表了大猩猩–黑猩猩群体中的一个属。它确实表明颌骨和面部在发育过程中发生了细微的变化，这在相应年龄的幼年大猩猩和黑猩猩中都是不曾见到过的。这些方面它确实显示出了与人类相像的特点。它具有非常狭窄的头部也是真实的，而其他非洲类人猿都是阔头型的——但是我们发现具有密切亲缘关系的人类种族也存在同样的差异。达特教授声称该标本的脑显示了某些确定的人类特征。这依赖于他是否正确地辨认了某一大脑裂缝——平行裂缝的位置。在温布利的陈列柜里，一幅画与“脑模型”一起陈列在那里；但是当我们观察脑模型，查看图画上展示的裂缝所在位置时，我们发现只有一处裂开的表面，在这里，鉴定变成了一种猜测。

In every essential respect the Taungs skull is that of a young anthropoid ape, possessing a brain which, in point of size, is actually smaller than that of a gorilla of a corresponding age. Only in the lesser development of teeth, jaws, and bony structures connected with mastication can it claim a greater degree of humanity than the gorilla. Its first permanent molar teeth which have just cut are only slightly smaller than those of the gorilla, while the preparations which are being made in the face for the upper permanent canines show that these teeth were to be of the large anthropoid kind.

The other point on which we awaited information related to the geological age of the Taungs skull. Fortunately, Dr. Robert Broom (*Nature*, April 18, p. 569) has thrown a welcome light on this matter. The skull was blasted out of a cave which had become filled up by sand washed in from the Kalahari. The fossil baboons found in neighbouring caves differ in only minor structural details from baboons still living in South Africa. In Dr. Broom's opinion the Taungs skull is of recent geological date; it is not older than the Pleistocene; he thinks it probable that it may not be older than the fossil human skull found in a limestone cave at Broken Hills, Rhodesia. It is quite possible—nay, even probable—that the Taungs anthropoid and Rhodesian man were contemporaries. Students of man's evolution have sufficient evidence to justify them in supposing that the phylum of man had separated from that of anthropoid apes early in the Miocene period. The Taungs ape is much too late in the scale of time to have any place in man's ancestry.

In a large diagram, placed in the show-case at Wembley, Prof. Dart gives his final conception of the place occupied by the Taungs ape in the scale of man's evolution. He makes it the foundation stone of the human family tree. From the "African Ape Ancestors, typified by the Taungs Infant", Pithecanthropus, Piltdown man, Rhodesian man, and African races radiate off. A genealogist would make an identical mistake were he to claim a modern Sussex peasant as the ancestor of William the Conqueror.

In the show-case at Wembley plastic reconstructions are exhibited in order that visitors may form some conception of what the young Taungs Ape looked like in life. Although the skull is anthropoid it has been marked by a "make-up" into which there have been incorporated many human characters. It is true the ears are those of the chimpanzee, but the forehead is smooth and rounded, the hair of the scalp is sleek and parted; the bushy eyebrows are those of a man at fifty-five or sixty; the neck is fat, thick, and full—extending from chin to occiput. In modelling the nose, gorilla lines have been followed, whereas the nasal part of the skull imitates closely chimpanzee characters. The mouth is wide, with a smile at each corner.

Prof. Dart has made a discovery of great importance, and the last thing I want to do is to detract from it. He has shown that anthropoid apes had extended, during the Pleistocene period, right into South Africa—into a land where anthropoid apes could not gain a livelihood today. He has found an extinct relative of the chimpanzee and gorilla but one with more man-like features than are possessed by either of these. His discovery throws light on the history of anthropoid apes but not on that of man. Java-man (Pithecanthropus) still remains the only known link between man and ape, and this extinct type lies on the human side of the gap.

(**116**, 11; 1925)

汤恩头骨在各个关键方面，都显示出一只幼年类人猿的特点，它的脑在尺寸上确实比相应年龄的大猩猩的小。只在较欠发育的牙齿、颌骨和与咀嚼相关的骨质结构方面，可以将它称为一种比大猩猩更高等的似人动物。其刚刚萌出的第一恒臼齿只比大猩猩的略小，而面部针对上恒犬齿所作的准备则表明这些牙齿是属于大型类人猿那一类的。

我们等待的另外一个方面的资料与汤恩头骨的地质学年代有关。幸运的是，罗伯特·布鲁姆博士（《自然》，4 月 18 日，第 569 页）已经对这个问题进行了阐述，并且他的说法被很多人接受。这个头骨是从一个填满了来自卡拉哈里沙漠的沙子的山洞里炸出来的。在旁边山洞发现的狒狒化石与南非现存的狒狒只在微小的结构细节上存在差异。布鲁姆博士认为，汤恩头骨的地质学年代较近；应该不早于更新世；他认为一种可能的情况是，这具头骨的年代并不比在罗德西亚布罗肯希尔山的石灰石山洞中发现的人类头骨化石的时期早。非常可能——不，几乎可以肯定的是——汤恩类人猿和罗德西亚人是同时代的。研究人类进化的学者们有足够的证据证实他们所假设的人类这一支系是在中新世早期从类人猿中分离出来的。汤恩猿在时间尺度上太晚了，所以根本不可能是人类的祖先。

在温布利的陈列柜里摆放的一张大图表中，达特教授给出了他对汤恩猿在人类进化上的位置的最终想法。他把汤恩猿当作奠定人类家族树的基石。从“汤恩婴儿代表的非洲猿祖先”开始，爪哇猿人、皮尔当人、罗德西亚人和非洲人种呈辐射状散出。如果一位系谱专家宣称一位现代萨塞克斯的农民是征服者威廉的祖先，那他就是犯了同样的错误。

在温布利的陈列柜里，陈列着复原像的造型以便参观者可以对幼年汤恩猿在生活状态下看起来是什么样子有一定的概念。尽管这具头骨是类人猿的，但是许多人类特征被“捏造”出来表现在这件复原像上。确实其耳朵是黑猩猩的，但是前额平滑而圆润，头皮上的头发光滑而疏散；浓密的眉毛与 55~60 岁的人类很像；脖子粗、厚而丰满——从下巴一直延伸到枕部。在制作鼻子的模型时，参照了大猩猩的线条，然而这具头骨的鼻子部分模仿了与黑猩猩非常接近的特征。嘴部宽阔，嘴角带笑。

达特教授取得了非常重要的发现，但是我想做的最后一件事就是贬低它。他向我们展示了类人猿在更新世时期曾经一直扩展到南非——这是一片现在的类人猿无法生存的地域。他发现了黑猩猩和大猩猩的一种已灭绝的亲属，但是它又具有比起二者来更像人类的特征。他的发现对于类人猿历史有一定的昭示作用，但是对人类历史并不尽然。爪哇猿人仍然是唯一已知的人和猿之间的环节，这种已灭绝的类型处于缺口中人类这一边。

（刘皓芳 翻译；吴新智 审稿）

The Taungs Skull

Editor's Note

Here Raymond Dart responds to Arthur Keith's assertive criticism of Dart's claim to have identified an intermediate form between man and ape, called *Australopithecus africanus*. Dart refutes Keith's charges robustly at every point except one, which relates to the question of whether a cast of the skull should be made available to other anthropologists.

IN *Nature* of July 4, 1925, p. 11, Sir Arthur Keith has attempted to show first that I called the Taungs skull a "missing link", and secondly, that it is not a "missing link".

As a matter of fact, although I undoubtedly regard the description as an adequate one, I have not used the term "missing link." On the other hand, Sir Arthur Keith in an article entitled "The New Missing Link" in the *British Medical Journal* (February 14, 1925) pointed out that "it is not only a missing link but a very complete and important one". After stating his views so definitely in February, it seems strange that, in July, he should state that "this claim is preposterous".

Despite this reversal of opinion, Sir Arthur tells us that the skull "does show human-like traits in the refinement of its jaws and face which is not met with in young gorillas and chimpanzees at a corresponding age." He appears to have overlooked the fact that in addition to these and other facts brought forward by myself, the temporal bone, sutures, and deciduous and permanent teeth (according to Dr. Robert Broom) also show human-like traits. Moreover, as Prof. Sollas has so ably shown, the whole profile of the skull is entirely different from that in living anthropoids, thus indirectly confirming my discovery that the brain inside the skull-dome which caused this profound difference was very different from the brains inside the skulls of modern apes.

The fact that Sir Arthur was unable to find the *parallel sulcus* depression in the replica cast sent to Wembley illustrates how unsatisfactory the study of the replica can be in the absence of the original.

With reference to the question of endocranial volume, I would state with Prof. Sollas that this "is a matter of only secondary importance". Nothing could exemplify this matter better than the condition of affairs in the Boskop race, where the endocranial volume was in the vicinity of 1,950 c.c. (The average European's endocranial volume is 1,400–1,500 c.c.) Indeed, the world's record in human endocranial volume (2,000 c.c.) was discovered in a "boskopoid" skull by Prof. Drennan in a dissecting room subject at Capetown this year. It is well known that the elephant and the whale have brains much larger than those of

汤恩头骨

编者按

雷蒙德·达特此前声称发现并鉴定了一种介于人与古猿之间，被命名为南方古猿非洲种的中间类型，阿瑟·基思对此提出了批评。在这里，雷蒙德·达特对阿瑟·基思的批评作出了回应。除了是否应该将这具头骨的模型对其他人类学家开放的问题，达特对基思的每一条意见都进行了反驳。

在 1925 年 7 月 4 日的《自然》第 11 页，阿瑟·基思爵士首先试图表明我把汤恩头骨称为一个“缺失的环节”，其次，试图表明汤恩头骨不是一个“缺失的环节”。

事实上，尽管我毫不怀疑地认为这种描述是适当的，但是我并没有使用过“缺失的环节”这个短语。另一方面，阿瑟·基思爵士在发表于《英国医学杂志》（1925 年 2 月 14 日）上的一篇题为《新的缺失环节》的文章里指出“它不仅是一个缺失的环节，而且是一个非常完整而重要的环节”。在他 2 月份如此确定地陈述了自己的观点之后，在 7 月他又说“这种说法是荒谬可笑的”，这似乎有点奇怪。

尽管有这样矛盾的观点，阿瑟爵士还是告诉我们这个头骨“在颌骨和面部的细节上确实显示出了与人类相像的特征，这种特征在相应年龄的幼年大猩猩和黑猩猩中都是不曾见到过的。”看起来他似乎忽略了一个事实，那就是除了我本人提出来的这些和其他事实之外，颞骨、骨缝、乳齿和恒牙（根据罗伯特·布鲁姆）也显示出了与人类相像的特征。另外，索拉斯教授也非常巧妙地说明了该头骨的整个剖面与现存的类人猿完全不同，因此间接地证实了我的发现，即引起这种深刻差异的位于头骨之内的脑与现代猿的头骨内的脑差异很大。

阿瑟爵士未能发现送到温布利的复制品模型上的**平行沟**凹陷，这个事实显示了在缺乏原型时只是对复制品进行研究会多么令人不满。

关于颅腔容量的问题，我已经与索拉斯教授共同声明过，即这“只是第二重要的事情”。除了颅腔容量接近 1,950 毫升（欧洲人的颅腔容量平均值为 1,400～1,500 毫升）的博斯科普人（南非石器时代中期的一个人种）的情况之外，可能没有什么更好的可以用来证明这个问题的例子了。实际上，人类颅腔容量的世界纪录（2,000 毫升）是德雷南教授今年在开普敦的一间解剖室里发现的一具“类博斯科普人的”头骨的容量。众所周知，大象和鲸的脑比人类的大得多，但是没有人会从这一点得出它们

human beings, but no one has inferred from that that their intelligence is greater. It is fairly certain that size of brain has some relation to size of body, as Dubois has shown. It is highly probable that the australopithecid man-apes were relatively small as compared with the gorilla. It is not the quantity so much as the quality of the brain that is significant.

Sir Arthur is harrowed unduly lest the skull *may* be Pleistocene. It is significant in this connexion that Dr. Broom, who first directed attention to this possibility (of which I was aware before my original paper was sent away), regarded it nevertheless as "the forerunner of such a type as Eoanthropus". It should not need explanation that the Taungs infant, being an infant, was ancestral to nothing, but the family that he typified are the nearest to the prehuman ancestral type that we have.

In view of these facts, there is little justification for the attempted witticism that in making the "African ancestors typified by the Taungs infant" the "foundation stone of the human family tree"—whatever that may be—I am making "a mistake identical with that of claiming a Sussex peasant as the ancestor of William the Conqueror". This is merely a case of mistaken identity on the part of Sir Arthur. I have but translated into everyday English the genealogical table suggested by Dr. Robert Broom (*Nature*, April 18, 1925), with which I agree almost entirely. I take it, however, as a mark of his personal favour that Sir Arthur should have attacked my utterance and spared Dr. Broom's.

Sir Arthur need have no qualms lest his remarks detract from the importance of the Taungs discovery—criticism generally enhances rather than detracts. Three decades ago Huxley refused to accept Pithecanthropus as a link. Today Sir Arthur Keith regards Pithecanthropus as the only known link. There is no record that Huxley first accepted it, then retracted it, but history sometimes repeats itself.

Raymond A. Dart

* * *

Prof. Dart is under a misapprehension in supposing that I have in any way or at any time altered my opinion regarding the fossil ape discovered at Taungs. From the description and illustrations given by him (*Nature*, Feb. 7, 1925, p. 195) the conclusion was forced on me that Australopithecus was a member of "the same group or sub-family as the chimpanzee and gorilla" (*Nature*, Feb. 14, 1925, p. 234). In the same issue of *Nature*, Prof. G. Elliot Smith expressed a similar opinion, describing Australopithecus as "an unmistakable anthropoid ape that seems to be much on the same grade of development as the gorilla and chimpanzee without being identical with either."

All the information which has come home since Prof. Dart made his original announcement in *Nature* has gone to support the close affinity of the Taungs ape to the gorilla and to the chimpanzee—it is a member of that group. Prof. Bolk, of Amsterdam,

的智慧也比人类高的推论。正如杜波伊斯指出的，脑的尺寸肯定与身体的尺寸有一定的关系。有一种很大的可能性是，与大猩猩相比，南方古猿属的人猿相对较小。显然脑的量不如质那样具有重大意义。

阿瑟爵士因为唯恐该头骨**可能**是更新世时期的而过度苦恼了。在这个关系上，值得注意的是，首先将注意力放到这种可能性（我意识到这种可能性是在我的初稿送出去之前）上的人是布鲁姆博士，他认为其不过是“像曙人一样类型的先祖”。需要解释的不是汤恩婴儿作为一个婴儿能不能作为任何生物的祖先，而是他所代表的科与我们拥有的人类之前的祖先型的亲缘关系是否是最近的。

鉴于这些事实，企图抓住将“汤恩婴儿代表的非洲祖先”当作“人类家族树的基石”——无论可能是什么——这些话，说我正在犯着“与宣称一位萨塞克斯农民是征服者威廉的祖先同样的错误”这种风凉话，几乎是没有道理的。这只不过是阿瑟爵士自己搞错罢了。我只是将罗伯特·布鲁姆建议的家族树（《自然》，1925 年 4 月 18 日）翻译成日常用的英语，对于该家族树，我几乎完全赞成。阿瑟爵士攻击我的意见，却没有批评布鲁姆博士的意见，我将这当作是他个人偏爱的一个标志。

阿瑟爵士不必担心他的言论是否会贬低汤恩发现的重要性——批评通常会提高影响而不会贬低影响。大约 30 年前，赫胥黎拒绝接受爪哇猿人作为一个环节而存在。今天阿瑟·基思爵士又认为爪哇猿人是唯一一个已知的环节。虽然没有关于赫胥黎首先接受它、而后又摒弃它的记载，但是历史有时是会重演的。

雷蒙德·达特

* * *

达特教授认为我随随便便改变对汤恩发现的猿化石的观点，他的这种想法是一种误解。从他给出的描述和说明（《自然》，1925 年 2 月 7 日，第 195 页）来看，他是将如下结论强加于我，即南方古猿是“与黑猩猩和大猩猩一样的群体或亚科”的成员之一（《自然》，1925 年 2 月 14 日，第 234 页）。在同期的《自然》中，埃利奥特·史密斯教授表达了相似的观点，将南方古猿描述为“肯定是一只类人猿，该类人猿似乎处于与大猩猩和黑猩猩同样的发育阶段，但是与两者又都不相同。”

自从达特教授在《自然》上最初公布结果之后，这里的所有信息都趋向于支持汤恩猿与大猩猩和黑猩猩存在亲密的亲缘关系——认为它是那一群体的成员之一。阿姆斯特丹的博尔克教授和俄亥俄州克利夫兰的温盖特·托德教授都注意到了一个事

and Prof. Wingate Todd, of Cleveland, Ohio, have directed attention to the fact that the skulls of occasional gorillas show the same kind of narrowing and lengthening as has been observed in that of the Taungs ape. Prof. Arthur Robinson has shown that there is a wide variation in the size of jaws of young chimpanzees of approximately the same age, the smaller of the jaws approaching in size and shape to the development seen in the Taungs ape. The dimensions of the erupting first permanent molar of the Taungs ape and the form of its cusps point to the same conclusion—that Australopithecus must be classified with the chimpanzee and gorilla. It is, therefore, "preposterous" that Prof. Dart should propose to create "a new family of Homo-simiadae for the reception of the group of individuals which it (Australopithecus) represents". It is preposterous because the group to which this fossil ape belongs has been known and named since the time of Sir Richard Owen.

The position which Prof. Dart assigns to the Taungs ape in the genealogical tree of man and ape has no foundation in fact. A large diagram in the exhibition in Wembley, prepared by Prof. Dart, informs visitors that the Taungs ape represents the ancestor of all forms of mankind, ancient and modern. Before making such a claim one would have expected that due inquiry would first be made as to whether or not the geological evidence can justify such a claim. From his letter one infers that Prof. Dart does not set much store by geological evidence. Yet it has been customary, and I think necessary, to take the time element into account in constructing pedigrees of every kind. Dr. Robert Broom and, later, Prof. Dart's colleague, Prof. R. B. Young, have reviewed the evidence relating to the geological antiquity of the Taungs fossil skull, and on data supplied by them one can be certain that early and true forms of men were already in existence before the ape's skull described by Prof. Dart was entombed in a cave at Taungs. To make a claim for the Taungs ape as a human ancestor is therefore "preposterous".

Finally, Prof. Dart reminds me that whales and elephants have massive brains and that many large-headed men and women show no outstanding mental ability. Still the fact remains that every human being whose brain fails to reach 850 grams in weight has been found to be an idiot. Size as well as convolutionary pattern of brain have to be taken into account in fixing the position of every fossil type of being that has any claim to be in the line of human evolution—the Taungs brain cast at Wembley possesses no feature which lifts it above an anthropoid status.

Arthur Keith

(**116**, 462-463; 1925)

Raymond A. Dart: University of the Witwatersrand, Johannesburg.
Arthur Keith: Royal College of Surgeons of England, Lincoln's Inn Fields, London, W.C., September 5.

实,即个别的大猩猩头骨显示出与在汤恩猿中观察到的同样形式的变窄、拉长的现象。阿瑟·鲁宾逊教授也表明大约同样年龄的幼年黑猩猩的颌骨大小存在很广泛的变异范围，其中稍小的颌骨就与汤恩猿中见到的大小和形状的发育程度接近。汤恩猿正在萌出的第一颗恒臼齿的尺寸和其牙尖的形式指向同样的结论——南方古猿肯定与黑猩猩和大猩猩属于同类。因此达特教授建议创建“一个名为人猿科的新科来接纳标本（南方古猿）代表的生物类群”的想法是“荒谬可笑的”。因为这例化石猿所属的群体是已知的，早在理查德·欧文先生时期就已经被命名过了，所以说他是荒谬可笑的。

达特教授给汤恩猿在人类和猿类家族树上指定的位置事实上没有任何基础。陈列在温布利的一幅达特教授制定的大图表告诉参观者，汤恩猿代表了古代和现代所有人类类型的祖先。一个人在提出这样的主张之前,应该会想到先通过适当的调查,看看地质学证据是否支持这种主张。从达特教授的信中可以推断出他并没有对地质学证据投入太多精力。但是通常习惯性的做法是，在构建每一类谱系时将时间因素考虑在内，我觉得这是必需的。罗伯特·布鲁姆博士审视过与汤恩头骨化石的地质学年代相关的证据，后来达特教授的同事扬教授也对此作过调查，根据他们所提供的数据，可以肯定早期的真正的人类类型早在达特教授描述的猿类头骨埋入汤恩洞穴之前就已经存在了。因此声称汤恩猿是人类祖先是“荒谬可笑的”。

最后，达特教授提醒我，鲸和大象具有巨大的脑，并且许多大脑袋的男人和女人并没有过人的智力。但是事实仍然是，研究发现每个脑重量不足 850 克的人都是白痴。在确定每种与人类进化世系有所关联的生物化石类型的地位时，脑的大小和脑回式样都是必须考虑的因素——温布利的汤恩脑模型不具备将其提升到类人猿以上地位的任何特征。

阿瑟·基思

（刘皓芳 翻译；吴新智 审稿）

Tertiary Man in Asia: the Chou Kou Tien Discovery*

D. Black

Editor's Note

The fossil-bearing cave site of Chou Kou Tien (Zhoukoudian in China) had been discovered in 1921, and soon yielded bones of various fossil mammals from horses to bats. This note is a secondary account of discoveries made to date that included, notably, two teeth attributable to "*Homo ? sp.*", the first human remains known to science in mainland Asia. The reporter, Canadian-born Davidson Black, Anatomy Professor at Peking Union Medical College, would go on to make his name at Chou Kou Tien with discoveries of other remains that he would call Sinanthropus, now regarded as *Homo erectus*. Black died in his office of heart problems in 1934, the remains of "Peking Man" close by. He was 49.

A rich fossiliferous deposit at Chou Kou Tien, 70 li [about 40 kilometres] to the southwest of Peking, was first discovered in the summer of 1921 by Dr. J. G. Andersson and later surveyed and partially excavated by Dr. O. Zdansky. A preliminary report on the site was published by Dr. Andersson in March 1923 (*Mem. Geol. Surv. China*, Ser. A, No. 5, pp. 83–89), followed in October of that year by a brief description of his survey by Dr. Zdansky (*Bull. Geo. Surv. China*, No. 5, pp. 83–89). The material recovered from the Chou Kou Tien cave deposit has been prepared in Prof. Wiman's laboratory in Upsala and afterwards studied there by Dr. Zdansky. As a result of this research, Dr. Andersson has now announced that in addition to the mammalian groups already known from this site, there have also been identified representatives of the Cheiroptera, one cynopithecid, and finally two specimens of extraordinary interest, namely, one premolar and one molar tooth of a species which cannot otherwise be named than *Homo ? sp.*

Judging from the presence of a true horse and the absence of Hipparion, Dr. Andersson in his preliminary report considered that the Chou Kou Tien fauna was possibly of Upper Pliocene age, an opinion also expressed by Dr. Zdansky. It is possible, however, in the light of recent research, that the horizon represented by this site may be of Lower Pleistocene age. Whether it be of late Tertiary or of early Quaternary age, the outstanding fact remains that, for the first time on the Asiatic continent north of the Himalayas, archaic hominid fossil material has been recovered, accompanied by complete and certain geological data. The actual presence of early man in eastern Asia is therefore now no longer a matter of conjecture.

---

* Announcement of the Chou Kou Tien discovery was first made by Dr. J. G. Andersson on the occasion of a joint scientific meeting of the Geological Society of China, the Peking Natural History Society and the Peking Union Medical College held in Peking on October 22, 1926, in honour of H. R. H. the Crown Prince of Sweden.

亚洲的第三纪人：周口店的发现*

步达生

编者按

1921 年人们在中国发现了藏有化石的周口店洞穴遗址，很快地，从中发掘出了从马到蝙蝠等多种哺乳动物的化石标本。这篇文章间接地介绍了之前在周口店遗址取得的各种发现，其中包括两件著名的来自“人属（种名不确定）”的牙骨，以及对于科学界来说亚洲大陆的首例人类化石。文章作者是出生于加拿大的戴维森·步达生，当时他是北京协和医学院的解剖学教授，后来他因为在周口店发现了一些其他的化石（他称之为中国猿人，现在被称为直立人）而闻名于世。1934 年，49 岁的步达生因心脏病死于办公室中，当时“北京人”的化石就在他身旁。

1921 年夏天，安特生博士在北京西南 70 里 [大约 40 公里] 的周口店首次发现了一套富含化石的沉积层，日贾尔斯基博士随后进行了调查及部分发掘工作。1923 年 3 月，有关这个遗址的初步报告由安特生博士发表（《中国地质专报》，A 辑，第 5 期，第 83~89 页）。同年 10 月，日贾尔斯基博士发表了他的初步调查结果（《中国地质学会会志》，第 5 期，第 83~89 页）。从周口店洞穴沉积中发现的材料在乌普萨拉的威曼教授的实验室进行修整，之后日贾尔斯基博士在那里对这些材料进行了研究。安特生博士发布了这项研究的一个结果：在这个化石点，除了已知的哺乳动物类群，还发现了一些翼手目的代表性物种，一个猕猴的标本，以及两个非常有趣的标本，即一个前臼齿和一个臼齿，对应的物种好像只能称为“人属（种名不确定）”。

根据存在真正的马以及没有发现三趾马来判断，安特生博士在他初期的报道中认为周口店动物群属于上新世早期，日贾尔斯基博士也表达了同样的观点。最近的研究表明，这一化石点代表的层位更可能属于上新世晚期。无论它属于第三纪晚期还是第四纪早期，最引人注目的事实是，这是第一次在喜马拉雅山北部的亚洲大陆发现原始人类的化石材料，并且具有完整的相关地质学资料。从此，关于早期人类事实上存在于东亚的说法不再只是一种猜测。

* 1926 年 10 月 22 日，为欢迎瑞典王储，中国地质学会、北京自然历史学会和北京协和医学院联合举办了科学会议，会上安特生博士首次宣布了周口店的发现。

While a complete description of these very important specimens may shortly be expected in *Palaeontologia Sinica*, the following brief notes may be of interest here. One of the teeth recovered is a right upper molar, probably the third, the relatively unworn crown of which presents characters appearing from the photographs to be essentially human. The posterior moiety of the crown is narrow and the roots appear to be fused. The other tooth is probably a lower anterior premolar, of which the crown only is preserved. The latter also is practically unworn, and appears in the photograph to be essentially bicuspid in character, a condition usually to be correlated with a reduction of the upper canine.

The Chou Kou Tien molar tooth, though unworn, would seem to resemble in general features the specimen purchased by Haberer in a Peking native drug shop and afterwards described in 1903 by Schlosser. The latter tooth was a left upper third molar having a very much worn crown, extensively fused lateral roots, and from the nature of its fossilisation considered by Schlosser to be in all probability Tertiary in age. It was provisionally designated as *Homo? Anthropoide?* It is of more than passing interest to recall that Schlosser, in concluding his description of the tooth, pointed out that future investigators might expect to find in China a new fossil anthropoid, Tertiary man or ancient Pleistocene man. The Chou Kou Tien discovery thus constitutes a striking confirmation of that prediction.

It is now evident that at the close of Tertiary or the beginning of Quaternary time man or a very closely related anthropoid actually did exist in eastern Asia. This knowledge is of fundamental importance in the field of prehistoric anthropology; for about this time also there lived in Java, Pithecanthropus; at Piltdown, Eoanthropus; and, but very shortly after, at Mauer, the man of Heidelberg. All these forms were thus practically contemporaneous with one another and occupied regions equally far removed respectively to the east, to the south-east, and to the west from the central Asiatic plateau which, it has been shown elsewhere, most probably coincides with their common dispersal centre. The Chou Kou Tien discovery therefore furnishes one more link in the already strong chain of evidence supporting the hypothesis of the central Asiatic origin of the Hominidae.

(**118**, 733-734; 1926)

Davidson Black: Department of Anatomy, Peking Union Medical College, Peking, China.

关于这些非常重要的标本的完整描述在《中国古生物志》上发表之前，以下这些简要的描述可能是有价值的。发现的牙齿中有一个是右上臼齿，可能是第三个，它的照片显示，牙冠几乎没有磨损，所显示出的基本上是人类的特征。其牙冠后半部分狭窄，根部融合。另一个牙齿可能是一个前部的下前臼齿，只有牙冠保存了下来。这一个也几乎没有磨损，它的照片显示出基本的二尖齿特征，这种情况通常与上犬齿退化有关。

周口店臼齿，尽管没有磨损，其总体特征却和哈贝雷尔在北京当地的一个药店买到的标本（1903 年，施洛瑟描述了这一标本的特征）似乎相似。这一来自药店的牙齿是一颗左上第三臼齿，其牙冠磨蚀严重，外侧两齿根大部分合并。从它石化的性质考虑，施洛瑟认为它很可能是第三纪的。它被临时归属于人属？或者猿属？回想起来，非常有趣的是，当施洛瑟总结他对那颗牙齿所作的描述时曾指出，将来的调查者将有希望在中国发现一种新的类似人的化石第三纪人或古老更新世人。周口店的发现是对那个预测的一个惹人注目的确认。

现在很明显了，在第三纪结束或第四纪开始之际，在东亚的确存在人类或者说一种与人类关系密切的类人猿。这在史前人类学领域是一个重要的基本发现。在大约同一个时期，爪哇存在爪哇直立猿人，皮尔当存在曙人，之后不久在毛尔地区出现海德堡人。所有这些生物类型都是属于同一时期的，并且分别在东部、东南部和西部各自占据了距中亚平原相同距离的区域，而其他的证据表明，中亚平原极有可能是它们共同的扩散中心。周口店的发现，在已有的有关人类中亚起源假说的有力证据链中又增添了一环。

（刘晓辉 翻译；徐星 审稿）

Thermal Agitation of Electricity in Conductors

J. B. Johnson

Editor's Note

In the early days of electronics and telephone communications, engineers aimed to reduce the noise in their circuits to a minimum. But there are fundamental limits, as physicist and engineer John Johnson of Bell Telephone Laboratories here reports. Johnson shows that ordinary electrical conductors exhibit spontaneous, thermally induced fluctuations of voltage, even when not being driven by external currents. This voltage noise depends directly on the temperature of the sample, leading Johnson to surmise that it must arise from continual agitation and excitation by random thermal energy. Thus, he concludes, the minimum voltage that can be usefully amplified in electronic circuits is limited by the very matter from which they are built.

ORDINARY electric conductors are sources of spontaneous fluctuations of voltage which can be measured with sufficiently sensitive instruments. This property of conductors appears to be the result of thermal agitation of the electric charges in the material of the conductor.

The effect has been observed and measured for various conductors, in the form of resistance units, by means of a vacuum tube amplifier terminated in a thermocouple. It manifests itself as a part of the phenomenon which is commonly called "tube noise". The part of the effect originating in the resistance gives rise to a mean square voltage fluctuation V^2 which is proportional to the value R of that resistance. The ratio V^2/R is independent of the nature or shape of the conductor, being the same for resistances of metal wire, graphite, thin metallic films, films of drawing ink, and strong or weak electrolytes. It does, however, depend on temperature and is proportional to the absolute temperature of the resistance. This dependence on temperature demonstrates that the component of the noise which is proportional to R comes from the conductor and not from the vacuum tube.

A similar phenomenon appears to have been observed and correctly interpreted in connexion with *a current sensitive* instrument, the string galvanometer (W. Einthoven, W. F. Einthoven, W. van der Horst, and H. Hirschfeld, *Physica*, 5, 358–360, No. 11/12, 1925). What is being measured in these cases is the effect upon the measuring device of continual shock excitation resulting from the random interchange of thermal energy and energy of electric potential or current in the conductor. Since the effect is the same for different conductors, it is evidently not dependent on the specific mechanism of conduction.

The amount and character of the observed noise depend upon the frequency-characteristic of the amplifier, as would be expected from experience with the small-shot

导体中电的热扰动

约翰逊

编者按

在电子通讯和电话通信的早期，工程师们的目标是将电路中的噪声降至最小。但贝尔电话实验室的物理学家、工程师约翰·约翰逊认为，噪声的减少是有基本极限的。约翰逊指出：即使在没有加载外部电流的情况下，导电体也会表现出由自发热扰动导致的电压波动。这种电压噪声与样品的温度直接相关，这使约翰逊联想到它应该是由随机热能的连续激励和扰动造成的。因此他得出以下结论：在电路中能够被有效放大的最小电压受到组成电路的材料的限制。

普通的导电体是电压自发涨落的来源，而这种波动电压可以用足够灵敏的仪器进行测量。导体的这种性质是导体材料中电荷热扰动的结果。

在各种导体中都已经观测到这类效应，并以电阻的形式，通过终端连接温差电偶的真空管放大器对此进行了测量。它通常作为所谓的“电子管噪声”现象的一部分而表现出来。源于电阻波动的这部分效应造成了电压均方值 V^2 的波动，此波动与电阻值 R 成正比。比率 V^2/R 与导体的性质和形状无关，对金属丝、石墨、金属薄膜、绘图墨水薄膜及强电解质或弱电解质来说，该比率都是相同的。然而，该比率却与温度有关，并与电阻的绝对温度成正比。这种温度依赖性表明，与 R 成比例的噪声分量来自导体，而不是来自真空管。

目前已经有人用**对电流灵敏**的仪器——弦线电流计观测到类似现象并正确地进行了解释（老艾因特霍芬、小艾因特霍芬、范德霍斯特以及赫希菲尔德，《物理学》，第5卷，第358~360页，第11/12期，1925年）。在这些情况下测量到的是，导体中的热能与电势或电流能量的随机交换引起的持续冲击激发对测量装置的影响。因为该效应对不同导体是相同的，所以显然它与传导的特定机制无关。

正如从散粒效应的经验可以预期的那样，观测到的噪声的量及其特性与放大器的频率特性有关。在室温下，来自电阻的表观输入功率在 10^{-18} 瓦的量级。至少在声

effect. The apparent input power originating in the resistance is of the order 10^{-18} watt at room temperature. The corresponding output power is proportional to the area under the graph of *power amplification–frequency*, at least in the range of audio frequencies. The magnitude of the "initial noise", when the quietest tubes are used without input resistance, is about the same as that produced by a resistance of 5,000 ohms at room temperature in the input circuit. For the technique of amplification, therefore, the effect means that the limit to the smallness of voltage which can be usefully amplified is often set, not by the vacuum tube, but by the very matter of which electrical circuits are built.

(**119**, 50-51; 1927)

J. B. Johnson: Bell Telephone Laboratories, Inc., New York, N. Y., Nov. 17.

频范围内，相应的输出功率与**功率放大–频率**关系图中曲线下的面积成正比。当采用无输入电阻的最平稳的真空管时，“初始噪声”的量级约与在室温下输入回路中5,000欧姆的电阻产生的噪声量级相同。因此，对于放大技术而言，这种效应意味着可以被有效放大的电压的最小值与构成电路的材料有关，而不由真空管决定。

（沈乃澂 翻译；赵见高 审稿）

The Scattering of Electrons by a Single Crystal of Nickel

C. Davisson and L. H. Germer

Editor's Note

In his PhD thesis of 1924, Louis de Broglie hypothesized that electrons and other massive particles might behave as waves, just as photons reveal the particle-like aspects of otherwise wave-like electromagnetic radiation. This implied that material particles should exhibit wave-like phenomena such as interference or diffraction. Here Clinton Davisson and Lester Germer of the Bell Telephone Laboratories verified this prediction by measuring the diffraction of an electron beam from a nickel crystal. Using the known spacing of atomic planes in nickel, they were able to calculate the expected angles where diffracted beams should occur if the electrons were indeed acting as waves in the manner de Broglie suggested. The experimental results agreed with these predictions, and they won Davission the 1937 Nobel Prize in physics.

IN a series of experiments now in progress, we are directing a narrow beam of electrons normally against a target cut from a single crystal of nickel, and are measuring the intensity of scattering (number of electrons per unit solid angle with speeds near that of the bombarding electrons) in various directions in front of the target. The experimental arrangement is such that the intensity of scattering can be measured in any latitude from the equator (plane of the target) to within 20° of the pole (incident beam) and in any azimuth.

The face of the target is cut parallel to a set of {111}-planes of the crystal lattice, and etching by vaporisation has been employed to develop its surface into {111}-facets. The bombardment covers an area of about 2 mm^2 and is normal to these facets.

As viewed along the incident beam the arrangement of atoms in the crystal exhibits a threefold symmetry. Three {100}-normals equally spaced in azimuth emerge from the crystal in latitude 35°, and, midway in azimuth between these, three {111}-normals emerge in latitude 20°. It will be convenient to refer to the azimuth of any one of the {100}-normals as a {100}-azimuth, and to that of any one of the {111}-normals as a {111}-azimuth. A third set of azimuths must also be specified; this bisects the dihedral angle between adjacent {100}- and {111}-azimuths and includes a {110}-normal lying in the plane of the target. There are six such azimuths, and any one of these will be referred to as a {110}-azimuth. It follows from considerations of symmetry that if the intensity of scattering exhibits a dependence upon azimuth as we pass from a {100}-azimuth to the next adjacent {111}-azimuth (60°), the same dependence must be exhibited in the reverse

镍单晶对电子的散射

戴维森，革末

编者按

1924 年，路易斯·德布罗意在他的博士论文中猜测，电子以及其他一些有质量的粒子的运动方式可能与波类似，就像光子除了具有电磁辐射的波动行为以外，还显示出与此不同的粒子行为。这一假说意味着，物质粒子会表现出干涉或者衍射这样的波动特征。在这篇文章中，贝尔电话实验室的克林顿·戴维森和莱斯特·革末通过对从镍晶中发出的电子束的衍射效应进行测量证实了这一假说。如果正如德布罗意猜想的那样电子具有像波一样的性质，那么根据镍晶中已知的原子平面之间的间距，他们就能计算出应该出现衍射束的预期角度。实验结果与这些预测是相吻合的。戴维森因此获得了 1937 年的诺贝尔物理学奖。

现在正在进行的一系列实验中，我们用一窄束电子垂直轰击一块从镍单晶上切下来的靶子，并测量了靶前方不同方向上的电子散射强度（与每单位立体角中速度和轰击电子的速度相似的电子数量）。我们的实验装置可以测量从赤道（靶平面）到与轴（入射束）成 20° 范围内在任意纬度处和在任意方位角上的散射强度。

被切割下来的靶面平行于镍单晶点阵的一组 {111} 面，切下来之后用蒸气进行蚀刻，以使其表面形成一系列 {111} 小晶面。轰击的电子束打在约 2 平方毫米的范围内，并与以上的小晶面垂直。

沿着入射电子束的方向观察，我们发现原子在晶体中的排列呈现三重对称。从晶体出发，在纬度为 35° 处有三条相隔同样方位角的 {100} 法线，并且在这些方位角中间，有三条 {111} 法线出现在纬度 20° 处。为了方便起见，任何一条 {100} 法线的方位角都可以被看作是一个 {100} 方位角，任何一条 {111} 法线的方位角也可以被看作是一个 {111} 方位角。我们还必须指定第三组方位角；它将平分相邻的 {100} 方位角和 {111} 方位角之间的二面角，并且包含一条位于靶平面上的 {110} 法线。有 6 个这样的方位角，它们当中的任意一个都可以被看作是一个 {110} 方位角。考虑到对称性的要求，当我们从 {100} 方位角转到下一个相邻的 {111} 方位角 (60°) 时，如果散射强度的变化依赖于方位角的话，那么当我们从 60° 转到下一个相邻的

order as we continue on through 60° to the next following {100}-azimuth. Dependence on azimuth must be an even function of period $2\pi/3$.

In general, if bombarding potential and azimuth are fixed and exploration is made in latitude, nothing very striking is observed. The intensity of scattering increases continuously and regularly from zero in the plane of the target to highest value in co-latitude 20°, the limit of observations. If bombarding potential and co-latitude are fixed and exploration is made in azimuth, a variation in the intensity of scattering of the type to be expected is always observed, but in general this variation is slight, amounting in some cases to not more than a few percent of the average intensity. This is the nature of the scattering for bombarding potentials in the range from 15 volts to near 40 volts.

At 40 volts a slight hump appears near 60° in the co-latitude curve for azimuth-{111}. This hump develops rapidly with increasing voltage into a strong spur, at the same time moving slowly upward toward the incident beam. It attains a maximum intensity in co-latitude 50° for a bombarding potential of 54 volts, then decreases in intensity, and disappears in co-latitude 45° at about 66 volts. The growth and decay of this spur are traced in Fig. 1.

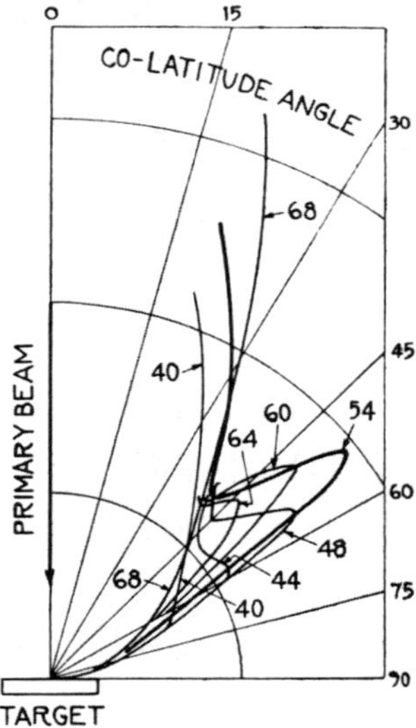

Fig. 1. Intensity of electron scattering vs. co-latitude angle for various bombarding voltages—azimuth-{111}-330°

A section in azimuth through this spur at its maximum (Fig. 2—Azimuth-330°) shows that it is sharp in azimuth as well as in latitude, and that it forms one of a set of three such spurs, as was to be expected. The width of these spurs both in latitude and in azimuth is almost completely accounted for by the low resolving power of the measuring device. *The spurs are due to beams of scattered electrons which are nearly if not quite as well defined as the primary*

{100} 方位角时，相同的依赖关系会按相反的次序再现。散射强度对方位角的依赖关系肯定是一个周期为 $2\pi/3$ 的偶函数。

一般来说，如果轰击电压和方位角固定，当沿着纬度测量的时候，不会观察到任何令人惊奇的现象。散射强度连续而有规律地从靶平面处的零增加到余纬 20° 时的最大值，余纬 20° 是我们的观测极限。如果轰击电压和余纬固定，而沿着方位角进行探测，则通常可以观察到预期之中的散射强度变化，但总的来说这种差别很小，在某些情况下总的变化量不会大于平均强度的百分之几。这就是散射在轰击电压从 15 伏特到接近 40 伏特之间变动时的特性。

当电压为 40 伏特时，在方位角为 {111} 的余纬曲线上接近 60° 处出现了一个小峰。随着电压的增加，这个峰迅速变大形成一个很强的尖峰，同时朝着入射电子束的方向缓慢向上提升。当轰击电压为 54 伏特，余纬为 50° 时，峰值达到最大，然后强度逐渐减小，在电压约为 66 伏特，余纬为 45° 时消失。图 1 描述了这个尖峰从增大到衰减的过程。

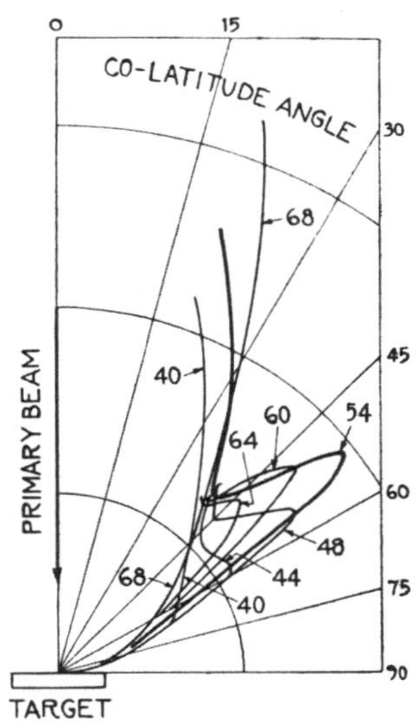

图 1. 不同轰击电压下电子散射强度随余纬角的变化——方位角{111}-330°

在最大值处穿过这个尖峰的方位角截面图（图 2 之方位角 330°）表明，它以方位角为变量的时候和以纬度为变量的时候一样尖锐，并且，如我们所料，它形成了 3 个尖峰组合中的一个。这些尖峰的宽度在随方位角变化和随纬度变化时都较大，几乎可以肯定地说这是由测量仪器的低分辨率造成的。**尖峰是由散射的电子束引起**

beam. The minor peaks occurring in the {100}-azimuth are sections of a similar set of spurs that attains its maximum development in co-latitude 44° for a bombarding potential of 65 volts.

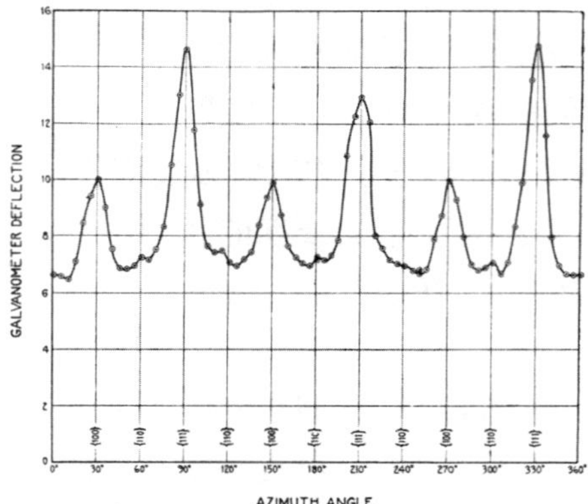

Fig. 2. Intensity of electron scattering vs. azimuth angle—54 volts, co-latitude 50°

Thirteen sets of beams similar to the one just described have been discovered in an exploration in the principal azimuths covering a voltage range from 15 volts to 200 volts. The data for these are set down on the left in Table I (columns 1–4). Small corrections have been applied to the observed co-latitude angles to allow for the variation with angle of the "background scattering", and for a small angular displacement of the normal to the facets from the incident beam.

Table I

	Electron Beams			X-ray Beams						
Azimuth	Bomb. Pot (volts)	Co-lat. θ	Intensity	Reflections	$\lambda\times10^8$ cm	Co-lat. θ	Co-lat. θ'	$v\times10^{-8}$ cm/sec	$n\lambda\times10^8$ cm	$n\left\{\frac{\lambda mv}{h}\right\}$
{111}	54	50°	0.5	{220}	2.03	70.5	52.7	4.36	1.65	0.99
	100	31	0.5	{331}	1.49	44.0	31.6	5.94	1.11	0.91
	174	21	0.9	{442}	1.13	31.6	22.4	7.84	0.77	0.83
	174	55	0.15	{440}	1.01	70.5	52.7	7.84	1.76	2(0.95)
{100}	65	44	0.5	{311}	1.84	59.0	43.2	4.79	1.49	0.98
	126	29	1.0	{422}	1.35	38.9	27.8	6.67	1.04	0.95
	190	20	1.0	{533}	1.04	28.8	20.4	8.19	0.74	0.83
	159	61	0.4	{511}	1.05	77.9	59.0	7.49	1.88	2(0.97)
{110}	138	59	0.07	{420}	1.22	78.5	59.5	6.98	1.06	1.02
	170	46	0.07	{531}	1.04	57.1	41.7	7.75	0.89	0.95
{111}	110	58	0.15	..	..	..	..	6.23	1.82	1.56
{100}	110	58	0.15	..	..	..	..	6.23	1.82	1.56
{110}	110	58	0.25	..	..	..	..	6.23	1.05	0.90

的，对这些散射电子束的界定即使不能像对原射线束的界定那样精确，至少也能相差不多。在 {100} 方位角上出现的次级峰是另一组类似的尖峰，它们在余纬 44°，轰击电压 65 伏特时达到最大值。

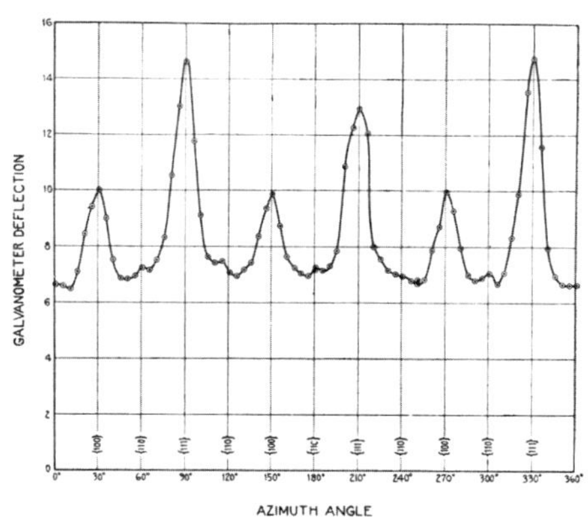

图 2. 电子散射强度随方位角的变化——54伏特，余纬50°

电压在从 15~200 伏特的范围内变动时，我们在主方位角方向发现了 13 组与上面所描述的电子束类似的射束。结果见表 I 左侧（第 1~4 列）。考虑到“背景散射”随角度的变化，以及晶面法线与入射电子束之间的角位移，我们对观察到的余纬角进行了微小的修正。

表 I

方位角	电子束			X 射线束						
	轰击电压（伏特）	余纬度 θ	强度	反射	$\lambda\times10^{8}$厘米	余纬度θ	余纬度θ'	$v\times10^{-8}$ 厘米/秒	$n\lambda\times10^{8}$ 厘米	$n\left\{\frac{\lambda mv}{h}\right\}$
{111}	54	50°	0.5	{220}	2.03	70.5	52.7	4.36	1.65	0.99
	100	31	0.5	{331}	1.49	44.0	31.6	5.94	1.11	0.91
	174	21	0.9	{442}	1.13	31.6	22.4	7.84	0.77	0.83
	174	55	0.15	{440}	1.01	70.5	52.7	7.84	1.76	2(0.95)
{100}	65	44	0.5	{311}	1.84	59.0	43.2	4.79	1.49	0.98
	126	29	1.0	{422}	1.35	38.9	27.8	6.67	1.04	0.95
	90	20	1.0	{533}	1.04	28.8	20.4	8.19	0.74	0.83
	159	61	0.4	{511}	1.05	77.9	59.0	7.49	1.88	2(0.97)
{110}	138	59	0.07	{420}	1.22	78.5	59.5	6.98	1.06	1.02
	170	46	0.07	{531}	1.04	57.1	41.7	7.75	0.89	0.95
{111}	110	58	0.15	..	..	..	..	6.23	1.82	1.56
{100}	110	58	0.15	..	..	..	..	6.23	1.82	1.56
{110}	110	58	0.25	..	..	..	..	6.23	1.05	0.90

If the incident electron beam were replaced by a beam of monochromatic X-rays of adjustable wave-length, very similar phenomena would, of course, be observed. At particular values of wave-length, sets of three or of six diffraction beams would emerge from the incident side of the target. On the right in Table I (columns 5, 6 and 7) are set down data for the ten sets of X-ray beams of longest wave-length which would occur within the angular range of our observations. Each of these first ten occurs in one of our three principal azimuths.

Several points of correlation will be noted between the two sets of data. Two points of difference will also be noted; the co-latitude angles of the electron beams are not those of the X-ray beams, and the three electron beams listed at the end of the Table appear to have no X-ray analogues.

The first of these differences is systematic and may be summarised quantitatively in a simple manner. If the crystal were contracted in the direction of the incident beam by a factor 0.7, the X-ray beams would be shifted to the smaller co-latitude angles θ' (column 8), and would then agree in position fairly well with the observed electron beams—the average difference being 1.7°. Associated in this way there is a set of electron beams for each of the first ten sets of X-ray beams occurring in the range of observations, the electron beams for 110 volts alone being unaccounted for.

These results are highly suggestive, of course, of the ideas underlying the theory of wave mechanics, and we naturally inquire if the wave-length of the X-ray beam which we thus associate with a beam of electrons is in fact the h/mv of L. de Broglie. The comparison may be made, as it happens, without assuming a particular correspondence between X-ray and electron beams, and without use of the contraction factor. Quite independently of this factor, the wave-lengths of all possible X-ray beams satisfy the optical grating formula $n\lambda = d\sin\theta$, where d is the distance between lines or rows of atoms in the surface of the crystal—these lines being normal to the azimuth plane of the beam considered. For azimuths-{111} and-{100}, $d = 2.15\times10^{-8}$ cm and for azimuth-{110}, $d = 1.24\times10^{-8}$ cm. We apply this formula to the electron beams without regard to the conditions which determine their distribution in co-latitude angle. The correlation obtained by this procedure between wave-length and electron speed v is set down in the last three columns of Table I.

In considering the computed values of $n(\lambda mv/h)$, listed in the last column, we should perhaps disregard those for the 110-volt beams at the bottom of the Table, as we have had reason already to regard these beams as in some way anomalous. The values for the other beams do, indeed, show a strong bias toward small integers, quite in agreement with the type of phenomenon suggested by the theory of wave mechanics. These integers, one and two, occur just as predicted upon the basis of the correlation between electron beams and X-ray beams obtained by use of the contraction factor. The systematic character of the departures from integers may be significant. We believe, however, that this results from

如果入射电子束被一束波长可调的单色 X 射线取代，我们当然可以观察到非常类似的现象。当波长为一些特定值时，在靶的入射面一侧就会出现 3 条一组或 6 条一组的衍射光束。表 I 右侧（第 5、6、7 列）列出了在我们能观测到的角度范围内由波长最长的 10 组 X 射线束得到的数据。前 10 组中每组数据都来自我们先前设定的 3 个主方位角中的一个。

在以上表中关于电子束和 X 射线束的两部分数据中有几点关系值得我们注意。同时还要注意两个区别：电子束的余纬角不是 X 射线束的余纬角；对于表格底部列出的 3 行电子束数据，没有 X 射线束数据与之对应。

第一个区别是系统上的，可以用一种简单的方式定量地总结出来。如果晶体沿着入射束方向被压缩至 70%，即压缩因子为 0.7，那么 X 射线束将移至更小的余纬角 θ'（第 8 列），这样它的位置将与观测到的电子束的情况精确地吻合——平均差异为 1.7°。同理，对于我们能观察到的前 10 组 X 射线束中的每一个，都存在一组电子束与之对应，只有 110 伏特的电子束不能应用这种方法。

当然，这些结果完全证实了波动力学理论的基本观点，我们自然很想知道，与电子束相关联的 X 射线束的波长是否就是德布罗意所说的 h/mv。也许不用假设 X 射线束和电子束之间的特定关系，也不用使用压缩因子，我们就可以作出比较。所有的 X 射线束都满足光栅公式 $n\lambda = d\sin\theta$，这与压缩因子完全不相干，在这个公式中，d 是位于晶体表面的原子行或列的间距——这些行列都垂直于我们粒子束所在的方位角平面。对于方位角 {111} 和方位角 {100}，$d = 2.15 \times 10^{-8}$ 厘米；对于方位角 {110}，$d = 1.24 \times 10^{-8}$ 厘米。我们可以把这个公式应用于电子束中而无需考虑支配电子束沿余纬度分布情况的因素。用这种方式得到的波长和电子速度 v 之间的关系列于表 I 的最后 3 列中。

至于在最后一列中列出的计算值 $n(\lambda mv/h)$，我们可能应该不考虑表格底部 110 伏特的电子束数据，因为我们已经有理由把这些电子束看作是异常情况。对于其他电子束，$n(\lambda mv/h)$ 值趋向于一个比较小的整数，与波动力学理论预言的现象十分吻合。使用压缩因子对电子束和 X 射线束之间的差别进行校正之后，也得到了预想的结果，即整数 1 和 2。对整数的偏离可能主要是由系统特性造成的。我们认为误差来自入

imperfect alignment of the incident beam, or from other structural deficiencies in the apparatus. The greatest departures are for beams lying near the limit of our co-latitude range. The data for these are the least trustworthy.

(**119**, 558-560; 1927)

C. Davission, L. H. Germer: Bell Telephone Laboratories, Inc., New York, N.Y., Mar. 3.

射束不严格准直或者仪器的其他结构缺陷。最大的偏离出现在我们能观察到的余纬度极限附近，那里的数据是最不可靠的。

（王静 翻译；赵见高 审稿）

The Continuous Spectrum of β-Rays

C. D. Ellis and W. A. Wooster

Editor's Note

Since the discovery of radioactivity in 1895, three different mechanisms had been identified. In some cases an atom would disintegrate by shedding an α-particle (the nucleus of a helium atom), in others by emitting a γ-ray (a high-frequency X-ray). In both these mechanisms, the amount of energy lost in the disintegration was found to be the same for each kind of disintegrating atom. In the third mechanism of radioactive decay, however, in which the particle shed during the disintegration is an electron, the energy carried away by the particle is indeterminate, ranging from zero to a certain maximum characteristic of the atom concerned. This raises problems referred to in the letter below.

THE continuous spectrum of the β-rays arising from radio-active bodies is a matter of great importance in the study of their disintegration. Two opposite views have been held about the origin of this continuous spectrum. It has been suggested that, as in the α-ray case, the nucleus, at each disintegration, emits an electron having a fixed characteristic energy, and that this process is identical for different atoms of the same body. The continuous spectrum given by these disintegration electrons is then explained as being due to secondary effects, into the nature of which we need not enter here. The alternative theory supposes that the process of emission of the electron is not the same for different atoms, and that the continuous spectrum is a fundamental characteristic of the type of atom disintegrating. Discussion of these views has hitherto been concerned with the problem of whether or not certain specified secondary effects could produce the observed heterogenity, and although no satisfactory explanation has yet been given by the assumption of secondary effects, it was most important to clear up the problem by a direct method.

There is a ready means of distinguishing between the two views, since in one case a given quantity of energy would be emitted at each disintegration equal to or greater than the maximum energy observed in the electrons escaping from the atom, whereas in the second case the average energy per disintegration would be expected to equal the average energy of the particles emitted. If we were to measure the total energy given out by a known amount of material, as, for example, by enclosing it in a thick-walled calorimeter, then in the first case the heating effect should lead to an average energy per disintegration equal to or greater than the fastest electron emitted, no matter in what way this energy was afterwards split up by secondary effects. Since on the second hypothesis no secondary effects are presumed to be present, the heating effect should correspond simply to the average kinetic energy of the particles forming the continuous spectrum.

To avoid complications due to α-rays or to γ-rays from parent or successive atoms, we

β 射线的连续谱

埃里斯，伍斯特

编者按

自从 1895 年发现放射性以来，人们已经确认放射性存在三种不同的机制。在某些情况下，原子会通过放射出一个 α 粒子（氦原子的原子核）而衰变，而在另一些情况下，原子会通过放射 γ 射线（一种高频 X 射线）而衰变。在这两种机制中，每一种衰变原子在衰变过程中的能量损失都是相同的。然而在放射性衰变的第三种机制中，原子衰变时放射出的粒子是电子，这个电子携带的能量是不确定的，其范围是从零到相关原子的最大特征能量。这就引发了下文中提到的问题。

在物质衰变的研究中，放射性物质产生的 β 射线连续谱是一个非常重要的问题。关于这种连续谱的起源有两种相反的观点。一种观点认为，和放射出 α 射线的衰变一样，在每一次衰变中原子核都发射出一个具有固定的特征能量的电子，并且对同一物体的不同原子这个过程都是相同的。于是，这些衰变电子的连续谱就被归因为二次效应，而有关二次效应的本质我们不必在这里进行讨论。另外一种观点认为，发射电子的过程对不同原子是不同的，并且连续谱是衰变原子类型的基本特征。关于这两种观点的讨论至今停留在是否有某种具体的二次效应能够产生可观测的连续谱这一问题上，尽管二次效应的假设还没能给出一个令人满意的解释，但最重要的是，我们应该用一种直接的方法来解释以上的问题。

有一个简易的方法可以区分以上两种观点，因为从第一个观点来看，每次衰变中放出的特定能量应该等于或者大于观察到的从原子中逃逸出的电子的最大能量，而从第二种观点来看，每次衰变的平均能量应该等于发射的电子的平均能量。如果我们测量一定量的物质发射出的电子的总能量，例如将这些物质放在一个厚壁量热计中，那么按照第一种观点，由热效应就能得出每次衰变的平均能量，这个值应该是等于或大于发射出的最快的电子的能量，不管这个能量随后以什么方式被二次效应分解。而在第二种观点中，由于不存在二次效应，所以热效应应该简单地对应于形成连续谱的电子的平均动能。

为了避免衰变中母原子或后续原子产生的 α 射线或者 γ 射线的干扰，我们在

measured the heating effect in a thick-walled calorimeter of a known quantity of radium E. This measurement proved difficult because of the small rate of evolution of heat, but by taking special precautions it has been possible to show that the average energy emitted at each disintegration of radium E is 340,000 ± 30,000 volts. This result is a striking confirmation of the hypothesis that the continuous spectrum is emitted as such from the nucleus, since the average energy of the particles as determined by ionisation measurements over the whole spectrum gives a value about 390,000 volts, whereas if the energy emitted per disintegration were equal to that of the fastest β-rays, the corresponding value of the heating would be three times as large—in fact, 1,050,000 volts.

Many interesting points are raised by the question of how a nucleus, otherwise quantised, can emit electrons with velocities varying over a wide range, but consideration of these will be deferred until the publication of the full results.

(**119**, 563-564; 1927)

C. D. Ellis, W. A. Wooster: Cavendish Laboratory, Cambridge, Mar.23.

厚壁量热计中测量了一定量的镭 E 的热效应。由于放热速度很慢，这个测量很难进行，但是通过采用特别的措施，我们得到镭 E 每次衰变辐射出的平均能量是 340,000 ± 30,000 电子伏。这个结果对于连续谱来源于原子核中的发射这一假设是个有力的确证。因为通过电离测量整个能谱而得到的电子平均能量大约是 390,000 电子伏，如果每次衰变放出的能量等于最快的 β 射线的能量，那么相应的热效应的值就应该是当前测量值的 3 倍，即 1,050,000 电子伏。

一个量子化的原子核怎么能发射速度变化范围如此广泛的电子？这个问题引发了很多有趣的观点，不过还是等到详细的结果发表之后再来考虑这些。

（王锋 翻译；江丕栋 审稿）

A New Type of Secondary Radiation

C. V. Raman and K. S. Krishnan

Editor's Note

Physics in the early twentieth century was dominated by scientists in Europe and the United States. Yet a landmark discovery in quantum physics is reported here by two Indian physicists in Calcutta. As hitherto understood, light scattering from a stationary material object should preserve its frequency. But Chandrasekhara Venkata Raman and Kariamanickam Srinivasa Krishnan demonstrate that a small part of the scattered light can significantly change frequency. This "Raman effect" involves an exchange of energy between the scattered photons and the internal degrees of freedom of atoms or molecules. The effect is used today to probe molecular structure and motion, and the chemical nature of materials. For his discovery, Raman was awarded the 1930 Nobel Prize in physics.

IF we assume that the X-ray scattering of the "unmodified" type observed by Prof. Compton corresponds to the normal or average state of the atoms and molecules, while the "modified" scattering of altered wave-length corresponds to their fluctuations from that state, it would follow that we should expect also in the case of ordinary light two types of scattering, one determined by the normal optical properties of the atoms or molecules, and another representing the effect of their fluctuations from their normal state. It accordingly becomes necessary to test whether this is actually the case. The experiments we have made have confirmed this anticipation, and shown that in every case in which light is scattered by the molecules in dust-free liquids or gases, the diffuse radiation of the ordinary kind, having the same wave-length as the incident beam, is accompanied by a modified scattered radiation of degraded frequency.

The new type of light scattering discovered by us naturally requires very powerful illumination for its observation. In our experiments, a beam of sunlight was converged successively by a telescope objective of 18 cm aperture and 230 cm focal length, and by a second lens of 5 cm focal length. At the focus of the second lens was placed the scattering material, which is either a liquid (carefully purified by repeated distillation *in vacuo*) or its dust-free vapour. To detect the presence of a modified scattered radiation, the method of complementary light-filters was used. A blue-violet filter, when coupled with a yellow-green filter and placed in the incident light, completely extinguished the track of the light through the liquid or vapour. The reappearance of the track when the yellow filter is transferred to a place between it and the observer's eye is proof of the existence of a modified scattered radiation. Spectroscopic confirmation is also available.

Some sixty different common liquids have been examined in this way, and every one of them showed the effect in greater or less degree. That the effect is a true scattering and not a fluorescence is indicated in the first place by its feebleness in comparison with the

一种新型的二次辐射

拉曼，克里希南

编者按

20 世纪早期的物理学主要是由欧美科学家主导的，然而，一项在量子物理方面具有里程碑意义的发现却是由两名印度物理学家在加尔各答作出的。就当时人们所知，从一个静止实物散射出来的光线应该保持频率不变，这一点大家都可以理解，但钱德拉塞卡拉·文卡塔·拉曼和卡瑞马尼卡姆·斯里尼瓦桑·克里希南却用实验证实，有一小部分散射光频率变化很大。“拉曼效应”包含散射光子与原子或分子的内自由度之间能量的交换过程。现在利用这个效应可以检测分子结构和分子运动以及材料的化学性质。拉曼因发现了这个效应而获得了 1930 年的诺贝尔物理学奖。

如果我们假定，康普顿教授观察到的“不变”的 X 射线散射对应于原子和分子的正常态或平均态，而波长发生改变的“变”散射对应于原子和分子相对于正常态或平均态的涨落，那么我们就可以预测，普通光的散射应该也存在两种类型，一种取决于原子或分子的正常光学性质，另一种则代表了它们相对于正常态的涨落效应。因此有必要检验真实的情况是否确实如此。我们的实验证实了上述预测。实验表明，由任何一种无尘的液体或气体分子造成的光散射，都不仅包含了与入射光波长相同的正常漫射辐射，同时也伴随频率发生变化的变散射辐射。

要观察到我们发现的这种新型光散射，自然就需要非常强的光照。在我们的实验中，一束太阳光依次通过口径为 18 厘米、焦距为 230 厘米的望远镜物镜和一个焦距为 5 厘米的透镜而被会聚。在第二个透镜的焦点处放置散射材料，这些材料或者是在真空中反复蒸馏得到的非常纯净的液体，或者是无尘的蒸气。为了探测变散射辐射的存在，我们使用了互补的滤光片。当把一个蓝紫色滤光片和一个黄绿色滤光片一起放置在入射光处时，透过液体或蒸气的光路会完全消失。而当把入射光处的黄色滤光片移置到散射材料和观测者的眼睛之间时，透过散射材料的光路就会重新出现。这就证实了变散射辐射的存在。光谱分析的结果也证实了这一点。

我们用这种方法检测了六十多种不同的常见液体，所有结果中都或多或少地出现了这种效应。这是一种真正的散射效应而不是一种荧光现象，因为与普通的散射

ordinary scattering, and secondly by its polarisation, which is in many cases quite strong and comparable with the polarisation of the ordinary scattering. The investigation is naturally much more difficult in the case of gases and vapours, owing to the excessive feebleness of the effect. Nevertheless, when the vapour is of sufficient density, for example with ether or amylene, the modified scattering is readily demonstrable.

(**121**, 501-502; 1928)

C. V. Raman, K. S. Krishnan: 210 Bowbazar Street, Calcutta, India, Feb. 16.

相比它的强度非常微弱，而且在很多情况下它具有与普通的散射相当的非常强的偏振性。这种效应的强度非常微弱，因此要在气体和蒸气中开展这项研究自然是非常困难的。不过，当蒸气浓度足够大时，例如乙醚或戊烯的蒸气，还是很容易观察到变散射的。

（王锋 翻译；李芝芬 审稿）

Anomalous Groups in the Periodic System of Elements

E. Fermi

Editor's Note

Enrico Fermi was one of the outstanding physicists of the 1930s and 1940s. He is credited with the construction of the first atomic reactor at the University of Chicago in 1941. In the early 1930s he was more concerned with the structure of atoms, and this brief note is a summary of a paper afterwards published in the proceedings of the Accademia dei Lincei, an Italian scientific academy. The message Fermi wished to emphasise was that the first group of transition elements in the periodic table could be accounted for by theoretical calculations of the behaviour of electrons in atoms whose atomic weight exceeded 21. Fermi received a Nobel Prize in physics in 1938.

IN a paper which will shortly appear in the *Rend. Accad. Lincei*, I have calculated the distribution of the electrons in a heavy atom. The electrons were considered as forming an atmosphere of *completely degenerated* gas held in proximity to the nucleus by the attraction of the nuclear charge screened by the electrons. Formulae were given for the density of the electrons and the potential as functions of the distance r from the nucleus.

In continuation of the previous work, I have applied the same method to the study of the formation of anomalous groups in the periodic system of elements. From the density of the electrons and their velocity distribution, one can easily calculate how many electrons have a given angular momentum in their motion about the nucleus, that is, how many electrons have a given azimuthal quantum number k.

It is known, for example, that the formation of the group of the rare earths corresponds to the bounding of electrons in 4_4 orbits, that is, to the presence in the atom of electrons with $k = 4$. Now it follows from the theory that electrons with $k = 4$ exist in the normal state only for atoms with atomic number $z \geq 55$. This agrees well with the empirical result that the group of the rare earths begins at $z = 58$ (cerium).

Similarly, the bounding of 3_3 electrons with $k = 3$ corresponds to the anomaly of the first great period beginning at $z = 21$ (scandium); according to the theory, electrons with $k = 3$ should appear in the atom just at $z = 21$.

Further details will be published later.

(**121**, 502; 1928)

E. Fermi: Physical Institute of the University, Rome.

元素周期系中的反常族

费米

编者按

恩里科·费米是20世纪三四十年代最杰出的物理学家之一。他享誉于1941年在芝加哥大学建成了第一个原子反应堆。20世纪30年代初，他更关注于原子的结构，这篇短文是他随后发表在《林琴科学院院刊》（林琴科学院是意大利的一个科学院）上的一篇论文的摘要。在这篇短文中费米强调，对原子量超过21的原子中电子的行为的理论计算可以用来解释周期表中的第一组过渡元素。费米于1938年获得了诺贝尔物理学奖。

在不久将会发表于《林琴科学院院刊》的一篇短文中，我计算了重原子中电子的分布。电子被看作是一团**完全简并**的气体，它们被由电子屏蔽了的核电荷吸引而围绕在核附近。另外我也给出了计算电子密度和势能的公式，它们都是与核之间的距离 r 的函数。

作为此前工作的延续，我使用了同样的方法来研究元素周期系中反常族的形成。根据电子密度和它们的速率分布，可以很容易地计算出相对于核的运动角动量为某一指定值的电子的数目，也就是说，角量子数 k 为某一指定值的电子的数目。

例如，已经知道稀土族的形成对应于电子被束缚在 4_4 轨道上，也就是说，对应于原子内存在 $k = 4$ 的电子。而根据理论可以得出，只有在那些原子序数 $z \geq 55$ 的原子中 $k = 4$ 的电子才能以正常状态存在。这与稀土元素族从 $z = 58$（铈）开始这一经验结果吻合得很好。

类似的，$k = 3$ 的电子被束缚在 3_3 轨道上，对应于从 $z = 21$（钪）开始的第一个长周期的反常。而根据理论，$k = 3$ 的电子刚好应该出现在 $z = 21$ 的原子之中。

进一步的细节将随后发表。

（王耀杨 翻译；李芝芬 审稿）

Wave Mechanics and Radioactive Disintegration

R. W. Gurney and E. U. Condon

Editor's Note

The "wave mechanics" description of quantum theory arose from the work of Erwin Schrödinger in the 1920s, which built on the suggestion of Louis de Broglie in 1924 that matter can possess wave-like properties. Schrödinger's description of the behaviour of quantum particles was formulated purely in terms of waves (or wavefunctions) whose amplitude in different parts of space specified the probability of the particle being there. Here Ronald Gurney and Edward Condon perceptively states what this implies for the decay of radioactive atomic nuclei by emission of alpha particles. They points out that the escape of the alpha particle can be regarded simply in terms of overlap between wavefunctions inside and outside the nucleus. This is a form of quantum "tunnelling" through an energy barrier.

AFTER the exponential law in radioactive decay had been discovered in 1902, it soon became clear that the time of disintegration of an atom was independent of the previous history of the atom and depended solely on chance. Since a nuclear particle must be held in the nucleus by an attractive field, we must, in order to explain its ejection, arrange for a spontaneous change from an attractive to a repulsive field. It has hitherto been necessary to postulate some special arbitrary "instability" of the nucleus; but in the following note it is pointed out that disintegration is a natural consequence of the laws of quantum mechanics without any special hypothesis.

It is well known that the failure of classical mechanics in molecular events is due to the fact that the wave-length associated with the particles is not small compared with molecular dimensions. The wave-length associated with α-particles is some 10^5 smaller, but since the nuclear dimensions are smaller than atomic in about the same ratio, the applicability of the wave mechanics would seem to be ensured.

In the classical mechanics, the orbit of a moving particle is entirely confined to those parts of space for which its potential energy is less than its total energy. If a ball be moving in a valley of potential energy and have not enough energy to get over a mountain on one side of the valley, it must certainly stay in the valley for all time, unless it acquire the deficiency in energy somehow. But this is not so on the quantum mechanics. It will always have a small but finite chance of slipping through the mountain and escaping from the valley.

In the diagram (Fig.1), let O represent the centre of a nucleus, and let $ABCDEFG$ represent a simplified one-dimensional plot of the potential energy. The parts ABC and GHK represent the Coulomb field of repulsion outside the nucleus, and the internal part $CDEFG$ represents the attractive field which holds α-particles in their orbits. Let DF be an allowed orbit the energy of which, say 4 million volts, is given by the height of DF above

波动力学和放射性衰变

格尼，康登

编者按

20 世纪 20 年代，埃尔温·薛定谔在路易斯·德布罗意于 1924 年提出的物质可能具有波动性这个观点的基础上进行了一系列研究工作，开启了用“波动力学”描述量子理论的先河。薛定谔完全从波（或者说是波函数）的角度系统地描述了量子化粒子的行为，空间中不同位置上波的振幅代表了粒子在该处出现的概率。在这篇文章中，罗纳德·格尼和爱德华·康登敏锐地指出了薛定谔的描述对于理解放射性原子核发射出 α 粒子的衰变的意义。他们认为，α 粒子的逃逸可以简单地理解成核内外波函数的重叠。这正是一种穿透能量壁垒的量子“隧道效应”。

在 1902 年放射性衰变的指数律被发现后，人们很快认识到一个原子衰变的时间与这个原子以前的历史无关，而完全是由几率决定的。因为核子一定是被引力场束缚在原子核中的，所以为了解释核子的发射，我们必须假设一个从引力场到斥力场的自发转变。目前，有必要假设原子核具有某种特别的任意“不稳定性”，但在下文中，我们会指出衰变是量子力学原理的固有结果，并不需要任何特别的假设。

众所周知，经典力学在分子事件中失效是由于与分子尺度相比粒子的波长较大。α 粒子的波长约为分子尺度的 $1/10^5$，但因为原子核的尺度比原子尺度小差不多同样的比率，波动力学的适用性似乎是有保证的。

在经典力学中，一个运动粒子的轨道被严格限制在势能小于总能量的区域内。如果一个球在势阱中运动，并且其能量不足以使它翻过势垒，那么它将一直待在势阱中，除非它以某种方式获得足以翻越势垒的能量。但在量子力学中情况并非如此，球总是有一个很小但不为零的机会可以穿透势垒从势阱中逃逸出来。

如图 1 所示，设 O 表示一个原子核的中心，$ABCDEFG$ 表示一条简化了的一维势能曲线。其中，ABC 段和 GHK 段表示核外的库仑排斥场；中间的 $CDEFG$ 段表示将 α 粒子束缚在其轨道上的引力场。设 DF 是一个容许轨道，它的能量，比如说 400 万电子伏，由 OX 到 DF 的高度表示。我们可以近似地说，这个轨道对应的波函

OX. Approximately, we can say that with this orbit will be associated a wave-function which will die away exponentially from *D* to *B*. Again, corresponding to motion outside the nucleus along *BM*, there will be a wave-function which will die away exponentially from *B* to *D*. The fact that these two functions overlap in the region *BD* means that there is a small but finite probability that the particle in the orbit *DF* will escape from the nucleus along *BM*, acquiring kinetic energy equal to the height of *DFBM* above *OX*, say 4 million volts. This occurrence will be spontaneous and governed solely by chance.

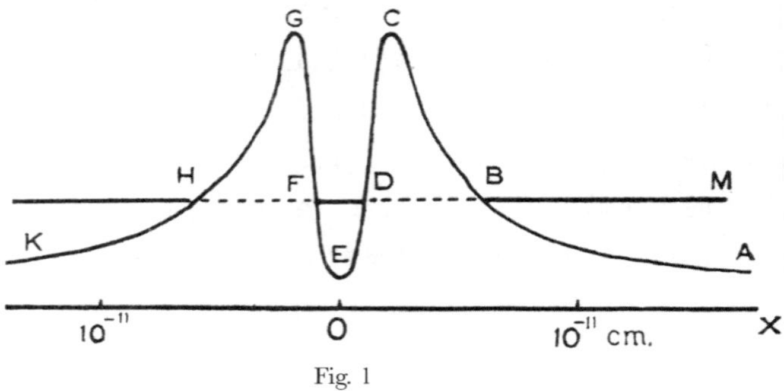

Fig. 1

The rate of disintegration, that is, the probability of escape, depends on the amount of overlapping of the wave-functions in the regions *DB* and *FH*, and this is extremely sensitive to the height to which the potential curve at *C* rises above *BDF*. By varying this height through a small range we can obtain all periods of radioactive decay from a fraction of a second, through the 10^9 years of uranium, to practical stability. (In considering the transmutation of a molecule into its isomer, Hund found a similar vast range of transformation periods, *Zeit. f. P.*, 43, 810; 1927) If the potential curves for the interaction of an α-particle with the various radioactive nuclei are similar, we can obtain a qualitative understanding of the Geiger-Nuttall relation between the rate of disintegration and the range of the emitted α-particles. For the α-particles of high energy the wave function for outside motion will overlap that for the inside motion more, and the rate of disintegration will be greater.

Besides obtaining a general idea of the mysterious instability of the nucleus, we can visualise in this way one of the most puzzling results of recent experimental work. An α-particle having the same range (2.7 cm) as those emitted by uranium should, if fired directly at the uranium nucleus, penetrate its structure; while faster α-particles should do so, even when not fired directly at the nucleus. It was therefore disconcerting when, on examining the scattering of fast α-particles fired at uranium, Rutherford and Chadwick (*Phil. Mag.*, 50, 904; 1925) could find no indication of any departure from the inverse square laws. But from the model outlined above, this is what would be expected. For if the height of *BM* above *OX* represents the energy of the uranium α-particles, then a faster particle fired at the nucleus will simply run part way up the hill *ABC* and return without having encountered any change in the repulsive field or any nuclear particles (which are describing orbits within the region *GEC*).

数从 *D* 到 *B* 呈指数衰减。此外，相应于核外运动的 *BM* 段中，粒子运动的波函数从 *B* 到 *D* 也呈指数衰减。这两个波函数在 *BD* 区域交叠的事实意味着，存在一个很小但不为零的几率，使得在 *DF* 轨道上的粒子能够沿着 *BM* 逃逸出原子核，同时获得 *OX* 到 *DFBM* 的高度所表示的动能，比如说 400 万电子伏。这一事件是自发的并且只受几率的控制。

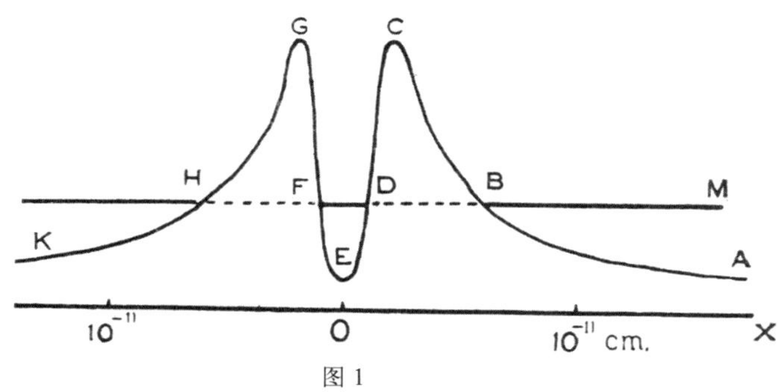

图 1

衰变速率，即逃逸几率，依赖于 *DB* 和 *FH* 区域中波函数的交叠量，并且对势能曲线中 *C* 点到 *BDF* 的高度十分敏感。通过在小范围内改变这个高度，我们可以获得所有放射性衰变元素的衰变周期，从几分之一秒、10^9 年（铀）到实质上是稳定的，各种情况都存在。（在考察分子的异构体之间的转化时，洪德发现了一个与此类似的转变周期的巨大范围，见《物理学杂志》，第 43 卷，第 810 页，1927 年）如果 α 粒子与各种放射性核相互作用的势能曲线是类似的，我们就能定性地理解衰变速率和所发射 α 粒子的射程之间的盖革–努塔耳关系。对于高能 α 粒子，核外运动波函数与核内运动波函数交叠得更多，从而衰变速率将会更大。

除了获得关于原子核神秘不稳定性的一般观点以外，以此方法我们还可以解释最近实验工作中最令人迷惑的一些结果中的一个。如果一个与铀发射出的 α 粒子射程相同（2.7 厘米）的 α 粒子直接射在铀核上，它将穿透铀核；而更快的 α 粒子即使不是直接射在铀核上，也会穿透铀核。因此，令人不解的是，在快 α 粒子射在铀上的散射实验中，卢瑟福和查德威克（《哲学杂志》，第 50 卷，第 904 页，1925 年）没能发现任何偏离库仑平方反比定律的迹象。但是从以上概述的模型来看，这应该是预期的结果。因为如果 *OX* 到 *BM* 的高度表示铀发射的 α 粒子的能量，那么射在核上的更快的粒子将仅仅沿着势垒 *ABC* 向上爬一段然后返回，其间没有经历排斥场的任何变化也没有遇到任何核子（核子在 *GEC* 区域内运动）。

The peculiar property of the wave mechanical equations which finds application here has also been applied to the theory of the emission of electrons from cold metals under the action of intense fields (Oppenheimer, *Proc. Nat. Acad. Sci.*, 14, 363; 1928; and Fowler and Nordheim, *Proc. Roy. Soc.*, A, 119, 173; 1928). Ordinarily, an atom does not lose its electrons because the attractive field of the atom remains attractive to all distances. But when an intense field is applied, then the attractive field is reversed in sign a short distance from the atom. This makes the resultant potential energy curve similar to that in the diagram, and so the atoms begin to shed their electrons.

Much has been written of the explosive violence with which the α-particle is hurled from its place in the nucleus. But from the process pictured above, one would rather say that the α-particle slips away almost unnoticed.

(**122**, 439; 1928)

Ronald W. Gurney, Edw. U. Condon: Palmer Physical Laboratory, Princeton University, July 30.

这里应用到的波动力学方程的特殊性质也曾被应用于强场作用下冷金属发射电子的理论中（奥本海默，《美国科学院院刊》，第 14 卷，第 363 页，1928 年；福勒和诺德海姆，《皇家学会学报》，A 辑，第 119 卷，第 173 页，1928 年）。一般情况下，原子不会失去它的电子，因为不论相隔多远的距离，原子的引力场都吸引着电子。然而，当施加一个强场时，原子附近一个短距离范围之内，原子引力场将被颠倒。这使得合成的势能曲线类似于图 1 中的曲线，因此原子开始发射它们的电子。

很多文献都提到了将 α 粒子从原子核内抛出的一种爆炸性的力量。然而，从以上描绘的过程来看，人们宁愿说 α 粒子几乎是神不知鬼不觉地溜出原子核的。

（王锋 翻译；李军刚 审稿）

Sterilisation as a Practical Eugenic Policy

E. W. MacBride

Editor's Note

Perhaps the most chilling aspect of this discussion of eugenics by zoologist Ernest William MacBride, reviewing of a book on eugenic policies in the United States, is its ignorance and prejudice several decades after Darwin's evolutionary theory sparked calls for controls on breeding in human society. MacBride does not even respect that theory: he was a notorious late advocate of Lamarckian inheritance of acquired characteristics, evident here in how he thinks people who acquire mental defects might pass them on to offspring. There is also an elision from sterilization of "mental defectives" to the control of fertility in people whom MacBride deems merely "stupid" or feckless—apparently, many of the poor. It is a stark reminder of how ideas about genetic heredity were confused and abused.

Sterilisation for Human Betterment: a Summary of Results of 6,000 Operations in California, 1909–1929. By E. S. Gosney and Dr. Paul Popenoe (A Publication of the Human Betterment Foundation.) pp. xviii+202. (New York: The Macmillan Co., 1929.) 8*s*. 6*d*. net.

THIS little book is a storehouse of information on the efforts which have been made in the United States to improve the human stock by sterilising the feeble-minded and the insane. It appears that although *more Americano* laws have been passed in about twenty states of the Union providing for the legal sterilisation of sexual perverts, and imbecile and insane patients in public institutions, these laws have been put into practical operation only in the State of California, so that in the book discussion is mainly concerned with the results obtained in that State.

The justification for these attempts to aid Nature in eliminating the unfit is set forth in the introduction. Amongst our unsentimental forefathers, no efforts were made to keep alive weakly and diseased children, and hence the race was propagated only from its most vigorous members; but nowadays, when unreflecting humanitarian sentiment is in fashion, all babies are kept alive so far as medical science can avail, and this science is paid for by levying tribute on the thrifty and self-supporting. The result is that this section of society limits its offspring, and future generations are likely to be recruited not from the fit but from the unfit.

How drastically and efficiently natural selection operated amongst the young in England during the eighteenth century may be gathered from figures given by Miss Buer in her book, "Health, Wealth, and Population in the Early Days of the Industrial Revolution". In 1730, out of all babies born in London, 74 percent died before they were five years of age; in 1750, 63 percent died; in 1770 the percentage was 50, and it did not sink to 30 until

作为一项实用优生学政策的绝育术

麦克布赖德

编者按

达尔文在进化论中曾提出应该对人类的生育有所控制，几十年后，动物学家欧内斯特·威廉·麦克布赖德仍然对此缺乏理解并存有偏见，他对一本关于美国优生政策的书所作的评论令人心寒。麦克布赖德甚至并不推崇达尔文的进化论。他作为拉马克获得性遗传假说的支持者之一而广为人知，从这篇文章中可以很明显地看出，他认为精神病患者可能会把他们的疾病基因传给后代。其实，不管是对“精神病患者”的绝育，还是对在麦克布赖德看来只是略显“愚钝”的无能之辈的生育控制，都是没有必要的，后者显然是指穷人。这篇文章是提醒人们关注遗传学的观点是如何被混淆和滥用的绝好实例。

《用于人类改良的绝育术：1909~1929 年加利福尼亚 6,000 例手术结果的概要》
戈斯尼和保罗·波普诺博士著（人类改良基金会的一部著作）
xviii + 202 页。（纽约：麦克米伦公司，1929 年。）8 先令 6 便士

这本小册子是一座知识宝库，蕴藏着美国通过使低能者和精神病患者绝育的方式来提高人口质量的过程中取得的成就。尽管美国联邦约 20 个州已经通过了**越来越多的美式**法律，为性异常者、智能低下者及精神病患者在公共机构进行绝育手术提供合法依据，但是看起来只有加利福尼亚州真正实施了这些法律，因此本书中主要是关于加利福尼亚州取得的成果的讨论。

为了帮助造物主减少不适于生存者而采取这些绝育措施的合理性，在本书的引言里已经有所陈述。我们那些无情的祖先，没有为使虚弱和得病的孩子能够活下来而进行过任何努力，因此种系都是从其最强壮的成员中繁衍而来的；但是如今，在这个浅薄的人道主义情感泛滥的时代，医学在其力所能及的范围内力求使所有婴儿都能活下来，并且通过向节俭的自食其力者征税来支付这些费用。这样的结果就是，社会上的这部分自食其力者限制自己的后代个数，未来的孩子们更可能是从那些不适者，而非适者中繁衍而来。

自然选择在 18 世纪英国的年轻人中所起的作用是多么明显有效，这可以从比埃小姐书中给出的数字推断出来，那本书的书名是《工业革命早期的健康、财富和人口》。1730 年伦敦出生的所有婴儿中有 74% 死于 5 岁之前，1750 年这一比例为 63%，1770 年为 50%，直到 1833 年这一比例才降到 30%。这一比例在该国的其他

1833. The percentage was probably even higher in other parts of the country. The help given by hospitals, and later by the State, to indigent mothers has all grown up in the last century, so that the argument that because we have maintained a vigorous, enterprising, fighting race in these islands for eight hundred years since the Norman Conquest, we shall continue to do so, is not one for which there is any sound basis.

It is, however, not practical politics to suggest a return to the old plan of *laissez-faire*. How then shall the elimination of the unfit be promoted? The authors of this book suggest "by legalised sterilisation". The method of sterilisation advocated is cutting the ducts (vasa deferentia in the male, and Fallopian tubes in the female) which convey the germ cells to the exterior. The authors point out that more than six thousand operations of this sort have been already performed in California, and that only seven failures are recorded (three in males, four in females). The operation does not interfere with sexual desire or the performance of the sexual act. The genital organs in man, as in Vertebrata generally, have two functions, namely: (1) to produce the germ cells; (2) to produce a hormone which diffuses through the system and maintains youth and vigour. In a man the spermatozoa forms a minor part of the sexual discharge, the main portion of which is constituted by the prostatic secretion, and some authorities hold that this secretion when absorbed by the female has an invigorating effect on the constitution. As to a woman, when it becomes necessary on account of tumours to remove the uterus, if a portion of one of the ovaries is preserved and sewn to the abdominal wall, this will prevent the premature onset of the menopause and maintain in the patient all the qualities of a young woman.

But are insanity and mental defect hereditary? Some British authorities hold that in many cases they are not. So far as insanity is concerned, however, there is general agreement, as our authors point out, that the condition known as "dementia praecox" is the result of an inborn weakened constitution, and that it is a mere question of time when it will manifest itself in the life of the unfortunate individual who has inherited this constitution. As to mental defect, the argument that it is sometimes not of hereditary origin, overlooks the consideration that all "mutations", of which mental defect is one, must ultimately have been produced by some external cause, and there is nothing to show that an "accidental" mental defective will not propagate mentally defective children. In any event, even if a defective should produce healthy children, such a person would make the worst possible parent to carry out the duty of caring for and training the children; and it is a little too much to ask the State to allow a defective to go on having children on the chance of some of them being normal, if the State has to support them all.

Our authors urge that sterilisation should not be regarded as a punishment but as a hygienic measure; that defectives confined in asylums might be allowed out on condition of their consenting to this operation. But whilst we agree that this argument is good so far as it goes, a little reflection will show that it only touches the fringe of the problem. The defectives most dangerous to society are those who are never confined in institutions at all! The high-grade defectives are just able to support themselves in the lowest paid and most unskilled occupations, and no civilised government would take the responsibility of

地方可能更高。上个世纪，医院给予这些贫困母亲的帮助（后来国家也给予了帮助）已经有所增加。因此有一种观点认为，诺曼征服以来的800年间，这些岛上的人们一直保持着一个精力充沛的、有进取心的、好战的民族应有的素质，所以应该继续保持这种做法，这一论点是缺乏可靠根据的。

然而，重新起用**自由放任**的旧政策并不切实可行。那么怎样才能促进不适于生存者数量的减少呢？本书的作者建议采取“合法绝育”的手段。这里提倡的绝育方法，是指切断向外输送生殖细胞的管道（男性是输精管，女性是输卵管）。作者指出加利福尼亚已经实施了六千多例此类手术，根据记录，只有7例失败（3例男性绝育术，4例女性绝育术）。该手术不会影响性欲和性行为的进行。与脊椎动物门其他成员一样，男性生殖器有两个功能，即：(1)产生生殖细胞；(2)产生散布于全身系统并维持机体青春与活力的激素。在男性分泌物中精子仅仅占很小部分，大部分是前列腺分泌液。一些权威人士认为，这些分泌液在被女性吸收后，会起到令女性的体质充满生气与活力的作用。对于女性，当她们因为肿瘤而必须摘除子宫时，如果保留一个卵巢的一部分，并将其与腹壁缝合在一起，就可以延缓更年期的到来并使病人保有年轻女人的所有特质。

但是精神病和心智缺陷是可遗传的吗？英国的一些权威人士认为，许多情况下它们并不遗传。然而就精神病而言，人们普遍认可的观点正如本书作者指出的那样，通常所说的“早发性痴呆”就是先天体质虚弱造成的，遗传了这一体质的不幸个体，发病只是时间问题。对于心智缺陷，有观点认为，它有时不是由于遗传。这一看法忽略了一点，那就是包括心智缺陷在内的所有“突变”最终都是由某一外因引起的，而没有任何证据表明“偶然的”心智缺陷者不会生育出心智缺陷的孩子。无论如何，即使一个有缺陷的个体可以生出健康的孩子，作为负有照顾和培养孩子的责任的父母来说，他们也是最不该成为父母的人；如果国家不得不供养所有有缺陷的个体的话，那么怀着“有缺陷的人也可能生出正常孩子”的投机心理而请求国家允许他们生育，这未免太过分了。

本书作者认为不应该将绝育视为一种惩罚，而应该看作一项卫生措施；对于那些被关在收容所的有缺陷的个体，在他们同意接受该手术的情况下，可以允许他们出去。就目前状况来看，我们承认这一想法很好，但是稍加考虑就会发现，这种做法只是刚刚触及这一问题的边缘而已。对社会最具危险性的有缺陷的个体是那些从来没有受到公共机构限制的人！高级的有缺陷的个体能够凭最低收入和最不需技能的工作来养活自己，并且任何文明社会的政府都不会承担约束他们的责任，所以

confining them, and so they go on propagating large families as stupid as themselves. As Mr. Lidbetter has shown*, it is from the ranks of just these classes that in the last hundred years the majority of paupers and criminals of London has been recruited.

It seems to us that in the last resort compulsory sterilisation will have to be inflicted as a penalty for the economic sin of producing more children than the parents can support. Whether a man has a large or a small family is—given a healthy wife—a matter of taste, so long as he provides for his own children; but when he comes to the State and demands that it—that is to say, his neighbours—should support these children, then the State can say, "Very well—we shall help you with the family which you have, but if after this you have any more children you shall be sterilised."

Before, however, such an alternative is presented to any citizen, he may justly claim that he should receive instruction from the State in the means of birth-control. It is obviously unfair that such knowledge should be denied to the poor whilst it is easily accessible to the rich. It is often said, and with justice, that the great objection to birth-control is that the wrong people practise it. But this knowledge once attained cannot be taken away; the middle classes possess it and cannot be prevented from putting it into practice. If, however, the knowledge and practice of birth-control were widely spread among the working-class, there would be created such a resentment against the reckless production of children that the movement to establish compulsory sterilisation of the unfit would prove irresistible.

(**125**, 40-42; 1930)

* "Pauperism and Heredity", by E. J. Lidbetter, *The Eugenics Review*, vol. 14, p. 152; 1923.

他们就会继续繁衍出和他们自己一样有缺陷的大家族。正如李德贝特先生所说*，最近的 100 年间，伦敦的大部分乞丐和罪犯正是从这类人中产生的。

对我们而言，强制性的绝育术似乎是对那些生育了很多孩子而无力供养的父母们所犯经济罪进行惩罚的最后手段。如果一个男人拥有一个健康的妻子，那么他选择拥有大家庭还是小家庭完全是个人喜好的问题，只要他能养得起自己的孩子；但是如果他要求国家（其实也就是他的左邻右舍们）帮助抚养他的孩子们，那么国家就可以说："没问题，我们会帮你养家，但是从现在起如果你再生孩子的话，那么你将会被绝育。"

不过，在向市民公布这种可供选择的方案之前，人们可能会理直气壮地声称自己有权利了解国家的生育控制政策。富人们可以很容易地获悉这些信息，而穷人们的知情权可能被剥夺了，这显然是非常不公平的。人们常常说，生育控制的最大问题在于是错误的人在实施政策，这么说也是公平的。但是，这些知识一旦为人们获得就不可能再拿走；中产阶级们知悉了这些，就不可避免会将其付诸实践。然而，假如节育的知识和实践在工人阶级中广泛传播，那么可能会出现一种对不计后果进行生育的怨恨，到那时对不适于生存者实施强制性绝育术的行动可能就无法遏止了。

（刘皓芳 翻译；刘京国 审稿）

---

*《贫困与遗传》，作者李德贝特，《优生学评论》，第 14 卷，第 152 页，1923 年。

The "Wave Band" Theory of Wireless Transmission

A. Fleming

Editor's Note

Ambrose Fleming here takes issue with a way of understanding wave communications, for telephone or television. In both technologies, devices encode signals as amplitude modulations of a carrier wave. In terms of Fourier analysis, one can view the resulting wave as occupying a "band" of frequencies around the carrier frequency, and it had become common to consider how these bands should be apportioned, which bands were allowed and so forth. But Fleming argues that talk of bands obscures the role of the amplitude. Too large an amplitude could cause interference between different transmissions, much as speaking too loudly at the theatre can be disruptive. Focusing on amplitude rather than bands, Fleming suggests, will help avoid unnecessary restrictions on the new technologies.

IN scientific history we meet with many examples of scientific theories or explanations which have been widely adopted and employed, not because they can be proved to be true but because they provide a simple, easily grasped, plausible explanation of certain scientific phenomena. The majority of persons are not able to see their way through complicated phenomena and so thankfully adopt any short-cut to a supposed comprehension of them without objection.

Ease of comprehension is not, however, a primary quality of Nature, and it does not follow that because we can imagine a mechanism capable of explaining some natural phenomenon it is therefore accomplished in that way. There is a widely diffused belief in a certain theory of wireless telephonic transmission, and also of television, that for securing good effects it is necessary to restrict or include operations within a certain width of "wave band". But although this view has been very much adopted there is good reason to think that it is merely a kind of mathematical fiction and does not correspond to any reality in Nature.

Let us consider how it has arisen. We send out from all wireless telephone transmitters an electromagnetic radiation of a certain definite and constant frequency expressed in kilocycles. Thus 2LO London broadcasts on 842 kilocycles. This means that it sends out 842,000 electric vibrations or waves per second. Every broadcasting station has allotted to it a certain frequency of oscillation and it is not allowed to depart from it.

It is like a lighthouse which sends out rays of light of one pure colour or an organ which emits a single pure musical note. For most broadcasting stations this peculiar and individual frequency lies somewhere between a million and half a million per second, though for the long wave stations like Daventry it is so low as 193,000 or 193 kilocycles.

无线传输的“波带”理论

弗莱明

编者按

在本文中安布罗斯·弗莱明对应用于无线电话和电视的波通信有不同的理解。在这两项技术中，装置把信号编译成振幅调制的载波。在傅立叶分析中，我们可以认为得到的波占据了载波频率附近的一个频“带”。大家通常要考虑的是这些频带如何分配，哪些频带能够被允许等等。但弗莱明认为人们对频带的过分关注掩盖了振幅的作用。太大的振幅可能会导致不同传输过程之间的干扰，就像在戏院中大声讲话带来的麻烦一样。弗莱明指出，对振幅而不是频带的关注将会帮助我们避免在应用这项新技术时遇到不必要的麻烦或限制。

在科学史上，我们遇到的许多科学理论或解释被广泛接受并应用的情况，并不是因为它们能够被证明是真理，而是因为它们为某些科学现象提供了简单、易于被理解并且似乎合理的解释。大多数人并不能透过复杂现象发现真理，因此也就乐于不加质疑地接受某种能够便捷地解释复杂现象的假说。

然而，简单的理解并不是自然界的基本特征，也不能随即得到，即不会因为我们想象一种能够解释某些自然现象的机制，它就以那种方式来实现。在关于无线电话传输以及电视的某种理论中，存在一种广为流传的认识，即为了达到可靠的良好效果，必须限制在具有确定宽度的“波带”内操作。这种观点虽然已被广泛接受，但我们有理由认为，这只是一类数学虚构，并不能与自然界的任何现实相对应。

让我们考虑这是如何产生的。我们从所有的无线电话发射台发出一个具有确定的恒定频率（以千周表示）的电磁辐射。伦敦 2LO 电台按照这种方式以 842 千周的频率进行广播。这意味着它每秒内发出 842,000 次电子振荡或 842,000 个波。每个广播站已分配到一个确定的振荡频率，而且不允许偏离此频率。

这就像一座发出单色光的灯塔或一个发出单一音符的风琴。对于大多数广播站而言，这类特有的专用频率位于每秒 50 万~100 万之间，然而对于类似达文特里这样的长波站，它的频率低至 193,000，或 193 千周。

When we speak or sing or cause music to affect the microphone at a broadcasting studio the result is to cause the emitted vibrations, which are called the *carrier waves*, to fluctuate in height or wave amplitude, but does not alter the number of waves sent out per second. It is like altering the height or size of the waves on the surface of the sea without altering the distance from crest to crest which is called the wave-length.

Suppose the broadcasting station emits a carrier wave of frequency n and let $p = 2\pi n$. Then we may express the amplitude a of this wave at any time t by the function $a = A \sin pt$ where A is the maximum amplitude. If on this we impose a low frequency oscillation due to a musical note of frequency m and let $2\pi m = q$, then we can express the modulated vibration by the function

$$a = A \cos qt \sin pt$$

But by a well-known trigonometrical theorem this is equal to

$$\frac{A}{2}\{\sin(p+q)t + \sin(p-q)t\} ,$$

and thence may be supposed to be equivalent to the simultaneous emission of two carrier waves of frequency $n + m$ and $n - m$.

If the imposed note or acoustic vibration is very complex in form, then in virtue of Fourier's theorem it may be resolved into the sum of a number of simple harmonic terms of form $\cos qt$, and each of these may be considered to be equivalent to a pair of co-existent carrier waves. Hence the complex modulation of a single frequency carrier wave might be imitated by the emission of a whole spectrum or multitude of simultaneous carrier waves of frequencies ranging between the limits $n + N$ and $n - N$, where n is the fundamental carrier frequency and N is the maximum acoustic frequency occurring and $2N$ is the width of the wave band. This, however, is a purely mathematical analysis, and this band of multiple frequencies does not exist, but only a carrier wave of one single frequency which is modulated in amplitude regularly or irregularly.

If the sounds made to the microphone at the broadcasting station are very complex, such as those due to instrumental music or speech, then in virtue of this mathematical theorem the very irregular fluctuations in amplitude of the single carrier wave can be imitated if we suppose the station to send out simultaneously a vast number of carrier waves of various frequencies lying between certain limits called the "width of the wave band".

This, however, is merely a mathematical artifice similar to that employed when we resolve a single force or velocity in imagination into two or more component forces. Thus, if we consider a ball rolling down an inclined plane and desire to know how far it will roll in one second, we can resolve the single vertical gravitational force on the ball into two components, one along the plane and one perpendicular to it. But this is merely an ideal division for convenience of solution of the problem; the actual force is one single force acting vertically downwards. Similar reasoning is true with regard to wireless telephony. What happens, as a matter of fact, is that the carrier wave of one single constant

当我们在播音室里对着麦克风说话、唱歌或放出音乐，都会导致发射振荡的产生，这被称作**载波**，它会使波的高度或波幅出现起伏，但并不改变每秒钟发出的波数。这类似于只改变海面上波浪的高度或大小，而不改变被称作波长的从波峰到波峰的距离。

假定广播站发射一个频率为 n 的载波，并令 $p = 2\pi n$。我们通过函数 $a = A \sin pt$ 表示在任意时刻 t 这个波的振幅 a，其中 A 是最大振幅。如按这种方式，我们施加一个由频率为 m 的音符引起的低频振荡，并令 $2\pi m = q$，则我们可以通过下面的函数方程表示调制振荡

$$a = A \cos qt \sin pt$$

通过人们熟知的三角定理，上式等于

$$\frac{A}{2}\{\sin(p+q)t + \sin(p-q)t\} ,$$

因此可以认为其等效于同时发射频率为 $n + m$ 和 $n - m$ 的两个载波。

如果施加的音律或声音振荡在形式上很复杂，那么利用傅立叶定理，我们可将其分解为如 $\cos qt$ 形式的许多简谐项之和，可以等效地认为其中每一项都是一对共存的载波。因此单频载波的复杂调制可以用整个谱的发射或大量频率在 $n + N$ 到 $n - N$ 之间的同步载波来模拟，其中 n 是基本载波频率，而 N 是出现的最大声波频率，$2N$ 是波带的宽度。然而，这是纯数学分析，这个多频波带并不存在，而只存在单频的载波，可以对其振幅进行规则或不规则的调制。

如果向广播站里的麦克风发出的声音非常复杂，例如乐器发出的声音或讲话的声音，那么根据这个数学原理，我们可以假定广播站里同时发出大量频率被限定在“波带宽度”之内的载波来模拟单个振幅不规则波动的载波。

然而，这仅是一个数学技巧，类似于我们在假想中将单个力或速度分解为两个或更多个分量。因此，如果我们考虑一个沿倾斜平面滚下的球，并要求知道球在一秒钟内滚动多远，我们就可以将球受到的单一的垂直引力分解为两个分量，分别与平面平行和垂直。但这只是便于解决问题的一个理想的分解；实际的力仍只是作用方向垂直向下的单个力。对于无线电话也存在着相似的思考过程。事实上，发生的情况是，一个单一恒定频率的载波按照某种规则或不规则的规律在振幅上发生变

frequency suffers a variation in amplitude according to a certain regular or irregular law. There are no multiple wave-lengths or wave bands at all.

The receiver absorbs this radiation of fluctuating amplitude and causes the direct current through the loud speaker to vary in accordance with the fluctuations of amplitude of the carrier wave; the carrier wave vibrations being rectified by the detector valve.

The same thing takes place in the case of wireless transmission in television. The scanning spot passes over the object and the reflected light falls on the photoelectric cells and creates in them a direct current which varies exactly in proportion to the intensity of the reflected light. This photoelectric current is employed to modulate the amplitude of a carrier wave, and the neon lamp at the receiving end translates back these variations of carrier wave amplitude into variations in the cathode light of the neon tube.

There is neither in wireless telephony nor in television any question of various bands of wavelength. There is nothing but a carrier wave of one single frequency which experiences change of amplitude. The whole question at issue then is, what range in amplitude is admissible?

In the case of television it is usual for critics of present achievements to say that good or satisfactory television cannot be achieved within the limits of the nine kilocycle band allowed. But there is in reality no wave band involved at all. It is merely a question of what change in amplitude in a given carrier wave can be permitted without creating a nuisance.

It is something like the question: How loud can you whisper to your next neighbour at a concert or theatre without being considered to be a nuisance? People do whisper in this way, and provided not too loudly, it is passed over. But if anyone is so ill-mannered as to speak too loudly he is quickly called to order, or turned out.

It is, however, not an easy thing to define a limit to wave amplitudes. They are measured in microvolts per metre and are difficult to measure. But a wave-length is easy to define in kilocycles or in metres, and hence the method has been adopted of limiting emission to an imaginary band of wavelengths which, however, do not exist.

The definition is imperfect or elusive. It is something like the old-fashioned definition of metaphysics as "a blind man in a dark room groping for a black cat which isn't there". Similarly, the supposed wave band is not there. All that is there is a change, gradual or sudden, in the amplitude of the carrier wave. It is clear, then, that sooner or later we shall have to modify our code of wireless laws.

We have no reason for limiting the output of our broadcasting stations to some imaginary wave band of a certain width, say nine kilocycles or whatever may be the limiting width, but we have reason for limiting the range of amplitude of the carrier waves sent out.

化。根本不存在多重的波长或波带。

接收器吸收了这类振幅振荡变化的辐射，并使通过扬声器的直流按照载波振幅的振荡变化而变化；载波的振动通过检测器的电子管进行整流。

对于电视的无线传输来说也存在相同的情况。扫描点在物体上扫描，而反射光落在光电元件上，并在其中产生了与反射光强度精确成正比变化的直流。这个光电流用于载波振幅的调制，在接收端的氖灯会将这些载波振幅的变化转换回氖管的阴极光的变化。

在无线电话和电视中均不存在各种不同波带的问题。只是存在振幅被调制的单频载波。那么争论的全部问题就是，可以允许振幅在什么范围内变化呢?

对于电视而言，批评现有成果的评论者们通常会提出，在允许的 9 千周的限制范围内,性能良好的或令人满意的电视不可能实现。但实际上根本不存在所谓的波带。这仅仅是一个在不产生干扰的情况下可以允许给定的载波中振幅如何变化的问题。

这就类似于如下的问题：在音乐厅或剧场中你可以用多大的声音悄悄对邻座说话而不至于影响其他人呢？人们以这种方式耳语，并保持声音不是很大，这样可以不被注意。但是如果任何人非常不礼貌地用很大的声音说话，他很快就会被要求保持安静或被逐出会场。

然而，确定波幅的调制范围并不是件容易的事情。它们以每米微伏的数量级来进行计量，并且是很难被测定的。但波长很容易用千周或米来定义，因此现已采用的方法是将发射限定到虚构的波带范围内，然而，这个波带实际上并不存在。

这样的定义是不完善的，或者说是难以理解的。这有点像形而上学的老式定义，如“一个在暗室里摸索一只并不存在的黑猫的盲人”。类似地，我们假设的波带并不存在。存在的只是载波中振幅逐步或突然的变化。显然，总有一天我们将不得不修改我们的无线通信的编码规则。

我们没有理由将我们的广播电台的输出限制在某种确定宽度的假想波带内，比如说 9 千周或其他任何限制范围，但我们有理由限定输出载波振幅的范围。

Some easily applied method will have to be found of defining and measuring the maximum permissible amplitude of the carrier waves as affected by the microphone or other variational appliance. It may perhaps be thought that an unnecessary fuss is here being made on what may be regarded as simply a way of explaining things, but experience in other arts shows how invention may be greatly retarded by unessential official restrictions. Consider, for example, the manner in which mechanical traction was retarded in Great Britain for years by ridiculous regulations limiting the speed of such vehicles on highway roads. The only restrictions that should be imposed are those absolutely necessary in the interests of public safety or convenience, and all else tend to throttle and retard invention and progress.

(**125**, 92-93; 1930)

我们必将找到某些易于应用的方法，用来定义及测量被麦克风或其他有变化的设备影响的载波的最大允许振幅。也许有人会认为，这只是事物的一种解释方式，没有必要像我们这样小题大做，但其他技术中的经验表明，不必要的官方限定是如何使可能的重大发明被推迟的。考虑一下这样的事例，机械牵引在英国的发展多年受阻，正是由于制定了荒谬的法规限制公路上的这类车辆的速度而导致的。唯一应该加以限制的，是那些出于公共安全性或便利性的考虑绝对必要的方面，而其他所有方面的限制往往会扼杀并妨碍发明和进步。

（沈乃澂 翻译；李军刚 审稿）

Electrons and Protons

Editor's Note

Paul Dirac was one of the most creative physicists of the early twentieth century. In 1932 he was appointed Lucasian professor at Cambridge University, and his theory unifying quantum mechanics with Einstein's theory of special relativity won him the 1933 Nobel Prize in physics. Here *Nature* reports on one of the implications of this theory, as Dirac outlined in a paper in the *Proceedings of the Royal Society*. His equations predicted "electrons of negative energy", which meant, of positive charge. The report echoes Dirac's initial suspicion that these predicted positively charged particles would behave like protons, but they soon proved to be "positive electrons" or positrons, made of antimatter.

A theory of positive electricity has been put forward by Dr. P. A. M. Dirac in the January number of the *Proceedings of the Royal Society*. The relativity quantum theory of an electron leads to a wave equation which possesses solutions corresponding to negative energies—the energy of the electron of ordinary experiment being reckoned as positive—and although there are serious difficulties encountered in any immediate attempt to associate these negative states with protons, the existence of positive electricity can be predicted by a fairly direct line of argument. Since the stable states of an electron are those of lowest energy, all the electrons would tend to fall into the negative energy states—with emission of radiation—were it not for the Pauli exclusion principle, which prevents more than one electron from going to any one state. If, however, it is assumed that "there are so many electrons in the world that... all the states of negative energy are occupied except perhaps a few...", it may be supposed that the infinite number of electrons present in any volume will remain undetectable if uniformly distributed, and only the few "holes", or missing states of negative energy will be amenable to observation. The step is then made of regarding these "holes" as "*things of positive energy*" which are identified with the protons. A difficulty now arises in ordinary electromagnetic theory which apparently has to cope with the presence of negative electricity of infinite density; this is met by supposing that for ordinary purposes volume-charges must be measured by departures from a "normal state of electrification", which is "the one where every electronic state of negative energy and none of positive energy is occupied." The problem of the large mass of the proton, as compared with that of the electron, is not discussed in detail, but a possible line of attack is indicated. Dr. Dirac has included the minimum of mathematical analysis in this paper, which can be followed in all essential points by anyone acquainted with the principles of the quantum theory.

(**125**, 182; 1930)

电子和质子

编者按

保罗·狄拉克是20世纪初最有建树的物理学家之一。1932年，他被聘任为剑桥大学的卢卡斯教授。他因提出能将量子力学与爱因斯坦的狭义相对论统一起来的理论而获得了1933年的诺贝尔物理学奖。在发表于《皇家学会学报》的一篇论文中，狄拉克对他的理论所蕴含的一个推断进行了概述,《自然》的这篇文章报道的正是这些。狄拉克的方程预测出“具有负能量的电子”，即带正电荷的电子。狄拉克最初认为，这些被预测到的带正电荷的粒子的行为方式与质子类似，这篇报道再次复述了这一观点，但不久之后，这些粒子就被证明是组成反物质的“正电子”。

在1月的《皇家学会学报》上，狄拉克博士提出了正电子理论。他通过电子的相对论量子理论导出了一个波动方程，而这个波动方程包含相应于负能量的解（普通实验中电子的能量被认为是正的）。尽管将这种负能态与质子联系起来的尝试遇到了很多困难，但是，通过非常直接的论证可以预言正电子的存在。因为电子的稳定态是能量最低的态，所以伴随着发射辐射，所有电子都将落入负能态，但是这与泡利不相容原理不同，这一原理规定不可能有多于一个的电子处于任何一个相同的状态。然而，如果假设“世界上有非常多的电子以至于除了极少的负能态以外，几乎所有的负能态都被占据了”，那么可以推测，这些存在于负能态体系中的无数个电子将不会被探测到（若电子是均匀分布的），而仅有极少的“空穴”或遗漏的负能态可以很容易地被观察到。接下来他把这些“空穴”视为与质子相同的“**具有正能量的物质**”。密度无限大的负电的出现使普通电磁学理论遇到了困难。不过，当假定一般情况下必须通过“带电的正常状态”的偏离来测量体电荷时，便可以解决这个难题，所谓带电的正常状态是指“每一个具有负能量的电子态均被占据而没有一个正能态被占据的状态”。在这篇文章中，他并没有详细讨论质子质量比电子质量大这一问题，但是却提出了一种可能的解决思路。狄拉克博士的论文仅涉及少量数学分析，这就使任何一个学习过量子理论基本原理的人都能看懂这些数学分析的要点。

（王锋 翻译；李淼 审稿）

The Connexion of Mass with Luminosity for Stars

J. Larmor

Editor's Note

Here the physicist Joseph Larmor writes to *Nature* to discuss important new work of Edward Milne on the topic of stellar interiors, and the relationship of stellar structure to luminosity. Given that we know very little about matter at the extremely high densities likely in stellar interiors, Larmor notes, Milne's work is a laudable attempt to work from basic principles, such as those of thermodynamics. The results suggest that a star's surface properties must be strongly constrained by the physics deep inside the interior, thereby offering a possible explanation for the empirical fact that the luminosity of a star depends for the most part only on its mass.

VERY remarkable and fruitful correlations have in recent years been detected, mainly at Mount Wilson, between the magnitudes of stars and their spectroscopic characteristics. The interpretation that would naturally present itself is that magnitude can enter into relation with the radiative phenomena of the surface atmosphere only through the intensity of gravity at the surface, which when great flattens down a steady atmosphere far more than proportionately. But if, following Eddington's empirical relation, total radiation of a star is a function of its mass alone, there must be more than this involved; for the radius of the star persists in this relation when expressed in terms of intensities of surface radiation and of gravity, the former determining the temperature roughly by itself. Modern hypothesis, which treats confidently of an "electron gas" with an atomic weight, as Ramsay boldly and prematurely proposed long ago, and subject to the Maxwell-Boltzmann exponential energy formula for statistics of distribution, and to its consequences for the theory of dissociation of mixed gases in relation to pressure and temperature, has on the initiative mainly of Saha led to promising applications to stellar atmospheres, which are held to be of densities low enough at any rate not to forbid this mode of treatment.

It would seem then to be necessary to conclude that these empirical spectroscopic relations on the surface require that the stellar atmosphere must be dominated to some degree by the remote steady interior of the star. Accordingly, tentative theories of the internal constitution of the stars and their flux of radiation have been developed in much detail. With Eddington the stars are perfect gases right down to the centre, though the density may there be hundreds of times that of platinum, as has apparently been verified for the case of the companion of Sirius—the high density involving the view at one time not unfamiliar that two atoms can occupy the same space, if the picturesque conception of atoms "stripped" irrevocably to the bone is to be avoided; and the energy emitted as radiation would come from a dissociation or destruction of matter according to a law involving temperature. On the other hand, it is insisted on by Jeans that the necessary radioactivity for the very long evolutions that are contemplated must be of constant and

恒星的质量与发光度之间的关系

拉莫尔

编者按

物理学家约瑟夫·拉莫尔在《自然》上发表的这篇文章，论述了爱德华·米耳恩在恒星内部结构以及内部结构与发光度之间的关系这一问题上的新成果。拉莫尔指出：因为我们对恒星内部处于超高密度的物质状态所知甚少，所以米耳恩尝试用一些基本原理（如热力学定律）进行的研究是值得称道的。米耳恩的研究结果表明，恒星的表面性质与它的内部物理状态有很强的相关性，这为解释恒星的发光度在很大程度上只取决于恒星质量这一经验事实提供了可能。

近几年的天文观测发现，恒星的绝对星等与它们的光谱特征之间存在很多明显的关联，这些观测主要是在威尔逊山进行的。对于这一点很自然的解释是，恒星的绝对星等只有通过位于表面处的引力强度才能和表面大气的辐射现象联系起来，引力强度在稳定的大气中下降很快，远不是成比例的关系。但是如果根据爱丁顿的经验关系式，一个恒星的总辐射量只是其质量的函数，则这其中一定还有另外的影响因素；因为在表面辐射强度和引力强度的关系式中含有恒星半径这个量，而凭借表面辐射强度本身就能大致确定恒星的温度。主要由萨哈倡导的一个最新的假设已经成功地解决了恒星大气方面的许多问题，一般认为恒星大气的密度很低，绝对不会影响这种处理方式的应用。该假设大胆地采用了拉姆齐在很久以前提出的冒险且欠成熟的假设——给“电子气”赋予一个原子量，其统计分布遵从指数形式的麦克斯韦–玻尔兹曼能量公式，并由此得到了混合气体的离解与压强和温度有关的理论。

接下来我们有必要假设，若要得到恒星表面的这些经验性的光谱关系就要求深藏在下面的稳定的恒星内部结构在某些程度上支配着恒星的大气。根据这一点，人们已经将关于恒星内部结构和辐射通量的试验性理论发展到很精细的程度了。根据爱丁顿的观点，恒星从表面到中心都可以看作是理想气体，尽管中心区的密度也许能达到铂的几百倍。这一点可以在天狼星的伴星上得到很好的证明——为了避免出现原子被不可逆地“离解”到只剩下骨架的独特概念，人们曾经用两个原子共同占据同一个空间这个并不陌生的观点来解释高密度的存在；根据一个与温度有关的定律，以辐射形式放出的能量来自于物质的离解或毁灭过程。另一方面，琼斯坚持认为，在预期的漫长演化过程中，那些必要辐射的强度肯定是稳定而确切的，否则恒星将

absolute intensity, else the star would explode: and he has essayed to regard the star as "liquid" in his investigations, apparently, however, implying a very imperfect gas rather than a special phase with its surface of sharp transition. There are other theories of less statical type.

A determined effort to shed off all such special hypotheses has been published very recently by Milne (*Monthly Notices R.A.S.* for November, pp. 17–53), which accordingly invites close attention and scrutiny. The procedure is the natural one, to try to make continuity between the gases of the atmosphere subject to laws more or less already formulated, and a dense interior about which as little is to be assumed as can be helped. He holds that it suffices merely to consider laws of internal density that are in mechanical equilibrium radially under internal pressure P, of which the fraction $(1-\beta)P$ is pressure of the internal field of radiation. He does not find it necessary to consider how this field of radiation of pressure $(1-\beta)P$ is sustained against loss by outward flux: for if he can arrive at results in terms of surface values that are valid for all such equilibrated densities whether otherwise possible or not, they must hold good for the one that follows the actual law of distribution whatever it may be.

The essential feature, so far as a reader can extract the gist from the complication of formulas that seems to be inherent in these discussions, appears to be that the coefficient β, while increasing rapidly downward in the atmosphere in a manner which can be regarded as known, suddenly rises when a photospheric level is reached, altering with steep gradient until a nearly constant value of β is soon attained for the interior of the star: and the same must apply only in less degree to the density ρ. The condition of mere mechanical equilibrium of the interior is found to express the pressure at the interface between atmosphere and photosphere in terms of values at the centre and one quantity C arising from an integral along the radius involving the arbitrarily assumed law of density. The expression for the atmospheric pressure at the interface involves the same constants in such way that on equating the pressures on the two sides of the interface they divide out of the result and only C remains. This C is held, in the light unforeseen of comparison with facts, to be in some degree a characteristic constant for all the stars, and thus may be the new element beyond surface values, and without assuming anything about their interiors, that the law as formulated requires.

This seems to be right enough in a general way, were it not that the formula for C involves the gradient of density within the star close to the interface, and thus its value must be very substantially changed, in absence of some verification to the contrary, by a very slight radial displacement of the surface which is chosen for that interface. For inside the photosphere ρ is as θ^3, while P which is continuous across the interface is as $\theta^4\phi(\theta)$: so C^{-1} is as the value of $P^{-1}(d\rho/dr)^2$ in which the second factor is the internal gradient, at the surface. If this consideration be correct it would appear that it is not legitimate to connect the chromosphere with the interior across a sharp boundary surface, as if they were different phases of matter like a liquid and its vapour. This conclusion would involve that the formula itself for C cannot be well founded: and the reason can be assigned, that

会发生爆炸。他在研究中试图假设恒星是“液体”的，这显然意味着恒星是一种非常不理想的气体，而不是处于一种在表面处变化很突然的特殊状态。另外还有一些其他的弱静态型理论。

米尔恩在最近发表的一篇文章（《皇家天文学会月刊》，11 月，第 17~53 页）中，坚决地剔除了所有这些特殊的假说，因而引发了人们的密切关注和深入思考。他很自然地试图让大气层中的气体与致密的内部之间保持连续性，其中，大气层中气体遵守的定律已经在前面或多或少地提到过了，而致密的内层则几乎没有什么理论可以参照。他提出，如果仅考虑内压为 P 时沿半径方向处于力学平衡的内部密度的变化规律，内压的一部分 $(1-\beta)P$ 源自内部辐射场，则这种连续性是可能存在的。米尔恩认为，没有必要考虑这个压强为 $(1-\beta)P$ 的辐射场如何克服向外的通量损失来维持自身，因为不管是否存在其他可能性，只要他能够得到对所有平衡态密度都有效的表面值，这些结果都会满足一个真实的分布规律，无论其形式如何。

只要读者能够从以上讨论中似乎不可避免的复杂公式中抽取出要点，就会看到基本特征是，系数 β 的值以一种我们已经知道的方式在恒星大气层中由外向内迅速增长，当到达光球层时突然增加，然后以很陡的梯度发生变化，直到很快在恒星内部达到一个基本恒定的值；密度 ρ 的情况肯定也是如此，只是在程度上会弱一些。内部的力学平衡条件可以表示为大气层和光球层间分界面上的压强，这个压强是中心物理量和 C 值的函数，其中 C 是由任意假定的密度分布沿半径方向积分得到的。大气压强在界面处的表达式中也包含同样的常数，在化简结果以使界面两侧的压强相等之后，其他常数都消失了，结果中只有 C 保留了下来。C 在某种意义上可以被认为是所有恒星的一个特征常数，这是在与实际情况的对比中没有料到的。因此它可能是除表面值之外的新要素，上述原理需要这些要素而不必考虑它们的内涵。

从一般意义上说上述做法已经足够完美了，如果不是因为 C 的表达式中包含恒星内部靠近分界面处的密度梯度，则在没有相反证据的前提下，只要所选的分界面沿半径方向有微小的位移，C 的数值必然会发生很大的变化。如果把光球层内的 ρ 看作 θ^3，把在分界面处数值连续变化的 P 看作 $\theta^4\phi(\theta)$，则 C^{-1} 就等于 $P^{-1}(d\rho/dr)^2$，其中第二个因子是在表面处的内部梯度。如果上述结果是正确的，那么将色球层与相隔一个突变边界面的恒星内部看成一个整体就显得不太合理，它们就像同一种物质的液态和气态一样属于不同的相。由此可见关于 C 的表达式不可能是合理的，理

the transition from Milne's formula (21) to (22) is invalid because the interior gradient of ϕ at the interface is very large and cannot be neglected even when multiplied by θ. Apparently one can only assert that the mass of the star involves the value of dP/dr within the photosphere and other quantities relating to the centre of the star, and the luminosity involves the value of P outside it, while the pressure P is continuous across a transition but not dP/dr.

In any case, perhaps not much stress would be laid on the deduction. The formula is regarded probably by its author as essentially an empirical result. When the value of C had been adapted to two prominent stars, the sun and Capella, it turned out in his hands, as he relates, to his astonishment, that it was a universal constant the same for all stars, and if so perhaps not connected with their interior constitutions at all.

(**125**, 273-274; 1930)

Joseph Larmor: Cambridge, Jan. 18.

由是从米尔恩的公式（21）推不出公式（22），因为内部梯度 ϕ 在分界面处非常大，即便是乘以 θ 以后也不能被忽略。显然我们只能说恒星的质量包含在光球层内的 dP/dr 值以及其他与恒星中心相联系的量中，发光度包含光球层外的 P 值，而穿过交界处时保持连续的是压强 P 而不是 dP/dr。

不管怎么说，也许对推导过程没有给予足够的重视。作者可能认为这个公式基本上是依赖经验结果得到的。当 C 值被应用于太阳和五车二这两颗著名的恒星时，作者发现，出乎他的意料，C 是一个对于所有恒星都相同的普适常数。如果真是这样，那么也许 C 值与恒星的内部结构根本没有关联。

（史春晖 翻译；蒋世仰 审稿）

Discovery of a Trans-Neptunian Planet

A. C. D. Crommelin

Editor's Note

Nature reports that astronomers at the Lowell Observatory in Flagstaff Arizona have for seven weeks been observing an object which appears to be a planet in orbit beyond Neptune. Its behaviour agrees fairly closely with predictions based on anomalies in the orbit of Uranus, and its size, based on one visual observation, seems intermediate between that of the Earth and Uranus. The article notes that the gravitation of this object might also account for a deviation of several days in the arrival of Halley's comet in each of its last two returns. The planet, soon named Pluto, is today no longer considered to be a proper planet, but rather a "dwarf planet"—an asteroid-like object comprised of frozen material.

ON the evening of Mar. 13 (an appropriate date, being the anniversary of the discovery of Uranus in 1781, and Mar. 14 being the birthday of the late Prof. Percival Lowell) a message was received from Prof. Harlow Shapley, director of Harvard Observatory, announcing that the astronomers at the Lowell Observatory, Flagstaff, Arizona, had been observing for seven weeks an object of the fifteenth magnitude the motion of which conformed with that of a planet outside Neptune, and agreed fairly closely with that of one of the hypothetical planets the elements of which had been inferred by the late Prof. Percival Lowell from a study of the small residuals between theory and observation in the positions of Uranus. That planet was better suited than Neptune for the study, since the latter had not been observed long enough to obtain the unperturbed elements.

Lowell's hypothetical planet had mean distance 43.0, eccentricity 0.202, longitude of perihelion 204°, mass 6.5 times that of the earth, period 282 years, longitude 84° at the date 1914–1915. Its position at the present time would be in the middle of Gemini, agreeing well with the observed place, which on Mar. 12 at 3h U. T. was 7 seconds of time west of δ Geminorum; the position of the star was R.A. 7h 15m 57.33s, north decl. 22° 6′ 52.2″, longitude 107.5°. This star is only 11′ south of the ecliptic, making it likely that the new planet has a small inclination. As regards the size of the body, the message states that it is intermediate between the earth and Uranus, implying perhaps a diameter of some 16,000 miles. A lower albedo than that of Neptune seems probable, to account for the faintness of the body. It appears from a New York telegram that at least one visual observation of the planet has been obtained, from which the estimate of size may have been deduced.

Mention should also be made of the predictions of Prof. W. H. Pickering; one of these, made in 1919 (*Harvard Annals*, vol. 61), gives the following elements; Epoch 1920; longitude 97.8°; distance 55.1, period 409 years; mean annual motion 0.880°; longitude of perihelion 280°; perihelion passage 1720, eccentricity 0.31; perihelion distance 38, mass twice earth's, present

发现海外行星

克罗姆林

编者按

据《自然》报道，天文学家在亚利桑那州弗拉格斯塔夫市的洛威尔天文台已经对一个看似是在海王星轨道外运行的行星进行了7周的观测。这颗行星的特征非常接近于人们根据天王星轨道的反常现象所作的预测，它的目测大小似乎介于地球和天王星之间。这篇文章指出，该天体的引力可能是使哈雷彗星最近两次回归地球的时间产生数天偏差的原因。这颗行星很快被命名为冥王星，现在它已不再被人们看作是一颗行星，而是一颗“矮行星”——一个由超低温物质组成的小行星。

3月13日晚（此日恰为1781年发现天王星的纪念日，而3月14日是已故天文学家珀西瓦尔·洛威尔的诞辰）从哈佛大学天文台台长哈洛·沙普利那里收到了一则消息，亚利桑那州弗拉格斯塔夫市洛威尔天文台的天文学家已经对一个星等为15等的天体进行了长达7周的观测，结果发现该天体的运动与海王星外的一颗行星同步，且与一颗假想行星的情况相当吻合。通过研究天王星的理论位置和观测位置之间的细小差别，已故的珀西瓦尔·洛威尔教授推断出了假想行星的根素。鉴于尚未对海王星观测足够长的时间来获得无扰动根素，相比之下这颗行星会更适合进行研究。

洛威尔假想行星的平均距离为43.0个天文单位，偏心率为0.202，近日点的黄经为204°，质量是地球质量的6.5倍，周期为282年，1914~1915年的黄经为84°。当前时刻该行星的位置应在双子座的中间，而在3月12日世界时3点观测到该星位于双子座δ星西部7s赤经的位置，因而观测与预言相吻合。该星的赤经是7h 15m 57.33s, 赤纬是+ 22°6′52.2″，黄经是107.5°。该星位于黄道南部11′的位置，很可能是一颗有很小倾角的行星。消息中写道，该行星的大小介于地球和天王星之间，直径约为16,000英里。该星的反照率低于海王星，因而星体本身极为暗弱。从来自纽约的电报看，目前已经至少获得过一次关于该星的目视观测，从而可能已经推测出了该星的大小。

这里需要提及的是皮克林教授的预言；他在1919年的一个预言（《哈佛年鉴》，第61卷）中给出的根素如下：历元1920年；黄经97.8°；距离55.1个天文单位，周期409年；平均周年运动0.880°；近日点的黄经280°；1720年通过近日点，偏心

annual motion 0.489°. This prediction gives the longitude for 1930 as 103°, which is within five degrees of the truth; actually it was in longitude 108° at discovery. Prof. Pickering's later prediction is further from the truth, making the longitude about 131°.

Gaillot and Lau also made predictions; like the other computers they noted that there were two positions, about 180° apart, that would satisfy the residuals almost equally well. Taking the position nearest to the discovered body, Lau gave longitude 153°, distance 75, epoch 1900. Gaillot gave longitude 108°, distance 66, epoch 1900. The latter is not very far from the truth; with a circular orbit, the longitude in 1930 resulting from Gaillot's orbit would be 128°, some 20° too great. Gaillot performed the useful work of revising Le Verrier's theory of Uranus, thus giving more trustworthy residuals. Lowell pointed out that the residuals of Uranus that led to the discovery of Neptune amounted to 133″, while those available in the present research did not exceed 4.5″; yet even in the case of Neptune the elements of the true orbit differed widely from the predicted ones, though the direction of the disturbing body was given fairly well. He noted that in the present case it would be wholly unwarrantable to expect the precision of a rifle bullet; if that of a shot-gun is obtained, the computor has done his work well.

Another method of obtaining provisional distances of unknown planets is derived from periodic comets; the mean period of the comets of Neptune's family is 71 years; it is pointed out in the article on comets ("Encyc. Brit." 14th edition, vol. 6, p. 102) that there is a group of five comets the mean period of which is 137 years; as stated there, "This family gives some ground for suspecting the existence of an extra-Neptunian planet with period about 335 years, and distance 48.2 units." This seems to be in fair accord with the new discovery, but probably the distance is nearer 45 than 48. Comets also suggest another still more remote planet, with period about 1,000 years, a suggestion which has also been made by Prof. G. Forbes and by Prof. W. H. Pickering.

The question has been asked, "Does the new planet conform to Bode's law?" It is difficult to assign a definite meaning to this question, since Bode's law broke down badly in the case of Neptune; Neptune's predicted distance was 38.8, its actual distance 30.1. For Bode's law, each new distance ought to be almost double the preceding one; the constant term of the law becomes negligible when the distance is great. For the extension of the terms given by the law we might (1) ignore Neptune as an interloper and take the next distance as double that of Uranus, giving 38.5 units; (2) we might take the next distance as four times that of Uranus, which would give 77 units; or (3) we might take the next distance as double that of Neptune or 60 units; none of these values is good, but (1) is the nearest to what we suppose to be the distance. Probably the best course is to assume that after Uranus the law changes; each new distance is then 1.5 times the preceding one; on this assumption, the hypothetical planet with distance 100 and period 1,000 years would be the next but one after the Lowell planet.

The low albedo of the new planet might be explicable if its temperature were much lower than that of Neptune. Owing to its smaller size, it would have lost more of its primitive

率 0.31；近日点距离 38 个天文单位，质量是地球的 2 倍，目前的周年运动为 0.489°。该预言认为 1930 年行星的黄经为 103°，该值与真实值相差不到 5°；事实上，发现时它的黄经为 108°。皮克林教授后来的预言与真实值相差更远，他预言的黄经达到 131°。

加约和劳也作了一些预言；和其他天文学家一样，他们认为在远离 180° 的两个位置能够很好地满足残差。选取离被发现天体最近的位置，劳给出黄经 153°，距离 75 个天文单位，历元 1900 年。加约给出黄经 108 °，距离 66 个天文单位，历元 1900 年。后者与真实值相差不远；对于圆形轨道，由加约的轨道得到 1930 年该星的黄经是 128°，比真实值大 20°。加约修正了勒威耶关于天王星的理论工作，给出了更可靠的残差。洛威尔指出导致发现海王星的天王星残差是 133″，然而在目前的研究中能够得到的残差小于 4.5″；然而即使在海王星的情况中，虽然精确地给出了扰动天体的方向，真实轨道根素也远远不同于预言结果。他还指出在目前的情况下要达到来复枪子弹那样的精度是完全不可能的；如果有更好的观测仪器，那么计算机可以将他的工作处理得更好。

另一种获得未知行星临时距离的方法来自周期彗星；海王星家族的彗星平均周期是 71 年；一篇关于彗星的文章（《不列颠百科全书》，第 14 版，第 6 卷，第 102 页）中指出一个由 5 颗彗星组成的团组的平均周期是 137 年；文章中说“这个家族提供了某种基础，可以假定存在周期大约为 335 年，距离为 48.2 个天文单位的海外行星。”这似乎与目前的新发现非常吻合，但是距离是接近 45 个天文单位而不是 48 个天文单位。福布斯和皮克林教授提出，彗星暗示了其他更远的、周期约为 1,000 年的行星的存在。

曾有人问到“新的行星满足波得定律吗？”由于海王星已经严重打破了波得定律，因而这个问题已不再有意义。海王星的预言距离是 38.8 个天文单位，实际距离是 30.1 个天文单位。根据波得定律，每一个新的天体距离都必须是前一个天体距离的 2 倍；当距离很远时可以忽略定律的常数项。对于定律的延展项，(1) 我们可以把海王星作为闯入者而忽略，下一个距离取天王星距离的 2 倍，得到 38.5 个天文单位；(2) 我们也可以把下一个距离取天王星距离的 4 倍，得到 77 个天文单位；(3) 或者我们还可以把下一个距离取成海王星的 2 倍即 60 个天文单位。这里的任何一个值都不太好，但是 (1) 的取值最接近我们估计的距离。那么在天王星之后，波得定律不再满足，这可能是最可取的方法；而每一个新的距离是前一个天体距离的 1.5 倍；基于这种假定，距离为 100 个天文单位、周期为 1,000 年的天体将是继洛威尔行星后的又一颗行星。

如果新行星的温度远低于海王星，那么就可以解释这类行星的低反照率。由于该行星很小，它将失去更多自身的原始热量而且只能吸收来自太阳能量的一半；从

heat, and would only receive half as much from the sun; hence its gases might be reduced to a liquid form, with great reduction of their volume. This would result in a relatively smaller disc than the one that might be inferred from its mass.

Some further particulars of the discovery are given by the New York correspondent of the *Times* in the issue for Mar. 15. Quoting an announcement which had been received there from the Lowell Observatory, it is stated that the planet was discovered on Jan. 21 on a plate taken with the Lawrence Lowell telescope; it has since been carefully followed, having been observed photographically by Mr. C. O. Lampland with the large Lowell reflector, and visually with the 24-inch refractor by various members of the staff. The observers estimate the distance of the planet from the sun as 45 units, which would give a period of 302 years, and mean annual motion of 1.2 degrees.

At discovery, the planet was about a week past opposition, and retrograding at the rate of about 1′ per day; this has now declined to 0.5′ per day, and the planet will be stationary in April. It should be possible to follow it until the middle of May, when the sun will interfere with observation until the autumn.

The details of the Lowell Observatory positions have not yet come to hand; when they do, it will be possible to derive sufficiently good elements to deduce ephemerides for preceding years. There are many plates that may contain images of the planet; those taken by the late Mr. Franklin Adams in his chart of the heavens, those taken of the region round Jupiter some twelve years ago for the positions of the outer satellites, and those taken at Königstuhl and elsewhere in the search for minor planets; these all show objects down to magnitude 15. If early images should be found, they will accelerate the determination of good elements of the new planet; in the case of Uranus, observations were found going back nearly a century before discovery, and in that of Neptune they went back fifty-one years. In the present case, forty years is the most that can be hoped for, and probably very few photographs showing objects of magnitude 15 are available before the beginning of this century.

One of the most difficult problems will be to find the mass of the new body; in Neptune's case, Lassell discovered the satellite a few months after the planet was found, and the mass was thus determined. It is to be feared, however, that the new planet would not have any satellite brighter than magnitude 21. Stars of this magnitude have been photographed with the 100-inch reflector at Mount Wilson, but it is doubtful whether it could be done within a few seconds of arc of a much brighter body. Failing the detection of a satellite, the mass can only be deduced from a rediscussion of the residuals of Uranus and Neptune; new tables of these planets will ultimately be called for, but that task must wait until the orbit of the new body is known fairly exactly.

The perturbations of Halley's comet will also require revision; at each of the last two returns, there has been a discordance of two or three days between the predicted and observed dates of perihelion passage; it will be interesting to see whether the introduction

而行星上的气体将会变成液态，体积也大大减小。这样它的盘面将小于由质量推算出的盘面的大小。

《泰晤士报》的纽约通讯员在 3 月 15 日的一期上刊载了发现新行星的更进一步的细节。文章引用了洛威尔天文台的一段宣告，说行星是在 1 月 21 日劳伦斯·洛威尔望远镜拍摄的一张底片中发现的；此后天文学家即对该星进行跟踪，兰普兰德先生用大洛威尔反射望远镜对该星进行了照相观测，其他的工作人员用口径 24 英寸的折射望远镜进行目视观测。观测者们推断该星与太阳的距离是 45 个天文单位，由此周期应当是 302 年，平均周年运动 1.2°。

发现之初，该行星刚通过对冲大约一星期，每天退行 1′；目前退行速度减小为每天 0.5′，并将在 4 月保持静止（留）。在 5 月中旬之前可以对该星进行跟踪观测，此后直到秋天来临之前这段时间内，太阳都会影响观测。

洛威尔天文台还未对观测结果进行详细处理；一旦进行处理，那么将会获得足够精确的轨道根素进而推导出前几年的天文历表。有很多底片可能包含这个行星的图像，如已故的富兰克林·亚当斯先生的天空星图，12 年前拍摄木星周围卫星的底片，以及那些在柯尼希斯施图尔山和其他地方拍摄的搜寻小行星的底片；这些底片都能够显示暗到 15 等的天体。如果可以找到更早期的图像将可以加速新行星轨道根素的确定；对于天王星，观测资料可以上溯到天王星发现前近一个世纪，而对于海王星则可以追溯到 51 年前。对于目前这颗行星至多追溯到 40 年前，本世纪之前可能很少有能够记录到 15 等星的照相底片。

最困难的一个问题是确定新发现天体的质量，对于海王星，在发现该行星几个月后拉塞尔就发现了它的卫星，从而确定了海王星的质量。然而，这颗新的行星的任何一个卫星的星等可能都将暗于 21 等。在威尔逊山用 100 英寸的反射式望远镜曾经拍摄过 21 星等的星，但是并不能确定使用该望远镜能否在几弧秒的范围内拍摄到这样的天体。由于不能探测到卫星，那么该行星的质量只能从对天王星和海王星的残差的再讨论中推导出来。而要进行这种再讨论推导就必须有关于新行星的新的位置数据表，而要得到这个表就必须有这个新行星的精确的轨道根素。

哈雷彗星的扰动也需要修正；在最近两次回归中，通过近日点日期的预言值与观测值之间都有两三天的差别；研究结果是否会由于引入新天体的扰动而有所改进

of the perturbations of the new body effects an improvement. The late Mr. S. A. Saunder made the suggestion at the time of the last apparition of the comet that an unknown planet might be the cause of the discordance, but it was not then possible to carry the suggestion further. The discovery of a new planet therefore opens a large field of work for mathematical astronomers. It will also appeal to students of cosmogony; Sir James Jeans, in an article in the *Observer* for Mar. 16, suggests that it may represent the extreme tip of the cigar-shaped filament thrown off from the sun by the passage of another star close to it. It would have been the first planet to cool down and solidify; he says, "As a consequence of this, it will probably prove to be unattended by satellites."

(**125**, 450-451; 1930)

将是一个有趣的课题。在彗星最近一次出现时，已故的桑德先生提出未知行星的存在将可能导致这种差异，但是并未对这个想法进行更进一步的研究。一颗新行星的发现则为数学天文学家提供了更广阔的工作空间。这也将吸引更多研究宇宙演化的学者投入其中；詹姆斯·金斯先生在 3 月 16 日《观察家报》的一篇文章中写道：这个新行星可能代表离太阳较近的另外一个恒星通过时太阳抛射出来的雪茄状纤维丝的顶端。它将是冷却凝固形成的第一颗行星；他说“作为这个现象的结果，将可能证明这颗新的行星没有卫星。”

（王宏彬 翻译；蒋世仰 审稿）

Lowell's Prediction of a Trans-Neptunian Planet

J. Jackson

Editor's Note

John Jackson of the UK Royal Observatory here discusses the possibility that the orbit of the newly discovered planet had been accurately predicted by the astronomer and polymath Percival Lowell. If true, Jackson suggests, then Lowell deserves high admiration, for the problem, while in principle similar to the prediction of Neptune, is in detail immeasurably more difficult. Jackson reviews Lowell's methods, in which, by hypothesizing a planet of particular mass and orbit, he could reduce the unexplained motion of Uranus by some 70%. However, the brightness of the observed planet is about ten times higher than predicted. Astronomers since have determined that Lowell's calculations were in error, and that Pluto was discovered principally through a painstaking empirical search of the sky.

THE reported discovery of a planet exterior to Neptune naturally arouses the interest of the general public. It will be of importance in theories concerning the genesis of the solar system as to how far it falls into line with the other planets as regards distance, mass, eccentricity and inclination of orbit, and presence or absence of satellites. Its physical appearance will be beyond observation. To those interested in dynamical astronomy, it may be of some interest to consider the data which led to its discovery and to make some comparison with the corresponding facts relating to Neptune.

If the planet which has been reported approximately follows the orbit predicted by Dr. Percival Lowell, the prediction and the discovery will demand the highest admiration which we can bestow. It is true that the problem as regards its general form is a repetition of that solved by Leverrier, Adams, and Galle more than eighty years ago; but its practical difficulty is of quite a different order of magnitude. In short, this discovery, if it turns out to be actually Lowell's predicted planet, was extremely difficult—while Neptune was in fact crying out to be found. Let us look at the actual data.

Uranus was discovered in 1781 by Herschel. Scrutiny of old records showed that it had been observed about a score of times dating back to 1690. The fact that Lemonnier observed it eight times within a month, including four consecutive days, without detecting its character, should be a lesson to anyone who makes observations without examining them. In 1820 Bouvard found that the old and the new observations could not be reconciled, and in constructing his tables boldly rejected the early observations, but the tables rapidly went from bad to worse; the residuals amounted to 20″ in 1830, 90″ in 1840, and to 120″ in 1844. Adams used in his first approximation data up to 1840, Leverrier data up to 1845. Now Uranus had passed Neptune in 1822. As the relative motion is about 2° a year, it means that for most of the time covered by the prediscovery observations the perturbations were very small, while from the fact that the difference

洛威尔对海外行星的预言

杰克逊

编者按

英国皇家天文台的约翰·杰克逊在此讨论了这颗新发现的行星的轨道已经被天文学家、博学者珀西瓦尔·洛威尔精确预测的可能性。在杰克逊看来，如果事实真的如此，那么洛威尔理应得到很高的荣誉，因为虽然计算的原则类似于海王星，但具体过程更加困难。杰克逊介绍了洛威尔的方法，洛威尔通过假设另一个具有一定质量和轨道的行星解释了天王星中 70% 的不明运动。可是，这颗行星的实际亮度比预测值高 10 倍。天文学家们后来认为洛威尔的计算存在错误，而冥王星的发现主要是通过有经验的观测者在太空中艰苦搜索得到的。

已报道的海王星外行星的发现引起了公众广泛的兴趣。该行星对于与太阳系的形成相关的理论来说将是至关重要的，诸如该星在线距离、质量、偏心率和轨道倾斜度，以及是否存在卫星等方面是否与其他行星的规律相符合的问题。该星的物理外貌是不能观测到的。如果对动力天文学感兴趣，可以研究一下发现该行星的数据并将它与海王星的相关资料进行比较。

如果报道的行星大致符合珀西瓦尔·洛威尔博士预言的轨道，那么这个预言和发现将是令人瞩目的。的确，就其一般形式而言，这个问题确实是 80 年前已经由勒威耶、亚当斯和伽勒解决了的问题的重现，但是实际研究中存在的困难在于数量级的差异。简而言之，如果这次发现的新行星的确是洛威尔预言的行星，那么这一发现是极其困难的，而海王星的发现其实是非常容易的。下面让我们看看实际的数据资料。

天王星是赫歇尔在 1781 年发现的。详细审查原始记录会发现，早在 1690 年就已经观测到了这颗星。勒莫尼耶一个月内 8 次观测天王星并且有 4 天是连续观测，但并未探测到它的特征，每个进行观测但未进行详细检测的观测者都应该引以为戒。1820 年布瓦尔发现新旧观测数据并不一致，他在建数据表的时候大胆舍弃了早期的观测资料，但是数据表的结果更糟了；1830 年残差达 20″，1840 年达 90″，到 1844 年则达到 120″。而亚当斯直到 1840 年使用的还是他的第一个近似数据，勒威耶则直到 1845 年还在用。1822 年天王星越过海王星。由于每年的相对运动只有 2°，那么在发现前观测所覆盖的大部分时间内，扰动是非常小的。然而考虑到行星

between the heliocentric distances is much smaller than expected from Bode's law, the perturbations at the time of conjunction were relatively large. Consequently the prediction of the longitude of the disturbing body was very easy, while the determination of the other elements were correspondingly difficult. The fact was that the simple hypothesis of the existence of an exterior planet with any sort of guess as to size and shape of orbit would suffice to predict the longitude. In other words, most of the residuals could be closely satisfied provided that substantially correct values of the longitude of the planet and its attractive force $m\left(\frac{1}{\Delta^2}+\frac{1}{r^2}\right)$ were used. Both Leverrier and Adams easily found values of these quantities, and Galle had no difficulty in detecting the planet.

We now turn to Lowell's "Memoir on a Trans-Neptunian Planet", published in 1915. The observational basis is the outstanding residuals in the motion of Uranus during two centuries, that is, rather more than two revolutions of that planet round the sun, of somewhat less than two revolutions relative to the predicted planet and of about one relative to Neptune. The following are the values of the observed residuals of Leverrier's and of Gaillot's theories taken from Lowell's memoir.

	Leverrier	Gaillot		Leverrier	Gaillot
1709	. .	+2.14″	1855	. .	−0.50″
1753	+5.52″	+4.45	1858	+0.50″	−0.20
1769	+4.77	+2.47	1861	. .	−0.36
1783	−3.30	−0.96	1864	+0.25	+0.18
1787	−5.12	−1.20	1867	. .	+1.20
1792	−3.50	+0.10	1870	−0.50	+1.32
1796	−1.88	−0.69	1873	. .	+0.75
1803	+0.40	−1.19	1876	−1.65	−0.50
1812	+2.00	−0.77	1879	. .	+0.58
1817	+0.50	−0.60	1882	−2.88	+0.52
1820	−0.75	−2.37	1885	. .	−0.17
1827	−2.10	+2.00	1888	−4.22	−0.85
1837	−1.10	−1.22	1891	. .	−1.11
1840	+0.63	+0.78	1894	−5.63	−0.50
1843	. .	+0.74	1897	. .	+0.35
1846	+0.38	−1.40	1900	−4.32	+1.00
1849	. .	−0.25	1903	−3.00	+0.65
1852	−1.17	−0.95	1907	. .	+0.25
			1910	. .	+1.10

The residuals show remarkable differences between the two theories, but Lowell deduced that the residuals exceeded their probable errors four or five times. The problem was to find from these residuals corrections to the elements of the orbit and to find the mass and the elements of the disturbing body. It might almost appear hopeless when we consider that the residuals must be affected by errors in the accepted masses of the known planets. There can be no doubt, however, that the masses adopted by Gaillot for Jupiter, Saturn, and Neptune are very accurate. Lowell's procedure was to adopt a value of the semimajor axis of the unknown body, and a complete series of values for its longitude, and then select the value of the longitude for which the sum of the squares of the residuals was

的日心距比由波得定律估算出的值小得多，那么在合发生时扰动将相对大一些。这样就可以较容易地确定扰动天体的黄经，然而确定该天体的其他根素就相对困难一些。实际上简单的外行星存在的假定，不管对轨道的大小和形状作何种猜测，都足以预言行星的黄经。换句话说，只要充分地利用行星黄经的正确值以及它的引力 $m\left(\frac{1}{\Delta^2}+\frac{1}{r^2}\right)$，那么大部分残差都可以得到很好的满足。勒威耶和亚当斯都很容易地找到了这些值，伽勒也很容易地探测到了行星。

下面我们看一下洛威尔 1915 年发表的《海外行星回忆录》。观测的主要内容是两个世纪中在那颗行星绕太阳旋转两圈多的过程中天王星运动中出现的残差，这种运动相对于预言中的行星不到两圈，相对于海王星大约为一圈。下面是来自洛威尔回忆录中相对于勒威耶理论和加约理论的观测残差值。

	勒威耶	加约		勒威耶	加约
1709	. .	+2.14″	1855	. .	–0.50″
1753	+5.52″	+4.45	1858	+0.50″	–0.20
1769	+4.77	+2.47	1861	. .	–0.36
1783	–3.30	–0.96	1864	+0.25	+0.18
1787	–5.12	–1.20	1867	. .	+1.20
1792	–3.50	+0.10	1870	–0.50	+1.32
1796	–1.88	–0.69	1873	. .	+0.75
1803	+0.40	–1.19	1876	–1.65	–0.50
1812	+2.00	–0.77	1879	. .	+0.58
1817	+0.50	–0.60	1882	–2.88	+0.52
1820	–0.75	–2.37	1885	. .	–0.17
1827	–2.10	+2.00	1888	–4.22	–0.85
1837	–1.10	–1.22	1891	. .	–1.11
1840	+0.63	+0.78	1894	–5.63	–0.50
1843	. .	+0.74	1897	. .	+0.35
1846	+0.38	–1.40	1900	–4.32	+1.00
1849	. .	–0.25	1903	–3.00	+0.65
1852	–1.17	–0.95	1907	. .	+0.25
			1910	. .	+1.10

这两种理论得到的残差值有着明显的差别，但是洛威尔推导发现残差与他们估测的误差相比超出 4~5 倍。现在的问题是，从这些残差的改正中找到轨道根素和扰动天体的质量及根素。当我们考虑到残差必然会受到已知行星质量误差的影响时，那么这些问题的解决就几乎是不太可能的了。毫无疑问，加约引用的木星、土星和海王星的质量都是非常精确的。洛威尔的做法是，引入未知天体的一个半长轴值以及该天体一系列完整的黄经值，然后选择残差平方和最小的黄经。用各种平均距离值（即未知行星的半长轴值）重复进行这一过程，直到变量的取值使得残差最小为

a minimum. The process was repeated with various values of the mean distance until values of the variables were found giving minimum residuals. The process was of course very laborious, but Lowell carried it through with great perseverance. The following extract from his final summary may be quoted: "By the most rigorous method, that of least squares throughout, taking the perturbative action through the first powers of the eccentricities, the outstanding squares of the residuals from 1750 to 1903 have been reduced 71 percent by the admission of an outside disturbing body."

The inclusion of further terms, of additional years and of the squares of the eccentricity, do not alter the results by any substantial amount. Lowell considered that the remaining irregularities could be explained by errors of observation. No trustworthy results could be found from the residuals in latitude so that the inclination of the orbit to the ecliptic could not be deduced, but Lowell considered that it might be of the order of 10°.

As the solution really depends on the difference of the attraction of the unknown planet on Uranus and on the sun, there are two possible solutions in which the longitudes differ by about 180°. The following elements are for the solution satisfying most nearly the position of the newly found body.

Heliocentric longitude on 1914, July	84.0°
Semimajor axis	43.0
Mass in terms of the sun's mass	1/50,000
Eccentricity	0.202
Longitude of perihelion	203.8°

This gives the longitude at the present time as about 104° compared with 107° of the new planet. The predicted magnitude was 12 to 13 or about ten times brighter than the observed; and a disc of more than 1″ was predicted. This is a rather serious discordance.

The smallness of the residuals indicated that the forces were small. The mass given above is only 0.4 of the mass of Neptune. At mean conjunction, the attraction of the predicted planet on Uranus would be only one-fifteenth of the attraction of Neptune in a similar position, and in addition it would last for a shorter time on account of the more rapid relative motion.

The discovery of a minor planet of the fifteenth magnitude is an everyday occurrence. The planet reveals itself by a decided motion relative to the stars in the course of taking a photograph. For a planet in the predicted orbit, the motion shown (mostly due to the earth's motion) would in the most favourable circumstances not be more than 2″ or 3″ an hour, and it would probably need a trail of at least 5″ for the planet to be detected. On the other hand, photographs taken on successive days would show decided motion, but the labour of finding the planet in a region containing many thousands of stars from separate photographs would be very great. Probably the Lowell observers have come across several minor planets before they were rewarded by the discovery of the very distant planet.

止。这无疑是一项非常艰苦的工作，但是洛威尔却坚持不懈地完成了。下面是从他的文章中引出的一句话，“利用最严密的方法，即最小二乘法，引入外界的一个扰动天体计算偏心率一次方的扰动行为，已经把 1750~1903 年天体明显的残差平方减少了 71%”。

进一步引入其他因素，加长年代以及使用偏心率的平方，并没有使结果有任何实质上的改变。洛威尔认为剩下的不规则性可以用观测误差来解释。从黄纬的残差中不能找到可信的结果，因而不能推导出轨道与黄道的倾角，但是洛威尔认为这个倾角可能是 10°。

由于问题的解直接取决于未知行星对天王星和太阳的引力差别，所以可能有两种使黄经相差 180° 的解。下面这些根素就是满足新发现天体条件的最近的解。

1914年7月太阳中心黄经	84.0°
半长轴	43.0
以太阳质量为单位表示的质量	1/50,000
偏心率	0.202
近日点黄经	203.8°

与这一新行星的当前黄经 107° 相比，据此根素得出的其当前黄经大约是 104°。预测星等为 12~13 等，或者说比观测到的星亮 10 倍；而且还预测到了一个大于 1″ 的盘。这里产生了很严重的矛盾。

残差小意味着力小。上面给出的质量仅是海王星质量的 0.4 倍。在平均会合期，预测行星给天王星的引力仅是同一位置海王星所施加引力的 1/15，并且由于较快的相对运动，这一状态持续的时间很短。

发现一个 15 等的小行星是件很常见的事情。在拍照过程中行星以一种相对于恒星确定的运动显露出来。对于预测轨道上的行星，一般在最好的环境下显现出的运动每小时不超过 2″ 或 3″（主要是由于地球的运动），对于可以被探测到的行星，至少需要 5″ 的踪迹。另一方面，连续几天的拍摄可以得到行星确定的运动，但是从分立的照片中在包含数千颗恒星的区域里找到这颗行星将是一件困难的工作。恐怕洛威尔天文台的观测者们在发现这颗遥远行星之前已经偶遇了很多小行星。

Astronomers all the world over will naturally look forward with great interest to see how nearly the newly discovered body moves in the orbit predicted by Lowell, and are anxiously waiting for further details of the observations.

(**125**, 451-452; 1930)

全世界的天文学家对新发现的这个天体的运动离洛威尔预测的轨迹有多近都非常感兴趣，而且充满好奇地期待着关于这一观测结果更详细的进展。

（王宏彬 翻译；蒋世仰 审稿）

Age of the Earth

A. P. Coleman

Editor's Note

This letter from Canadian geologist Arthur Coleman at the Royal Ontario Museum in Toronto argued that the Earth must be much older than then commonly believed. His starting point was his reading of a popular book by Sir James Jeans, a British astronomer and physicist. The modern estimate of the age of the Earth is about 4.6 billion years. Estimates of when life first appeared on the Earth are only a little smaller—perhaps 4.2 billion years.

I have just finished reading a most interesting book on "The Universe around Us", by Sir James Jeans. It opens up a complex and abstruse subject with admirable clearness, so that even a geologist possessed of very little mathematics can find his way through it without too much difficulty. The ease with which in this brilliant book millions of millions of stars are marshalled and their history outlined for millions of millions of years inspires no little awe and a large amount of envy in the breast of a plodding geologist who keeps to the solid earth. If the book contained only the inspiring visions of an astronomer in regard to the origin and the fate of the universe around us a geologist might refrain from comment; but at several points the history of the earth and its inhabitants is touched upon, giving him a right to a word of criticism.

Sir James in a page or two suggests that the earth began about 2,000,000,000 years ago as a globe of intensely hot gas, which gradually cooled down, becoming first a liquid, then plastic, and finally an outer crust solidified, "rocks and mountains forming a permanent record of the irregularities of its earlier plastic form." Life probably began on the earth about 300,000,000 years before the present.

A generation ago, when Lord Kelvin laid down the law as to geological time, this allowance would have seemed very liberal; but the discovery that the age of certain rocks can be determined by an analysis of the radioactive minerals they contain has completely changed our point of view. 2,000,000,000 years is decidedly too short a time for Pre-Cambrian history if the earth began in a gaseous form; and life existed far earlier than 300,000,000 years ago.

These points are easily proved by a brief study of the Grenville Series of Ontario and Quebec and of the Laurentian granite and gneiss which have erupted through it, since radioactive minerals are found in pegmatite dikes connected with the granite.

地球的年龄

科尔曼

编者按

这是一篇来自多伦多皇家安大略博物馆的加拿大地质学家科尔曼的快报文章，文中他认为地球的年龄远比大家公认的古老得多。他的出发点得益于英国天文学家、物理学家詹姆斯·金斯爵士的名著。目前估计的地球年龄为46亿年，而地球上首次出现生命的时间稍晚，可能为42亿年。

我刚读完一本十分有趣的书《我们周围的宇宙》，作者是詹姆斯·金斯爵士。这本书以清晰得令人惊叹的文字讲述了一个复杂而深奥的主题，即使是一位几乎没有数学基础的地质学家要读懂它也不会感到很困难。这部精妙绝伦的著作轻松自如地列举了数以百万计的恒星以及它们在数百万年中的演化史，让局限于固体地球的古板的地质学家感到既敬畏又十分羡慕。如果这本书只是一位天文学家对我们周围的宇宙的起源和结局的美好展望，那地质学家也就用不着说三道四了；但作者在某些方面又提到了地球的历史以及它的居民，所以地质学家还是有权提出批评意见的。

詹姆斯爵士用了一两页的篇幅说明地球诞生在约2,000,000,000年以前，开始是一个炽热的气团，随后逐渐冷却，先变成液体，然后是塑性体，最后外壳硬化，“岩石和山脉是地球早期塑性形态不规则性的永久记录。”地球上的生命可能始于距今约300,000,000年前。

30年前，当开尔文勋爵建立地质年代的理论时，这种定量看似非常自由，但是在发现某些岩石的年龄可以用其所含的放射性矿物测定后，我们的观点彻底改变了。如果地球是从气态开始演变的，那么地球上生命存在的时间将远远超过300,000,000年，而寒武纪之前的地球仅有2,000,000,000年的历史显然是太短了。

针对安大略和魁北克的格伦维尔岩系以及侵入其中的劳伦琴花岗岩和片麻岩的初步研究的结果可以很容易地证明以上观点，因为在混有花岗岩的伟晶岩脉处发现了放射性的矿物。

Ages of Pegmatite in Ontario and Quebec

Localities	Determined by T. L. Walker[1]	Determined by H. V. Ellsworth[2]
Parry Sound	1,274,000,000 years	1,179,000,000 years
Villeneuve	1,293,000,000 „	1,189,000,000 „
Cardiff Tp.	1,239,000,000 „	1,299,000,000 „
Butt Tp.		1,130,000,000 „

The localities are scattered over 200 miles from west to east, and it will be observed that the age of the pegmatites does not differ greatly at the various points and that the two analysts are not far apart in their determinations. We may conclude that the uraninites of the pegmatites were formed about 1,230,000,000 years ago.

The pegmatite dikes are the latest phases of the molten granite which heaved up the rocks of the Grenvile Series into domes forming important mountain chains which covered many thousand square miles, and the Grenville Series must have been solid rock long before this took place.

The Grenville rocks are now crystalline limestones, gneisses, and quartzites, but were originally ordinary limestones, shales, and sandstones, which were deposited as limy material, mud, and sand in a shallow sea.

The series is very thick according to Dr. Adams, who measured 17,824 feet in one section and 94,406 feet in another; and to the age shown by the radioactive minerals must be added many millions of years for the deposit of a great geological formation, its consolidation, and the thrusting up of the widespread Laurentian mountains.

But we are still far from the beginning. Before the Grenville sediments could be laid down the earth's crust must have been firm and solid to form shores and sea bottoms of a permanent kind, and must have been cool enough to allow rain to fall and rivers to bring down mud and sand into the sea. If the earth ever passed through a stage of heat and plasticity, that was completely over before the beginning of the Grenville sea; and the water was cool enough to allow algae or other lowly plants to thrive, since in places the sediments contain several percent of carbon, now in the form of graphite.

Life had already appeared in the sea.

The Grenville rocks have been chosen for this study because they are well known and are dated by analyses of radioactive minerals, but they are probably somewhat surpassed in age by the Keewatin, which is mainly volcanic, but with important amounts of sediments, and the Coutchiching, which is wholly sedimentary. No uraninites have yet been found in the granites which penetrate them.

安大略和魁北克伟晶岩的年龄

采样点	沃克的测定结果 [1]	埃尔斯沃思的测定结果 [2]
帕里湾	1,274,000,000 年	1,179,000,000 年
维伦纽夫	1,293,000,000 年	1,189,000,000 年
加的夫区	1,239,000,000 年	1,299,000,000 年
巴特区		1,130,000,000 年

样品采集区域分布在从西到东的 200 多英里范围内。我们可以看到不同地点的伟晶岩年龄没有太大差别，两个分析者的测定结果也相近，可以认为伟晶岩中的沥青铀矿形成于约 1,230,000,000 年前。

伟晶岩脉是熔融花岗岩结晶的最后阶段，熔融花岗岩岩浆拱起格伦维尔岩系的岩石，使其变形成穹隆状，从而形成了绵延数千平方英里的大山系，而格伦维尔岩系肯定在该过程发生之前早已是固化的岩石了。

格伦维尔岩系的岩石目前是结晶质石灰岩、片麻岩和石英岩，而原来曾是普通的石灰岩、页岩和砂岩，由浅海中含石灰的物质和泥沙沉积而成。

根据亚当斯博士的测定结果，该岩系很厚，在一个剖面的厚度为 17,824 英尺，另一个为 94,406 英尺。由于一个巨型地质层组的累积和固化过程以及广阔的劳伦琴山脉的推进需要很长时间，因此其地质年代比根据放射性矿物测定的年龄早数百万年。

然而，我们离出发点还很远。在格伦维尔沉积物形成之前，地壳必须足够坚硬致密，以形成永久性的海岸和海底。同时，地球一定已经足够冷，使雨水能够降落，河流能够把泥沙带到海里。如果地球曾经经历了炽热和塑性阶段，那么该阶段一定在格伦维尔海形成之前已经彻底结束；海水也冷得可以使海藻和其他低等植物旺盛生长，致使一些地方的沉积物中含有百分之几的碳，现在以石墨形式存在。

生命已经在海洋中出现。

本次研究之所以选择格伦维尔岩系，一是由于它的知名度，二是因为可以通过分析放射性矿物测量它的年代。但是，格伦维尔岩系可能稍晚于以火山岩为主但带有大量沉积物的基瓦丁岩系和完全由沉积岩组成的库奇金岩系。在侵入其中的花岗岩中至今没有发现沥青铀矿。

Taking all of these formations together, we have a known area of 1,000,000 square miles of cold and rigid rocks with well-established lands and seas in North America 1,300,000,000 or 1,500,000,000 years ago, with no suggestion of physical conditions fundamentally different from the present.

The oldest rocks in Brazil, South Africa, Australia, India, Scandinavia, and Scotland, judging from what I have seen of them, include similar sediments to those of the Grenville and Keewatin in Canada, though not on so broad a scale. In Holmes's interesting discussion of the "Age of the Earth", the Lower Pre-Cambrian of West Australia is stated to be 1,260,000,000 years old, which fits well with the age of the Grenville.

There were, then, solid land surfaces not too warm for lakes or seas to exist in all the continents in the earliest times known to the geologist; and there is, in fact, no geological evidence that the world ever was molten. If our globe passed through intensely hot gaseous, liquid, and plastic stages, the cooling had run its course completely many millions of years before the pegmatite veins penetrated the Grenville sediments; and the cold continents had undergone at least one major mountain-building revolution at an earlier time than 1,230,000,000 years ago.

Since then the earth has not been cooling down, but has kept its surface temperature surprisingly uniform, though with minor variations, including several ice ages. The carbon and limestone in the earliest rocks suggest lowly plant life in the waters from the very beginning of known geological history, and the Pre-Cambrian geologist is inclined to be a uniformitarian, and to ask the astronomer if the first quarter of the world's history was really so wild and turbulent as he describes it, when the later three-quarters were so temperate and uniform.

May not the earth have been built up of cold particles such as now reach us by the million from space, and may it not have escaped entirely the white hot stage of the nebular theory? Is it not possible that the hot gases cooled rapidly into innumerable solid particles which later came together to make the earth? The tiny scattered asteroids and meteorites suggest some such process; and this would provide the cold earth which the Pre-Cambrian geologist requires.

If the astronomers cannot provide a cold process of world building, they must allow the geologist a much longer time than 2,000,000,000 years to condense the hot cloud of gas into a solid world with continents and basins cool enough for the Grenville sea with its algae.

(**125**, 668-669; 1930)

A. P. Coleman: Royal Ontario Museum, Toronto.

References:

1. Ages of Some Canadian Pegmatites, *Contribs. to Can. Mineralogy*, Univ., Toronto, 1924.

2. Radioactive Minerals as Geological Age Indicators, *American Journal of Science*, No. 50, p. 127, etc.

如果综合考虑所有这些地层，那么在北美我们就拥有一块面积达 1,000,000 平方英里的区域，该区域由冷而坚硬的岩石构成，根据这些岩石，在 1,300,000,000 或 1,500,000,000 年前已确定存在的海陆，其物质环境与现在没有根本的不同。

根据我的判断，巴西、南非、澳大利亚、印度、斯堪的纳维亚和苏格兰最古老的岩石中含有和加拿大格伦维尔岩系和基瓦丁岩系相似的沉积物，尽管其规模稍小一些。霍姆斯在关于《地球的年龄》的有趣讨论中指出，西澳大利亚的前寒武纪在距今约 1,260,000,000 年前，这和格伦维尔岩系的年龄十分吻合。

在地质学家了解的最早地质历史时期，各个大陆的固态陆地表面不是太热，可以维持湖泊或海洋的存在。而且，事实上没有地质上的证据可以证明地球曾经处于熔融状态。如果我们的星球曾经经历过炽热气体、液体和塑性阶段，那么冷却过程的彻底完成要比伟晶岩脉侵入格伦维尔沉积岩早至少几百万年，而且冷却后的大陆至少要在 1,230,000,000 年前经历一次较大的造山运动。

从那以后，地球没有继续变冷，相反它的表面温度均衡得令人惊讶，尽管有包括几个冰川期在内的一些小的起伏。最早期的岩石中含有碳和石灰石说明生活在水中的低等植物出现于普遍认可的地质历史的开始阶段。同时，对于前寒武纪，地质学家倾向于均变论，他们会问天文学家：如果地球历史的前 1/4 真的如他们描述的那样荒凉和动荡，那为什么后 3/4 的历史如此平静和均匀？

难道地球不可以由与如今从宇宙大批飞向我们的冷物质类似的物质组成吗？难道它不可能完全避免星云历史上的白热阶段吗？难道炽热的气体不可能迅速冷却成无数构成地球的固体颗粒吗？那些散布于宇宙中的小行星和陨石表明确实存在这样的过程，这将支持地质学家认为寒武纪之前地球是冷的这个观点。

如果天文学家不能提供地球形成的冷却过程，他们必须允许地质学家认为至少需要 2,000,000,000 年时间才能使气团浓缩成固体世界，才能使大陆和盆地足够冷以便海藻可以生活在格伦维尔海中。

（金成伟 翻译；张忠杰 审稿）

Artificial Disintegration by α-Particles

J. Chadwick and G. Gamow

Editor's Note

In 1919 Ernest Rutherford had shown that alpha particles fired into nitrogen gas could create hydrogen, apparently because the particles kick a proton out of the nucleus while being themselves captured. Rutherford called the process "artificial disintegration" (popularly, splitting) of the atom. Here James Chadwick and George Gamow suggest that protons might also be produced without the alpha particle entering the nucleus. Gamow went on to propose that protons, with only half the charge of alpha particles, might approach the positively charged nucleus more easily. That idea stimulated work at Cambridge on a high-voltage device to accelerate protons to high energies, leading to the first induced splitting of a lithium nucleus by proton bombardment by John Cockroft and Ernest Walton.

IT is commonly assumed that the process of artificial disintegration of an atomic nucleus by collision of an α-particle is due to the penetration of the α-particle into the nuclear system; the α-particle is captured and a proton is emitted.

On general grounds it seems possible that another process may also occur, the ejection of a proton without the capture of the α-particle.

Consider a nucleus with a potential field of the type shown in Fig. 1, where the potential barrier for the α-particle is given by the full line and that for the proton by the dotted line. Let the stable level on which the proton exists in the nucleus be $-E_p^\circ$ and the level on which the α-particle remains after capture be $-E_\alpha^\circ$.

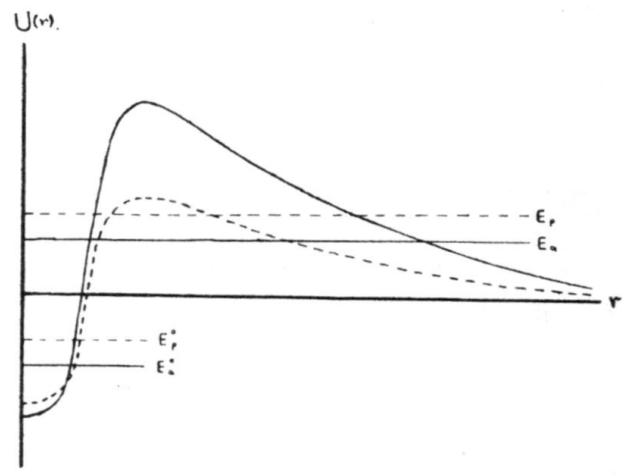

Fig. 1

α 粒子引发的人工衰变

查德威克，伽莫夫

编者按

1919 年，欧内斯特·卢瑟福发现 α 粒子射入氮气后可以产生氢，这显然是因为 α 粒子被俘获的同时把一个质子赶出了原子核。卢瑟福把这个过程称作原子的“人工衰变”（更通俗地说是裂变）。詹姆斯·查德威克和乔治·伽莫夫在本文中提出，在没有 α 粒子进入原子核的情况下我们也可以得到质子。伽莫夫还指出，质子所带电荷数只有 α 粒子的一半，它有可能更容易接近带正电的原子核。这种观点推动了剑桥大学的约翰·考克饶夫和欧内斯特·瓦耳顿的工作，他们用高压装置把质子加速到很高的能量，然后通过质子轰击首次实现了锂核的诱导裂变。

通常的假设是，一个 α 粒子与原子核发生碰撞的人工衰变过程，是由 α 粒子穿透到核系统内部导致的；在 α 粒子被俘获的同时，也会发射出一个质子。

按照通常的观点，另外一种过程也可能发生，即在没有俘获 α 粒子的情况下，释放出一个质子。

研究一个处于图 1 所示的势场条件下的核，其中实线代表 α 粒子的势垒，虚线代表质子的势垒。令原子核中质子的稳态能级为 $-E_p^o$，被俘获后的 α 粒子的能级为 $-E_\alpha^o$。

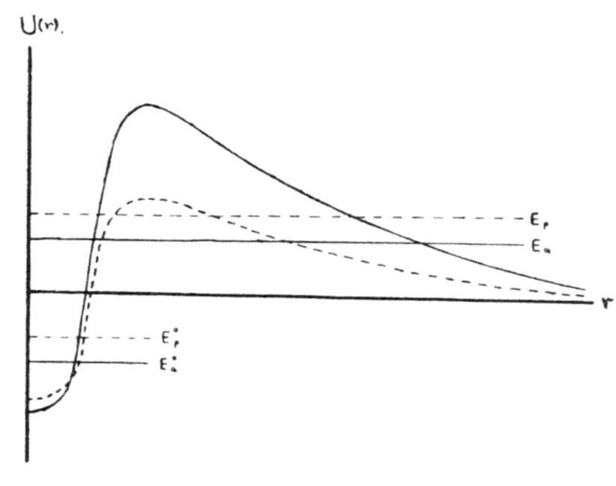

图 1

If an α-particle of kinetic energy E_α penetrates into this nucleus and is captured, the energy of the proton emitted in the disintegration will be $E_p = E_\alpha + E_\alpha^\circ - E_p^\circ$, neglecting the small kinetic energy of the recoiling nucleus. If the nucleus disintegrates without capture of the α-particle, the initial kinetic energy of the α-particle will be distributed between the emitted proton and the escaping α-particle (again neglecting the recoiling nucleus). The disintegration protons may have in this case any energy between $E_p = 0$ and $E_p = E_\alpha - E_p^\circ$.

Thus, if both these processes occur, the disintegration protons will consist of two groups: a continuous spectrum with a maximum energy less than that of the incident α-particles and a line spectrum with an energy greater or less than that of the original α-particles according as $E_\alpha^\circ > E_p^\circ$ or $E_\alpha^\circ < E_p^\circ$, but in either case considerably greater than the upper limit of the continuous spectrum (see Fig. 2).

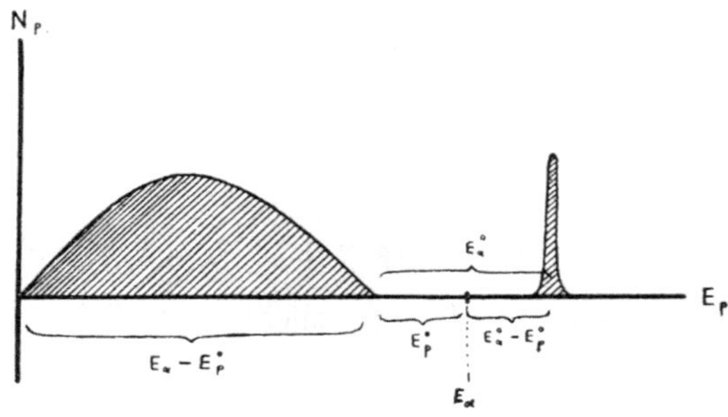

Fig. 2

In some experiments of one of us in collaboration with J. Constable and E. C. Pollard, the presence of these two groups of protons appears quite definitely in certain cases, for example, boron and aluminium. A full discussion of these and other cases of disintegration will be given elsewhere, but it may be noted that the existence of groups of protons has already been reported by Bothe and by Pose. In general the experimental results suggest that with incident α-particles of energy about 5×10^6 volts (α-particles of polonium) the process of non-capture is several times more frequent than the process of capture.

It is clear that, if our hypothesis is correct, accurate measurement of the upper limit of the continuous spectrum and of the line will allow us to estimate the values of the energy levels of the proton and α-particle in the nucleus. In the case of aluminium bombarded by the α-particles of polonium the protons in the continuous spectrum have a maximum range of 32 cm and those of the line spectrum a range of 64 cm. These measurements give the following approximate values for the energy levels:

$$E_p^\circ = 0.6 \times 10^6 \; e \text{ volts, and } E_\alpha^\circ = 2 \times 10^6 \; e \text{ volts.}$$

On the wave mechanics the probability of disintegration of both types is given by the square of the integral

$$W = \int f(r_{\alpha,p}) \cdot \psi_\alpha \cdot \psi_p \cdot \phi_\alpha \cdot \phi_p \cdot dV \cdot dV' \tag{1}$$

如果动能为 E_α 的 α 粒子穿透到这个核的内部并被俘获，那么衰变中发射出的质子的能量是 $E_p = E_\alpha + E_\alpha^o - E_p^o$，这里忽略了此过程中反冲核的微小动能。如果在核衰变的过程中没有俘获 α 粒子，α 粒子的初始动能将在发射出的质子与逃逸的 α 粒子之间分配（再次忽略反冲核）。在这种情况下，衰变质子的能量可能是 $E_p = 0$ 与 $E_p = E_\alpha - E_p^o$ 之间的任意值。

因此，如果这两种过程都存在，那么衰变的质子将由两部分组成：最大能量小于入射 α 粒子能量的连续谱和能量大于或小于初始 α 粒子能量的线状谱，相当于 $E_\alpha^o > E_p^o$ 或 $E_\alpha^o < E_p^o$，但对于这两种情况中的任何一种，能量都比连续谱的上限大很多（见图 2）。

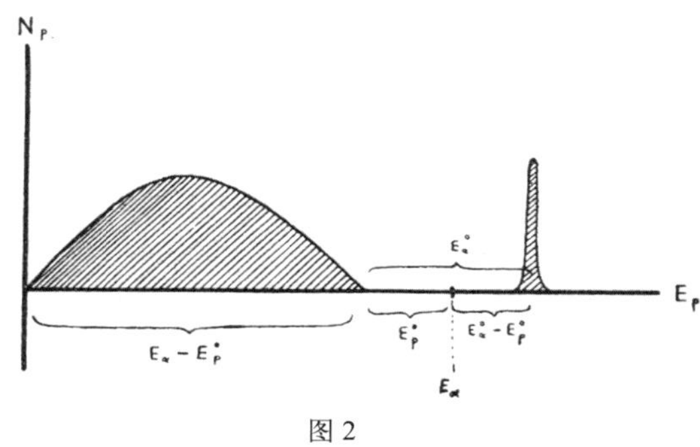

图 2

在我们中的一位与康斯特布尔和波拉德合作进行的一些实验中，在某些情况下这两组质子的存在会非常明确地表现出来，比如，硼和铝的情况。关于这些以及其他一些衰变的情况将会在别的地方给出全面的讨论，但可以注意到的是，博特和波泽已经报道了多组质子的存在。一般而言，实验结果显示，当入射 α 粒子（钋的 α 粒子）的能量约为 5×10^6 电子伏时，非俘获过程发生的频率比俘获过程高出几倍。

显然，如果我们的假设是正确的，那么对连续谱和线状谱上限的精确测量将使我们能对原子核中的质子和 α 粒子的能级数值进行估计。在用钋的 α 粒子轰击铝的情况下，在连续谱中质子的最大范围为 32 厘米，而在那些线状谱中的范围为 64 厘米。由这些测量得到的能级近似值如下：

$$E_p^o = 0.6 \times 10^6 \text{ 电子伏}，E_\alpha^o = 2 \times 10^6 \text{ 电子伏}。$$

根据波动力学，两种类型的衰变几率由积分的平方给出

$$W = \int f(r_{\alpha,p}) \cdot \psi_\alpha \cdot \psi_p \cdot \phi_\alpha \cdot \phi_p \cdot dV \cdot dV' \tag{1}$$

where $f(r_{\alpha,p})$ is the potential energy of an α-particle and a proton at the distance $r_{\alpha,p}$ apart, and the wave functions ψ_α, ψ_p represent the solutions for the α-particle and proton before and ϕ_α, ϕ_p after the disintegration. In calculating the integral (1) we must develop the incident plane wave of the α-particle into spherical harmonics corresponding to different azimuthal quantum numbers of the α-particle, and deal with each term separately.

In the case of capture of the α-particle the estimation of (1) can be carried out quite simply. It can be shown that the effect of the higher harmonics is very small, and that the disintegration is due almost entirely to the direct collisions. Thus we obtain for the probability of disintegration

$$W_1^2 = \frac{A}{v_\alpha^2} \cdot e^{-\frac{8\pi^2 e^2}{h} \cdot \frac{Z}{v_\alpha}} \cdot e^{-\frac{4\pi^2 e^2}{h} \cdot \frac{Z}{v_p}} \qquad (2)$$

where v_α and v_p are the velocities of the initial α-particle and the ejected proton respectively. Since only the first harmonic is important in disintegration of this type, it is to be expected that the protons will be distributed nearly uniformly in all directions.

When the α-particle is not captured the disintegrations will arise mainly from collisions in which the α-particle does not penetrate into the nucleus. For disintegration produced in this way the higher harmonics become of importance. The probability of disintegration can be roughly represented by the formula

$$W_2^2 = B \cdot e^{-\frac{8\pi^2 e^2}{h} \cdot Z\left(\frac{1}{v'_\alpha} - \frac{1}{v_\alpha}\right)} \cdot e^{-\frac{4\pi^2 e^2}{h} \cdot \frac{Z}{v_p}} \qquad (3)$$

where v'_α is the velocity of the α-particle after the collision, and B is a function of the angle of ejection of the proton. The protons of the continuous spectrum will not be emitted uniformly in all directions. According to the expression (3) the distribution with energy of the protons in the continuous spectrum will have a maximum value for an energy of ejection of about 0.3 of the upper limit, and will vanish for zero energy and at the upper limit.

More detailed accounts of the experimental results and of the theoretical calculations will be given shortly.

(**126**, 54-55; 1930)

J. Chadwick, G. Gamow: Cavendish Laboratory, Cambridge, June 18.

式中 $f(r_{\alpha,p})$ 是相距为 $r_{\alpha,p}$ 的一个 α 粒子和一个质子之间的势能，波函数 ψ_α 和 ψ_p 表示 α 粒子和质子在衰变前的解，ϕ_α 和 ϕ_p 表示它们在衰变后的解。在计算积分 (1) 时，我们必须将 α 粒子的入射平面波展开为对应于 α 粒子不同角量子数的球谐函数的形式，并对每一项分别处理。

在俘获 α 粒子的情况下，对 (1) 式进行估算非常简单。可以看到的是，高次谐波的效应很小，衰变几乎全部由直接碰撞产生。因此我们可以得到衰变几率

$$W_1^2 = \frac{A}{v_\alpha^2} \cdot e^{-\frac{8\pi^2 e^2}{h} \cdot \frac{Z}{v_\alpha}} \cdot e^{-\frac{4\pi^2 e^2}{h} \cdot \frac{Z}{v_p}} \tag{2}$$

式中 v_α 和 v_p 分别是初始 α 粒子和发射出的质子的速度。因为在这类衰变中，只有一次谐波起主要作用，所以可以认为质子在所有方向上都是均匀分布的。

当 α 粒子未被俘获时，衰变将主要通过 α 粒子并不穿透到核内的碰撞产生。在由这种方式产生的衰变过程中，高次谐波开始起主要作用。衰变几率可以通过下述公式近似地表示

$$W_2^2 = B \cdot e^{-\frac{8\pi^2 e^2}{h} \cdot Z\left(\frac{1}{v'_\alpha} - \frac{1}{v_\alpha}\right)} \cdot e^{-\frac{4\pi^2 e^2}{h} \cdot \frac{Z}{v_p}} \tag{3}$$

式中 v'_α 是碰撞后 α 粒子的速度，B 是质子发射角度的函数。连续谱中的质子在各个方向上的发射并不均匀。根据表达式 (3)，连续谱中质子的能量分布，在约为上限的 0.3 处其发射能量达到最大值，而在上限处突然变为零。

不久之后，我们将给出关于实验结果和理论计算的更详细的解释。

（沈乃澂 翻译；江丕栋 审稿）

A New Theory of Magnetic Storms

S. Chapman and V. C. A. Ferraro

Editor's Note

In the 1920s and 1930s, radio communications became an important public issue with the creation of broadcasting systems and the use of radio waves for telephones. These systems were sometimes disrupted by an unknown process apparently linked to changes in the Earth's magnetism, called magnetic storms. Here Sydney Chapman and Vincente Ferraro at Imperial College, both later leading figures of "space physics", provide the basis of the explanation for magnetic storms. Such events, they say, are caused by interactions of the geomagnetic field with charged particles streaming from the Sun (the solar wind).

AN attempt to infer the course of events when a neutral ionised stream of particles from the sun is directed towards the earth has now led to results which we believe indicate how magnetic storms are produced. A full discussion of the phenomena involves the solution of numerous intricate mathematical problems, many of which have not yet been attacked in detail; but it seems possible to outline the main sequence of events.

The motion of a neutral ionised stream in the earth's magnetic field was investigated by one of us in 1923[1], and it was concluded that the stream would be scarcely deflected by the field, though some slight convergence would occur within about one earth-diameter from the earth's centre *O* (Fig. 1).

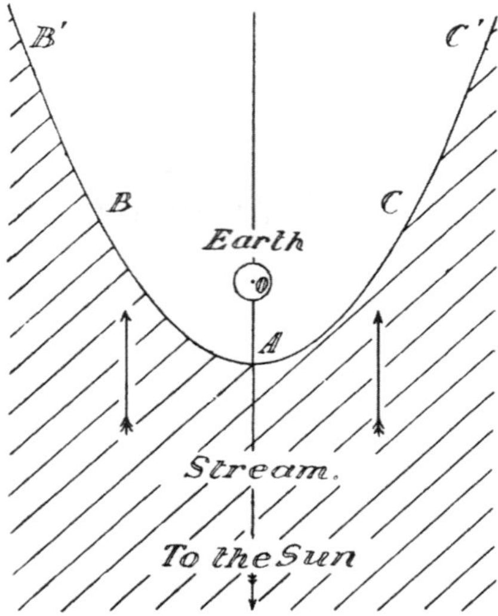

Fig. 1

一项关于磁暴的新理论

查普曼，费拉罗

编者按

20 世纪二三十年代，随着广播系统的发明和无线电波在电话上的应用，无线电通信成了一个重要的公共议题。这些无线电通信系统有时会被一种与地球磁场的变化有关的未知过程中断，这种未知过程叫做磁暴。在这篇文章中，帝国学院的辛迪·查普曼和文森特·费拉罗提出了磁暴现象的基本原理，他们两人后来都成了“空间物理学”领域的领军人物。他们认为，磁暴事件是由来自太阳的带电粒子流（太阳风）与地球磁场相互作用引起的。

我们尝试推断来自太阳的呈电中性的电离粒子流直接吹向地球时各事件的过程，如今有了结果，我们相信这一结果可以阐释磁暴是怎样产生的。完整地讨论这一过程需要求解大量复杂的数学问题，而且其中的许多问题尚未得到详细的研究。不过，概括出这些事件的大致顺序似乎还是可以做到的。

我们中的一位在 1923 年对呈电中性的电离粒子流在地球磁场中的运动作了研究[1]，结果表明，虽然在距地心 O 约一个地球直径的范围内会发生一些轻微的会聚，但是这种流几乎不会因地磁作用而偏转（如图 1）。

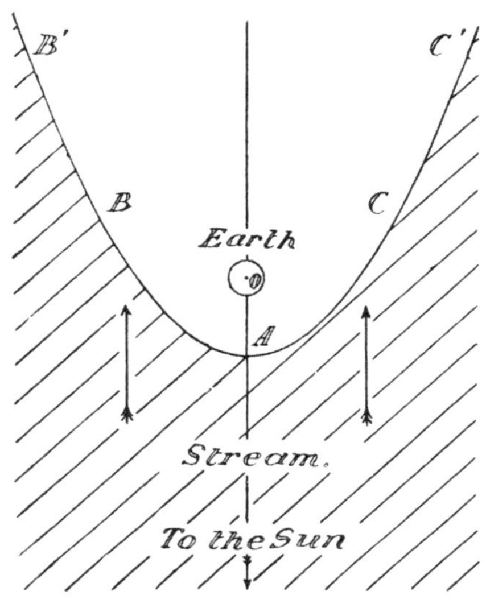

图 1

No indication as to how such a stream could produce magnetic storms and aurorae was obtained. It would seem that this failure was due to the assumption there made that the stream had enveloped the earth for a time long enough to enable a steady state to be set up, whereas it now appears that magnetic storms are essentially connected with the *approach* of the stream towards the earth. The important changes in the stream occur within a few earth radii from *O*, and beyond this distance the former conclusion that the stream travels almost without deflection remains valid.

The stream is in effect a highly conducting body, and as it enters the earth's field electric currents flowing parallel to its surface are induced in the surface layers, so that the interior of the stream is nearly shielded from the earth's field. Outside the stream the magnetic effect of the currents is roughly equivalent to that of an "image" magnetic doublet at a certain point inside the stream; in the equatorial plane the field between the earth and the stream is increased in intensity. It is as if the current layer, as it advances towards the earth, pushes forward and crowds together the earth's lines of force.

We identify this change with the observed increase in the earth's horizontal force during the first stage of a magnetic storm. Detailed examination shows that the magnitude of the effect, and its time scale, depend almost entirely on the kinetic energy of the stream per unit volume; if the velocity of the stream is of the order 1,000 km/s, the density requisite to explain the first phase of an average magnetic storm (taking account of the shielding effect of the Heaviside layer) is roughly of the order 10^{-22} gm/cc; this might be provided by about 1.5 calcium ions or 60 hydrogen ions per cc.

The magnetic energy of the field is increased during this phase at the expense of the kinetic energy of the stream; the retardation of the particles occurs in the current layer, which is continually increased in mass-density by the oncoming of particles from behind. The retardation is greatest at that part of the front of the stream (*A* in Fig. 1) which is moving along the direct line from the sun to the earth's centre *O*; on either side of *A* the stream will advance relatively to *A*, and the earth will become partly enclosed by the stream; the surface *B'BACC'* will continually close in, at a diminishing rate, upon the earth; whether it actually reaches the earth's atmosphere, in the equatorial plane, will depend on the density of the stream, and the length of time during which it is directed towards the earth (this is determined by the angular breadth of the stream viewed from the sun).

In the second (which is the main) phase of a magnetic storm the earth's horizontal force is decreased. We attribute this to the formation of a westerly current round the earth, due primarily to the flow of charges across the space "behind" the earth (viewed from the sun). Along the sides *BB'*, *CC'* of the enclosure there will be charged layers, *BB'* positive and *CC'* negative, due to the polarisation of the stream by the magnetic field. The charges in these layers will be subject to an outward electrostatic force, and the positive ions along *BB'* will cross over to *CC'*, partly guided by the earth's field. The electrons along *CC'* cannot flow along the reverse path because of the greater deflecting influence of the field upon them,

现在尚无法说明这样的流是如何形成磁暴和极光的。之所以无法说明，似乎是因为一项假设，即，认为粒子流包围地球的时间足够长，以至于建立起了一个平衡态。然而，在现在看来，磁暴在本质上与粒子流向地球**传输的过程**有关。在距离地心 O 几个地球半径的范围内，粒子流发生了重大变化。而在这个范围以外，前面的结论仍旧有效，即粒子流在运动过程中几乎没有发生偏转。

实际上，这种粒子流是一个良导体，当它进入地球磁场时，在其表面层会感应出平行于表面流动的电流，以至于粒子流内部几乎被地球磁场屏蔽。在粒子流外部，电流的磁效应则与粒子流内部某点的“镜像”磁偶极子的磁场大致相等；所以在赤道面上，地球与粒子流之间的磁场强度变大。当电流层向地球方向前进时，看起来就像是电流层推动地球磁力线向前并使之挤压在一起。

我们发现这一变化与磁暴初相时观测到的地球磁场水平分量的增加有关。进一步的研究表明，这种作用的数量级和持续时间几乎完全取决于单位体积粒子流的动能。假设粒子流速度的数量级是 1,000 km/s，那么一次中等强度磁暴的初相要求的粒子流密度应该大致在 10^{-22} gm/cc 这个数量级（考虑到亥维赛层的屏蔽效应），这大概相当于每立方厘米中有 1.5 个钙离子或 60 个氢离子。

在这一阶段，随着粒子流动能的消耗，磁场的磁能增大，电流层中的粒子出现阻滞现象，而其后相继而来的粒子使其质量密度不断增大。在沿太阳到地心 O 之间的直线上运动的粒子流的前端（图 1 中 A 点），阻滞现象最明显；A 两侧的粒子向 A 靠拢，进而把地球部分地包围起来。$B'BACC'$ 面将以持续减小的速度不断靠近地球；该面能否到达赤道面上空的大气层，取决于粒子流的密度和朝地球运动的时间（时间长短取决于粒子流以太阳为观测点的角宽度）。

在磁暴的第二阶段（即磁暴主相），地球磁场的水平分量开始减小。我们将这一现象归因于围绕地球的西向电流的形成，这种电流主要是由从地球“后面”（以太阳为观测点）穿过太空的电荷的流动引起的。由于磁场对带电粒子流的极化作用，沿封闭圈的 BB'、CC' 边将会形成电荷层，BB'边为正，CC'边为负。这些层上的电荷会受到一个向外的静电力。一定程度上由于地球磁场的作用，BB'边的正电离子将横渡到 CC'边。CC'边的电子不能向相反方向运动，因为地球磁场作用于其上的偏转力更

but negative charges from "above" and "below" the equatorial plane will travel along the earth's lines of force to neutralise the charge of the ions moving from BB' across the gap. The details of the process are not yet clear, but it appears likely that a westerly current can thus be set up round the earth. It can be shown that the current-ring, if formed, can persist in mechanical and electromagnetic equilibrium for some days after the cessation of the onward flow of particles from behind. The gradual dissipation of this ring current corresponds to the final phase of the storm.

One of the distinctive features of the theory here outlined is the distance from the earth within which the main electric currents flow, namely, a few earth radii; they are outside the earth's atmosphere (though secondary currents are induced therein), but they are much nearer the earth than the currents (in the equatorial plane) discussed by Birkeland, or the equatorial current proposed by Prof. Størmer and associated by him with the decrease of latitude of aurorae during magnetic storms.

We have not examined closely the extent to which the stream will cause inflow of ions and electrons into the earth's atmosphere in the polar regions, or how this inflow will give rise to the observed currents along the auroral zones; but it seems likely that present theories of the aurorae will need to be modified, because the particles of a neutral stream can approach much closer to the earth, in the equatorial plane, than the single charged particles hitherto considered. This must also have an important bearing on the theory of radio echoes, should it be proved that these are produced outside the earth's atmosphere.

(**126**, 129-130; 1930)

S. Chapman and V. C. A. Ferrapo: Imperial College of Science, South Kensington, London, S.W. 7, June 26.

Reference:

1. *Proc. Camb. Phil. Soc.*, **21**, 577; 1923.

大，但来自赤道面“上侧”和“下侧”的负电荷会沿地球磁力线运动，以中和从 BB' 穿过间隙迁移过来的正电离子的电荷。尽管对这一过程的具体细节还不清楚，但可以看出这样就能形成围着地球的西向电流。我们发现，该环电流一旦形成，在没有后继粒子补充时仍能保持力学和电磁平衡状态并持续好几天。该环电流的逐渐消失对应于磁暴的末相。

本文所述理论的一个与众不同的特征是主电流的作用区域与地球之间的距离，即几个地球半径。这些区域在地球的大气层之外（虽然次级电流是在大气层内感应产生的），但它们离地球的距离远比伯克兰所说的电流（位于赤道面上）以及斯托末教授提出的赤道电流（斯托末教授还指出赤道电流与磁暴期间极光出现的纬度降低有关）近得多。

我们还没有仔细研究粒子流造成正电离子和电子进入地球两极地区大气层的广度，以及带电粒子进入大气层的过程如何形成了我们沿极光带观测到的电流。但看来现今关于极光的理论很可能需要修正，因为，与目前设想的单电荷粒子相比，在赤道面上，呈电中性的带电粒子流中的粒子可以到达距地球更近的位置。如果能够证明这些过程是在地球大气层之外发生的，那么本研究在无线电回波理论方面也一定会产生重大的影响。

（齐红艳 翻译；张忠杰 审稿）

Deep Sea Investigations by Submarine Observation Chamber

Editor's Note

C. William Beebe was a naturalist with little formal training who financed some of his later ornithological expeditions with best-selling accounts of earlier ones. He also pioneered deep-sea exploration, collaborating with the independently wealthy Otis Barton, who made the first true bathysphere in 1928, a steel sphere with thick quartz windows. Together they made record-breaking descents in the Atlantic. This report describes one of the first: a descent to 1,426 ft off Bermuda. Reaching such depths was unprecedented at the time, but Beebe had doubled that figure by 1934.

ON June 11, 1930, in lat. 32° 16′ N., long. 64° 39′ W., in the Atlantic Ocean off Bermuda, Dr. William Beebe, accompanied by Mr. Otis Barton, descended to a depth of 1,426 feet below the surface of the sea.[1]

This announcement marks a new era in the exploration of the sea. All previous diving records shrink into insignificance compared with this depth; it was with no wish for record-making achievements that the descent was undertaken, but a real explorer's desire to see the animals beneath the waters as they live and not at second-hand from the collections of deep sea nets.

The construction of the chamber was financed by Mr. Barton, and he and Dr. Beebe, working from the New York Zoological Society's Oceanographic Expedition's headquarters at Nonsuch Island, have now made several descents, of which three were to a depth of 800 feet and one to 1,426 feet. The chamber is a steel sphere 57.3 in. in outside diameter and 1.5 in. thick. Observations could be made through a 6 in. diameter port fitted with a quartz window. Outside the window was hung a bag of decayed fish and some baited hooks, and a strong electric searchlight could be used to illumine the surrounding water. Telephonic communication was maintained with the ship above and a supply of oxygen carried.

One of the most striking phenomena was the "blue brilliance of the watery light to the naked eye, long after every particle of colour had been drained from the spectrum." The visual degeneration of the spectrum was observed, in connexion with an intensity metre. In Dr. Beebe's own words[2]: "The red had gone completely a few feet down ...; orange had been absorbed at sixty feet below the surface and yellow at less than 400. At our depth (800 feet) lavender, too, was non-existent, together with the two opposite ends of the

基于水下观测室的深海调查

编者按

威廉·毕比是一位没有接受过太多正规训练的博物学家。他早期研究的报告非常畅销，这为他后来进行鸟类学考察积累了资金。他还与独立而又富有的奥蒂斯·巴顿一起在深海探测方面做了许多开创性的工作。巴顿于 1928 年制作出了第一个真正的深海球形潜水器，这是一种带有厚石英窗的钢球。他们一起打破了多项大西洋中潜海深度的记录。这篇文章报道的就是其中比较早的一项：在百慕大下潜达到 1,426 英尺。在当时能达到如此深度是史无前例的，而到 1934 年时毕比的下潜深度又翻了一倍。

1930 年 6 月 11 日，在北纬 32° 16′，西经 64° 39′，距离百慕大不远的大西洋中，威廉 · 毕比博士在奥蒂斯·巴顿先生的陪同下潜到了距海面 1,426 英尺的地方。[1]

这一通告标志着海洋探测的一个新时代的到来。与这一深度相比，之前所有的潜水记录都黯然失色了。进行这次下潜活动并不是为了创造纪录，而是基于一个真正的探险者的渴望，为了亲眼看见处于原生状态的水下动物，而不是靠从深海拖网捕捞获得的二手资料来观察这些动物。

观测室是由巴顿先生出资建造的，他和毕比博士均就职于位于楠萨奇岛的纽约动物学会的海洋探险队总部。目前他们已完成了多次下潜任务，其中 3 次潜到了 800 英尺，1 次潜到 1,426 英尺。该观测室是一个钢球，外径 57.3 英寸，厚 1.5 英寸。观察者可以通过一个直径为 6 英寸的石英窗进行观察。在窗外悬挂着一袋烂鱼和一些装有鱼饵的鱼钩，以及一个大功率的探照灯，可用于照亮周围的海水。水面上方的船只与观测室之间保持电话通信，并向观测室提供氧气。

最令人吃惊的现象之一是，“在所有的颜色都从光谱中消失很久以后，裸眼看到的是海水发出的蓝色光芒。”他们利用一个强度计来观测光谱中各种光的衰减。用毕比博士自己的话说就是 [2]，“几英尺以下红光就完全消失了……；橙光在水面下 60 英尺处被吸收，而黄光在不到 400 英尺处被吸收。在我们所在的深度（800 英尺），淡紫色的光以及光谱两端的红外线和紫外线也都不复存在了，绿光依然存在，但

spectrum, infra-red and ultraviolet, while green still persisted, but greatly diluted. All that remained to our straining eyes were violet and blue, but blue such as no living man had ever seen."

It proved quite possible to observe pelagic animals drifting and swimming past the window, such as medusae, shrimps, and fish, and about a dozen true bathypelagic fish were identified. A very interesting result of these observations was the presence of certain species of fish and invertebrates in water layers well above the depth at which their occurrence is first indicated by net catches in the daytime.

Four descents have also been made in water up to 350 feet in depth along the shelving bottom of the Bermudian insular shelf as the vessel drifted seawards. Such exploration revealed a new fish fauna at these offshore depths, the recognisable shore fish also being of great size.

The observations will be continued another year, and it is to be hoped that this new weapon of marine research has come to stay and that similar submarine observation chambers may be built in time for a study of the floor of shallower seas and the habits of food fishes. Already shallow water diving has proved its scientific value. We shall await Dr. Beebe's and Mr. Barton's full reports with great interest.

(**126**, 220; 1930)

References:

1. *Science*, vol. 72, No. 1854, July 11, 1930, pp. 27-28. "A New Method of Deep Sea Observation First-hand". By Henry Fairfield Osborn.

2. *New York Times*, June 27, 1930.

也变淡了很多。我们睁大双眼只能看到紫光和蓝光，不过这种蓝是人类所不曾见过的。”

实践证明，极有可能会看到从窗前漂过或游过的浮游动物，如水母、虾类和鱼类，并且辨认出了约 12 种真正的深海鱼类。通过这些观察获得的一个非常有趣的结果是：在上层水体中发现的一些鱼类和无脊椎动物在这里也出现了，要知道它们在上层水体中的存在最早是通过白天用渔网捕捞而发现的。

当这艘船随海漂移时，沿着百慕大岛架的缓倾海底，在水中深达 350 英尺的地方又下潜过 4 次。这类探险活动在这些近海深度发现了一个新的鱼类区系，可识别的近海鱼类也达到了很大的数量。

明年水下观测还将继续进行，我们希望这一海洋研究的新装备能够得到普遍应用，并且能够及时建造出类似的水下观测室，用于研究浅海海底以及食用鱼类的生活习性。浅海潜水在科学上的价值已经得到了证明。让我们怀着极大的兴趣期待毕比博士和巴顿先生的详细报告吧。

（齐红艳 翻译；张泽渤 审稿）

Stellar Structure and the Origin of Stellar Energy

E. A. Milne

Editor's Note

Arthur Eddington had recently proposed a model for the structure of stellar interiors, which implied that the mass of a star determines its luminosity in a unique way. Here the English astronomer Edward Milne disputes Eddington's claim. Milne's alternative model predicts that stars should have an extremely dense core—the most likely setting, it was felt, for the processes giving stars their energy—surrounded by a gaseous body of lower density. The core temperatures in his theory could be as high as 10^{11} degrees, some 10,000 times higher than Eddington's theory suggests. Milne's theory made little use of the principles of quantum theory or emerging nuclear physics, but illustrated a growing interest in understanding stellar structure from first principles.

THE generally accepted theory of the internal conditions in stars, due to Sir A. S. Eddington, depends largely on a special solution of the fundamental equations, and according to this a definite calculable luminosity is associated with a given mass. If this were the only solution of the equations it would conflict, as I have repeatedly shown in recent papers, with the obvious physical considerations which show that we can build up a given mass in equilibrium so as to have an *arbitrary* luminosity (not too large) whatever the assumed physical properties of the material. I have recently noticed that the fundamental equations possess a whole family of solutions, corresponding to arbitrarily assigned luminosity for given mass. These solutions show immediately that Eddington's solution is a special solution and corresponds to an unstable distribution of mass. In the stable distributions the density and temperature tend to very high values as the centre is approached, theoretically becoming infinite if the classical gas laws held to unlimited compressibility.

The physical properties of the stable configurations can be described as follows. Suppose a star is built up according to Eddington's solution with his value of the rate of internal generation of energy. Let the rate of internal generation of energy diminish ever so slightly. Then the density distribution suffers a remarkable change. The mass suffers an intense concentration towards its centre, the external radius not necessarily being changed. The star tends to precipitate itself at its centre, to crystallise out so to speak, forming a core or nucleus of very dense material. The star tends to generate a kind of "white-dwarf" at its centre, surrounded of course by a gaseous distribution of more familiar type; the star is like a yolk in an egg. In this configuration the density and temperature are prevented from assuming infinite values by the failure of the classical gas laws, but they reach values incomparably higher than current estimates. For example, it seems probable (though the following estimates are subject to revision) that the central temperature exceeds 10^{11} degrees, in comparison with the current estimates of the order of 10^{7} degrees; and the density may run up to the maximum density of which ionised matter is capable.

恒星的结构和恒星能量的起源

米尔恩

编者按

阿瑟·爱丁顿最近提出了一种恒星内部结构的模型，该模型认为恒星的质量唯一地决定了恒星的光度。在这篇文章中，英国天文学家爱德华·米尔恩对爱丁顿的观点提出了异议。米尔恩提出的另一种模型预言，恒星应该具有一个非常致密的核心，在这个核心中极有可能进行着为恒星提供能量的反应，而核心的周围被低密度的气体包围。按照米尔恩的理论，恒星核心的温度可能会高达 10^{11} 度，这比根据爱丁顿的理论估计出来的值高出 10,000 倍。米尔恩的理论几乎没有用到任何量子理论和新兴的核物理学的原理，不过这表明了人们从基本原则出发解释恒星结构的兴趣正在增长。

目前，由爱丁顿爵士提出的恒星内部结构理论被大家普遍接受，该理论主要基于基本方程的特殊解，由此计算出的恒星的光度与恒星质量密切相关。然而，正如我在最近几篇文章中反复强调的那样，如果这是方程的唯一解，那么就会与一些显而易见的物理学原理相矛盾，比如，一颗质量一定的处于平衡态的恒星，可以具有**任意的**光度（只要不是太大），而且与物质的物理特性无关。我最近发现，基本方程可以得出一组解，对于同一质量，会解出任意的光度。这组解显而易见地说明了，爱丁顿的解只是一个特解，并且对应着质量的不稳定分布。在稳定分布的情况下，中心区域附近的密度和温度会变得非常高；若经典的气体定律在无限压缩的条件下仍然适用，则密度和温度在理论上可以达到无穷大。

这种稳定结构的物理特性如下所述：假如一颗恒星是按照爱丁顿的解和他给出的内部产能率而构造的，则只要该产能率略微减小，密度分布就会发生明显的变化，质量向中心紧密聚集，而外部半径并不一定会相应地发生变化。随着这颗恒星中的物质逐渐向中心沉积，就会形成一个致密的核，甚至会达到结晶。恒星在其中心形成了一种“白矮星”，其周围分布着常见的气态物质，这时整颗恒星就像一个鸡蛋中的蛋黄。在这种结构中，由于经典气体定律的失效，密度与温度不再是无穷大，但仍然远高于现行的估计值。比如，中心温度很可能（虽然下面的估计值曾被修正）高于 10^{11} 度，而目前的估计值为 10^{7} 度；而中心密度可能达到了电离物质所能具有的极限密度。

The unstable density distribution of Eddington's model (curve *A*) and the stable density distribution of actual stars (curves *B*) are indicated roughly in Fig. 1, which is not drawn to scale. It may be mentioned that the instability is of a radically different kind from that discussed by Sir James Jeans. He concluded that perfect-gas stars of Eddington's model were *vibrationally* unstable. In my investigations, the instability of Eddington's model arises from any slight departure of the rate of generation of energy below the critical value found by Eddington. The perfect-gas distribution of my solutions is perfectly stable, but the density necessarily increases until degeneracy or imperfect compressibility takes control.

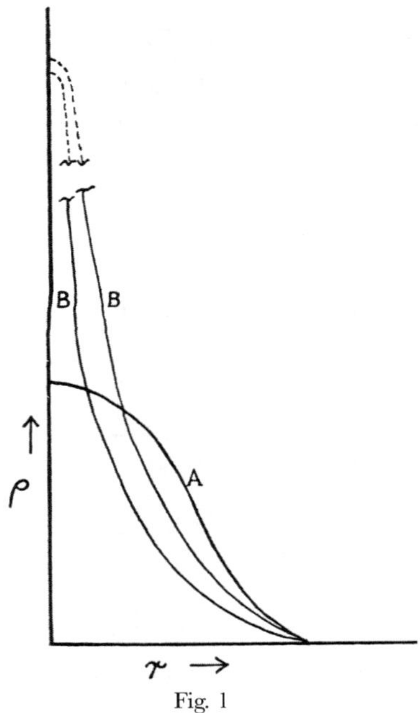

Fig. 1

The consequences amount to a complete revolution in our picture of the internal constitution of the stars. In the intensely hot, intensely dense nucleus, the temperatures and densities are high enough for the transformation of matter into radiation to take place with ease. It is to this nucleus that we must look for the origin of stellar energy, a nucleus the existence of which has previously been unsuspected. The difficulties previously felt as to stellar conditions being sufficiently drastic to permit the evolution of energy largely disappear. Many of the cherished results of current investigations of the interiors of stars must be abandoned; current estimates of central temperature, central density, the current theory of pulsating stars, the current view that high mass necessarily implies high radiation pressure, the supposed method of deducing opacity of stellar material from observed masses and luminosities, the supposed proof of the observed mass-luminosity correlation—all these require serious modification.

The new results are not a speculation. They are derived by taking the observed mass and luminosity of a star, and finding the restrictions these impose on the possible density

图 1 粗略地显示了爱丁顿模型中的非稳定密度分布（曲线 A），以及实际恒星的稳态密度分布（曲线 B），该示意图未标刻度。需要说明的是，这里的非稳定性完全不同于詹姆斯·金斯爵士的论述。詹姆斯·金斯认为，爱丁顿模型中的理想气体恒星处于非稳定的**振动**状态。而我的研究表明，爱丁顿模型的非稳定性源于产能率略低于爱丁顿给出的临界值。在我的解中，理想气体分布是完全稳定的，但是密度会不断增加，直到简并或非理想压缩起主导作用。

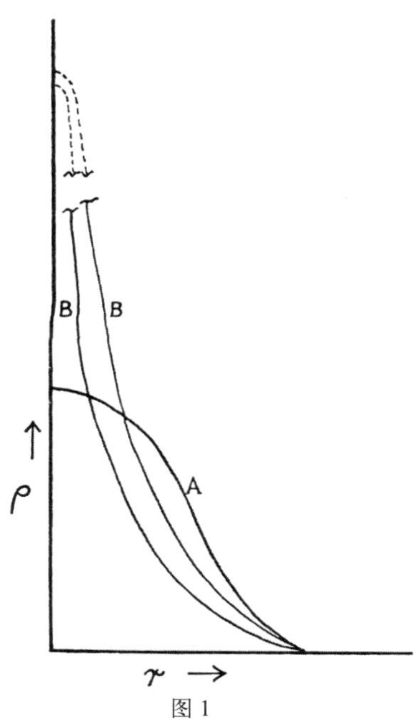

图 1

这些结果对于我们理解恒星内部结构具有革命性的意义。在极端炽热和致密的恒星核中，足够高的温度和压力可以轻易地使物质转化为辐射能。面对这样的核心，我们必须寻找恒星能量的来源，而核的存在以前并不为人所知。以前人们在考虑恒星内部必须存在极端条件才能完成能量演化时遇到的困难现在已经基本解决。然而，我们不得不放弃目前流行的许多关于恒星内部的珍贵研究结果：目前对于恒星中心温度、中心密度的估计，目前关于脉动星的理论，目前关于高质量必然导致高辐射压的观点，通过观测恒星的质量和光度得出恒星物质不透明度的方法，以及对观测到的质–光关系的证明——所有这些成果，都需要进行认真的修正。

这些新得到的结果不是猜测，而是通过测量恒星的质量和光度并给定一些限制条件后推导出来的，这些限制条件要求密度分布必须同恒星的质量和光度相容。通

distributions compatible with this mass and luminosity. By integrating the fundamental equations from the boundary inwards, we are inevitably led to high central temperatures and densities. So long as the classical gas laws persist, the solution is one of the family with a central singularity (infinities in ρ and T), and it is only the ultimate failure of the gas laws which rounds off the distribution with a finite though very large central ρ and T.

(**126**, 238; 1930)

E. A. Milne: Wadham College, Oxford, July 29.

过对基本方程从边界向内积分，我们必然会得到较高的中心温度和密度。只要经典气体定律仍然适用，就可以得到其中一个具有中心奇点的解（相应的密度 ρ 和温度 T 为无穷大）；而只有当气体定律最终不再适用时，才能够形成中心密度 ρ 和温度 T 虽然很高但不是无穷大的分布。

（金世超 翻译；何香涛 审稿）

Eugenic Sterilisation

Editor's Note

In the 1930s the potential benefits of eugenic sterilisation, first advocated by Darwin's cousin Francis Galton, were being considered in many European nations and in the United States. This editorial reacts to a proposal by the Eugenics Society of London for a legal change allowing sterilisation of the mentally impaired on their own consent, or that of a parent or legal guardian. The society had argued this would achieve a 17% reduction in mentally deficient individuals in one generation. The editorial counters that this figure is unreasonably optimistic, as it assumed permissions could be obtained for some 300,000 certifiable mental defectives in England and Wales. Yet like many scientists of the times, this editorial supports the principle of sterilisation.

SOME of the young people of Germany would have us believe that much of the time that can be spared from their more materially fruitful exploits is given over to singing a song which they call "Deutsche Jugend, heraus". Its language, borrowed from historical romanticism, permits, if it does not foster, a certain diversity of interpretation, and some lines with a frankly Christian significance may even be omitted at the discretion of the singer. Claim to popularity is thus made more catholic.

Wollt Ihr ein neues bauen
mit Händen stark und rein,
in gläubigem Vertrauen
lasst dies die Losung sein:
Den Feind in eigner Mitte
gefällt in ernstem Strauss....

Moralists, it is easy to see, may use these lines to assist them in focusing attention upon that enemy in their midst distinguished as the *beam*, while the nationalist may recognise more immediately its particular referability to the communistic *mote*.

We are assured, however, that the resiliency of this *credo* unites rather than divides, and such demonstrations as we have enjoyed tend to reinforce the assurance audibly. But we cannot help wondering what will happen in the world when the youth of one country or another not only present accessible enemies in their patriotic songs but also define them with scientific precision.

The real enemies of mankind are made, yearly, more and more accessible to attack by science, and if it were not for the protective screens, intangible and often fantastic, thrown up by the unscientific for whom nakedness, even the enemy's, still seems to possess terrible powers, mankind might subjugate very speedily its worst foes. But if, as Sir Walter Fletcher

优生绝育

编者按

20 世纪 30 年代，美国和很多欧洲国家都开始认识到，由达尔文表弟弗朗西斯·高尔顿最先提出的优生绝育将对社会产生积极的影响。这篇评论回应了伦敦优生学会的一项倡议，该倡议要求修改法律，以便在取得本人或父母中的一方或法定监护人同意的条件下，允许对心智缺陷者实施绝育。优生学会认为，这样做将使下一代中心智缺陷者的数量减少 17%。这篇评论认为这个数字过于乐观，因为它假设了英格兰和威尔士的 300,000 名确诊的心智缺陷者都会同意进行绝育。不过和当时的许多科学家一样，这篇评论的作者也是支持绝育原则的。

富有成效的物质文明建设使德国的年轻人有了更多的闲暇时间，但他们中的一些人把大部分多出来的时间都花费在吟唱一首被他们称为《德国青年》的歌曲上了。其带有历史浪漫主义色彩的歌词给整首歌赋予了多种解释，甚至是一些本来并不包含的意思。歌词中有些颇具虔诚基督教意义的部分可能会被歌手酌情删掉，从而可以更广泛地向大众普及。

你愿意建立一个新世界吗？
用强壮的双手建立一个纯净的世界，
坚定这一信念
并使它成为口号：
让处于中间阶层的敌人
就像虔诚的施特劳斯信徒一样……

很容易就能想到，道德家们可能会借用这些歌词来帮助他们将注意力集中在那些混杂于人民中间并以**国家栋梁**而著称的敌人，而民族主义者们通过这些歌词可能会更直接地意识到，相对于共产主义的**瑕疵**而言他们的主张具有的特殊借鉴意义。

尽管如此，我们还是确信这一**信条**的弹性有利于团结而非分裂，而且这种我们一直很喜欢的表述方式显然很有利于增强其可信性。但我们不禁要问，当一个国家或另一个国家的年轻人不仅在自己的爱国歌曲里提到触手可及的敌人，而且用科学精确的语言定义这些敌人的时候，这个世界将会发生什么呢？

一年一年过去，人类真正的敌人变得越来越容易被科学击倒，如果不是因为无知者甚至是反对者们非科学地抛出的一些无形的、通常是幻想出来的保护屏障看上去似乎拥有可怕力量的话，人类可能很快就能征服他们最大的敌人了。但是正如沃尔特 · 弗莱彻爵士最近指出的，如果对于一种纯粹的疾病（例如癌症），只有在破

has lately pointed out, a mere ailment, like cancer, has only been made accessible to scientific study through the lifting of foolish and superstitious taboos, how can we expect the direr social maladies to be approached courageously? A protective hedge of errors and superstitions hems them in on every side, so rank and poisonous that it seems that even science is infected and intimidated while it attacks.

How else is it possible to explain the demand just put forward* by a committee of the Eugenics Society for permissive legislation which would take a whole generation to achieve a reduction in the incidence of mental defect not of a hundred, not of fifty, not of twenty-five, but, problematically, of seventeen percent? Between our people and the realisation of this slender benefit stands "an ambiguity of the law" which the Society proposes to remove. A person may, with consent, be sterilised in the interests of his *own* health. In the interest of the public health, present or future, he may not be sterilised. By a curious legal inversion, the "willing mind" of the individual cannot take away the offence against the public even should he be prepared to save it from all possibility of contamination by his own progeny. The offence consists in a "maim" which deprives the individual, or so it may be contended, of martial courage, and the State of a vessel, however unsuitable otherwise, for this same virtue. To contentions of this sort, surely the monosyllabic genius of Mr. H. G. Wells's latest novel has supplied the only effective answer.

To meet the practical situation, the committee proposes a Bill legalising eugenic sterilisation. This would authorise the mental deficiency authority or superintendent of an institution to sterilise a mental defective, subject to the consent of the parent or guardian and of the Board of Control, and of the spouse if the defective is married. In the case of defectives deemed capable of giving consent, sterilisation would not be performed otherwise than with this consent. It would authorise the voluntary sterilisation of a person about to be discharged from a mental hospital for the insane as recovered, again with the added consent of Board and of parent, guardian, or spouse; and it would legalise voluntary sterilisation for the sole purpose of preventing the transmission of hereditary defect seriously impairing physical or mental health or efficiency.

Five members of the committee and another contributed to the *Lancet* for July 19 a letter defending this policy. The defence combats the assertion that if every certifiable mental defective had been sterilised twenty or thirty years ago it would have made little appreciable difference to the number of mental defectives existing today. It repeats a sentence of the committee's report urging that "if all the defectives in the community could be prevented from having children the effect would be even on the most unfavourable genetic assumptions with regard to defectiveness, to reduce the incidence of mental defect by as much as 17 percent in one generation".

---

* Committee for Legalising Eugenic Sterilisation. Eugenics Society, London, 1930.

除愚蠢迷信的忌讳后才能对其进行科学研究的话，那么我们又怎么能期望研究人员会大胆地去研究更加可怕的社会弊病呢？错误和迷信的保护罩将各种问题和弊病团团包围，当这些讨厌而又恶毒的保护罩发起进攻时，看起来似乎连科学都受到了影响和威胁。

优生学会的一个委员会最近提出，希望一项提案能获得立法通过*。他们认为提案的措施将会使下一代中心智缺陷的发生率下降，这种降低不是减少 100 个、50 个或者 25 个缺陷个体，而是使缺陷的发生率整体上降低 17%。对他们的此项要求，还有什么别的可能的解释吗？在这一微薄利益的实现和我们的民众意识之间，还存在着“法律上的模糊地带”，这正是该学会主张消除的。为了**自身**健康，一个人可能会同意做绝育手术。但如果是为了公众健康，那么无论是现在还是将来，他可能都不会去做绝育手术的。通过一种奇妙的立法转换，个人的“意愿”就不能再对公众利益有所冒犯，甚至他必须做好准备以免因为自己后代可能的缺陷而使公众利益受到损害。这里所说的冒犯包含在一种“伤害”中，这种伤害可能存在争议，它剥夺了个人的战斗勇气，剥夺了国家的命脉，然而同样的特点并不适合于其他情况。对于这种争论，单音节天才威尔斯先生的最后一部小说肯定已经提供了唯一有效的答案。

为了符合实际情况，委员会提出了一份使优生绝育合法化的议案。这将在心智缺陷者的父母或监护人和管理委员会及缺陷者的配偶（如果该缺陷者已婚）同意的情况下，赋予心智缺陷相关的权威机构或者机构管理人对心智缺陷者进行绝育的权利。如果缺陷者被认为具有作出决定的能力，则只有在本人同意的情况下才能对其进行绝育。自愿同意进行绝育的患者在康复后可以允许其离开精神病院，当然这需要来自委员会以及父母、监护人或配偶的同意；这将使把阻止遗传缺陷发生传递以防止其对生理或精神健康或功效造成严重损害作为唯一目的的自愿绝育合法化。

该委员会的 5 位委员和另外一人于 7 月 19 日向《柳叶刀》投稿捍卫这项政策。他们在辩护中反对了如下的断言：即使在二三十年前就对每一个确认具有心智缺陷的人进行了绝育，那也几乎不会对现在的心智缺陷者的数量带来多么显著的影响。文中重复了一个委员会报告中的句子，极力主张“如果能够使社会上所有有缺陷的个体不生育孩子的话，那么，即使是在最坏的遗传假设下，其影响也将使下一代中心智缺陷的发生率降低 17% 左右”。

* 优生绝育合法化委员会。优生学会，伦敦，1930年。

Obviously a 17 percent reduction in the incidence of mental deficiency is more desirable than a 17 percent increase. But do the committee's proposals ensure this reduction? Clearly, no. The words quoted promise at least that reduction if the fertility of *all* living mental defectives is prevented. The committee's proposals, with their emphasis upon the voluntary principle, by no means ensure that the 300,000 certifiable mental defectives in England and Wales would be sterilised. Who must consent? (1) The patient, if he is capable. (2) The parent or guardian. (3) The spouse if the patient is married. (4) The Board of Control. The calculation, it is true, is based on two assumptions "highly unfavourable" to the effectiveness of the proposals—that the genetic factor responsible for defectiveness (primary amentia) would be much "carried" and would only rarely produce manifest defectives, and that defectiveness is uniformly distributed throughout the community. (The fertility of defectives also is assumed to be that of the average of the population.)

How unfavourable, on the other hand, are the chances of permission? Nothing is gained by attempts to write off opposing assumptions. A figure is a figure, right or wrong.

Again, is a 17 percent reduction all that eugenic science can promise? Disregarding altogether those so-intelligent defectives who will strive to serve the country by seeking this minor mutilation, is it the institutional class that constitutes the chief danger to society? Prof. MacBride (*Nature*, Jan. 11, 1930, p. 40) says emphatically that this but touches the fringe of the problem. "The defectives most dangerous to society are those who are never confined in institutions at all! The high-grade defectives are just able to support themselves in the lowest paid and most unskilled occupations, and no civilised government would take the responsibility of confining them, and so they go on propagating large families as stupid as themselves." His idea of penal sterilisation, a punishment "for the economic sin of producing more children than the parents can support", is one which becomes more and more difficult to apply as more and more ways are devised by the State for screening the individual from biological estimation.

Is there not a real danger that the advocates of such legislation as here may mistake the assent of the political machine for victory? If assent were gained, would it not be much more accurately determined as the hall-mark of failure? It is not the assent of the State, but the initiative and creative power of the State, that is needed to secure essential progress, and that will not exist until our legislators of all parties or of any party derive their inspiration from the cultivation of natural knowledge.

(**126**, 301-302; 1930)

很显然，心智缺陷的发生率下降 17% 比增加 17% 更有利。但是该委员会的提案是否可以保证这一减少量的实现呢？当然不能。此处引用的文字认为，如果使**所有**活着的心智缺陷者不生育的话，那么至少可以保证实现这一减少量。按照委员会所强调的自愿原则，他们的提议根本不能保证英格兰和威尔士的 300,000 例确定具有心智缺陷的个体都被绝育。谁必须同意呢？（1）病人，如果他有自主能力的话。(2)父母或监护人。(3)配偶，如果病人已婚的话。(4)管理委员会。该计算是无误的，但它建立在两个“非常不支持”该提案有效性的假设之上。一个假设是：导致缺陷（先天痴愚）的遗传因子会在大量的后代中“被携带”，但仅仅在很少情况下才会产生明显的缺陷；另一个假设是：这种缺陷在整个社会中是均匀分布的。(缺陷者的生育能力也被假定为相当于整个人群的平均生育能力)。

另一方面，允许缺陷个体生育的话情况会有多么不利呢？通过试图取消对立的假设并没有取得任何成果。无论正误，数字就是数字。

再者，17% 的减少是否是优生学能够承诺的全部效果？如果完全不考虑那些非常聪明并尽力为国家作贡献的缺陷个体来判断最小损害的话，那么是不是公共机构就成了对社会的主要威胁呢？麦克布赖德教授（《自然》,1930 年 1 月 11 日,第 40 页）强调说这仅仅触碰到问题的边缘。“对社会最具危险性的有缺陷的个体是那些从来没有受到公共机构限制的人！高级的有缺陷的个体能够凭最低收入和最不需技能的工作来养活自己，并且任何文明社会的政府都不会承担约束他们的责任，所以他们就会继续繁衍出和他们自己一样有缺陷的大家族。”当国家设计出越来越多的方法通过生物评估来筛选缺陷个体时，他那将绝育术视为“对那些生育了很多孩子而无力供养的父母们所犯经济罪”的惩罚措施的观点就变得越来越难应用于实际了。

像这里提到的支持如此立法的倡导者们可能会将政治机构的同意误解为自己取得了胜利，这难道不是真正的危险吗？一旦得到同意，那么更准确地说，这更应该被视作失败的标志。因为这不是国家的同意，而是国家自发的创造性力量的同意，这种力量是用来保障最基本的发展的，除非所有政党的立法者从自然知识的熏陶中获得启发，否则这种力量将不会存在。

（刘皓芳 翻译；刘京国 审稿）

Fine Structure of α-Rays

G. Gamow

Editor's Note

George Gamow was a Russian scientist who left the Soviet Union in 1932 and worked at several Western European universities until moving to the United States before the Second World War. He made a powerful impression by his versatility as a scientist, his capacity to write clearly for the general public and his engagement in public causes such as advocacy of building nuclear weapons in the United States. This brief letter offers an explanation of why γ-rays emitted by radioactive atoms may have a variety of energies.

IT is usually assumed that the long range α-particles observed in C'-products of radioactive series correspond to different quantum levels of the α-particle in the nucleus. If after the preceding β-disintegration the nucleus is left in an excited state with the α-particle on one of the levels of higher energy, one of the two following processes can take place: either the α-particle will cross the potential barrier surrounding the nucleus and will fly away with the total energy of the excited level (long range α-particle), or it will fall down to the lowest level, emitting the rest of its energy in the form of electromagnetic radiation (γ-rays), and will later fly away as an ordinary α-particle of the element in question. Thus there must exist a correspondence between the different long range α-particles and the γ-rays of the preceding radioactive body. If p is the relative number of nuclei in the excited state, λ the corresponding decay constant, and θ the probability of transition of the nucleus from the excited state to one of the states of lower energy with emission of energy (in form of γ-quanta or an electron from the electronic shells of the atom), the relative number of long range α-particles must be $N = p\frac{\lambda}{\theta}$. Knowing the number of α-particles in each long range group and calculating, from the wave mechanical theory of radioactive disintegration, the corresponding values of λ, we can estimate for each group the value θ/p, giving a lower limit for the probability of γ-emission. For example, for thorium-C' possessing besides the ordinary α-particles also two groups of long range α-particles, we have for transition probabilities from two excited states to the normal state $\theta_1 < 0.4 \times 10^{12}$ sec^{-1} and $\theta_2 < 2 \times 10^{12}$ sec^{-1}, which is the right order of magnitude for the emission of light quanta of these energies. With decreasing energy λ decreases much more rapidly (exponentially) than θ, so that the number of long range α-particles from the lower excited levels will be very small. (From this point of view we can also easily understand why the long range α-particles were observed only for C'-products for which the energy of normal α-particles is already much greater than for any other known radioactive element.)

α 射线的精细结构

伽莫夫

编者按

乔治·伽莫夫是一位俄国科学家，他于 1932 年离开苏联，在几所西欧的大学里工作。第二次世界大战爆发前夕，他去了美国。他是一位多才多艺的科学家，为公众撰写的普及读物非常清晰易懂，他在美国投身于公众事业，支持原子武器，这些都给人们留下了深刻的印象。在这篇简短的快报文章中，他解释了为什么放射性原子发射出的 γ 射线可能具有不同的能量。

我们通常假设，C' 放射系列产物中观测到的长程 α 粒子对应于原子核中 α 粒子的不同量子能级。如果在 β 衰变之后，原子核处于 α 粒子占据某个更高能级的激发态，那么就可能发生下面两个过程中的一个：或者是 α 粒子穿过原子核周围的势垒，携带激发态的所有能量而逃逸（长程 α 粒子）；或者是 α 粒子降至最低能级，将剩余能量以电磁辐射（γ 射线）的形式发射出去，然后再以普通 α 粒子的形式逃逸出该原子核。这样，在长程 α 粒子和之前的放射体放出的 γ 射线之间就应该存在某种关联。如果 p 是处于激发态的原子核的相对数量，λ 是相应的衰减常数，θ 是原子核从激发态跃迁到某个低能态并辐射出能量（以 γ 量子或者是从该原子电子壳层中发射出的电子的形式）的几率，那么长程 α 粒子的相对数量就是 $N = p\dfrac{\lambda}{\theta}$。我们已经知道了每一个长程组内的 α 粒子数量，并且可以通过辐射衰变的波动力学理论计算出相应的 λ 值，那么我们就可以估算出每一组的 θ/p 值，得到一个 γ 辐射发生几率的下限。比如，在钍 C' 的衰变中，除释放普通的 α 粒子之外，还有两组长程 α 粒子，对于从这两个激发态跃迁到正常态的几率，我们知道：$\theta_1 < 0.4 \times 10^{12}$/ 秒，$\theta_2 < 2 \times 10^{12}$/ 秒，这个数量级对于辐射这些能量的光量子来说是合适的。当能量减小时，λ（呈指数减小）比 θ 减小的速度快很多，因此来自较低激发态的长程 α 粒子数量将会非常少。（这样看来，就不难理解为什么长程 α 粒子只有在 C' 的衰变产物中才会出现了，C' 过程产物中的正常 α 粒子的能量远远高于其他已知的任何放射性元素产生的能量。）

A difficulty arises with the recent experiments of S. Rosenblum (*C. R.*, p. 1,549; 1929; p. 1,124; 1930), who found that the α-rays of thorium-C consist of five different groups lying very close together. The energy differences and intensities of the different groups relative to the strongest one (α_0) are, according to Rosenblum:

$$E\alpha_1 - E\alpha_0 = +40.6 \text{ kv} \quad I\alpha_1 = 0.3$$
$$E\alpha_2 - E\alpha_0 = -287 \;\;,, \quad I\alpha_2 = 0.03$$
$$E\alpha_3 - E\alpha_0 = -442 \;\;,, \quad I\alpha_3 = 0.02$$
$$E\alpha_4 - E\alpha_0 = -421 \;\;,, \quad I\alpha_4 = 0.005$$

If we suppose that these groups are due to α-particles escaping from different excited quantum levels in the nucleus, we meet with very serious difficulties. The decay constant λ for the energy of thorium-C fine structure particles is very small ($\lambda \sim 10^{-2}$ sec^{-1}), and in order to explain the relatively great number of particles in different groups we must assume also very small transition probabilities. We must assume that thorium-C nucleus can stay in an excited state without emission of energy for a period of half an hour!

We can, however, obtain the explanation of these groups by assuming that we have here a process quite different from the emission of long range α-particles. Suppose that two (or more) α-particles stay on the normal level of the thorium-C nucleus. It can happen that after one of the α-particles has escaped the nucleus will remain in an excited state with the other particle on a certain level of higher energy. (In this case the energy of the escaping α-particle will be smaller than the normal level and obviously will not correspond to any quantum level inside the nucleus.) From the excited state the nucleus (thorium-C'' now) can afterwards jump down to the normal level, emitting the energy difference in form of a γ-quantum.

Thus the relative number of different groups will not depend on the probability of γ-emission but only on the transition integral:

$$W = \int f(r_{1,2})\, \psi E_0(\alpha_1)\, \psi E_0(\alpha_2)\, \psi E_n(\alpha_1) \psi E\alpha_n(\alpha_2)\, dv_1\, dv_2$$

where $f(r)$ is the interaction energy of two α-particles at a distance r apart, ψE_0 and ψE_n the eigenfunctions of an α-particle in the normal and n^{th} excited states, and ψE_a the eigenfunction of an escaping α-particle with the energy: $E_{\alpha_n} = E_0 - (E_n - E_0)$.

According to this scheme, the γ-rays corresponding to different fine structure groups of thorium-C must be observed as γ-rays of thorium-C (ejecting electrons from K, L, M, ... shells of the thorium-C$''$-atom) and not as the rays of thorium-B, as we would expect in the case of long range particle explanation. The level scheme of the thorium-C''-nucleus as given by fine structure energies is represented in Fig. 1.

罗森布拉姆在最近的实验中遇到了一些困难（《法国科学院院刊》，1929 年第 1,549 页，1930 年第 1,124 页），他发现钍 C 放射的 α 射线是由非常接近的 5 个不同的组组成的。罗森布拉姆给出了其余各组相对于最强的那一组（α_0）的能量差和强度：

$$E\alpha_1 - E\alpha_0 = +40.6 \text{ 千电子伏} \quad I\alpha_1 = 0.3$$
$$E\alpha_2 - E\alpha_0 = -287 \text{ 千电子伏} \quad I\alpha_2 = 0.03$$
$$E\alpha_3 - E\alpha_0 = -442 \text{ 千电子伏} \quad I\alpha_3 = 0.02$$
$$E\alpha_4 - E\alpha_0 = -421 \text{ 千电子伏} \quad I\alpha_4 = 0.005$$

如果我们假设这些组分是由于逃逸的 α 粒子曾处于原子核内的不同激发量子能级而形成的，那么我们将遇到很大的麻烦。钍 C 精细结构粒子的能量衰减常数 λ 非常小（λ 约为 10^{-2}/ 秒），而且为了解释各组中何以有相对那么大数量的粒子，我们必须同时假设跃迁几率非常小。我们必须假设钍 C 原子核可以停留在激发态而不辐射能量长达半个小时！

然而，如果我们设想一个完全不同于发射长程 α 粒子的过程，就可以解释这 5 组 α 粒子了。假设有两个（或者更多的）α 粒子处于钍 C 原子核的正常能级上。其中一个 α 粒子逃逸到原子核外后，原子核有可能还保持在激发态，因为剩下的 α 粒子可能处于某个能量较高的能级上。（在这种情况下，逃逸 α 粒子的能量低于正常能级，而且它显然与原子核内的任何量子能级都不相等。）随后原子核（这里是钍 C''）可以从激发态跃迁到正常态，放出一个 γ 量子以释放两态之间的能量差。

这样，不同组的相对数量将与 γ 辐射的几率无关，而只与跃迁积分相关：

$$W = \int f(r_{1,2})\,\psi E_0(\alpha_1)\,\psi E_0(\alpha_2)\,\psi E_n(\alpha_1)\,\psi E\alpha_n(\alpha_2)\,dv_1\,dv_2$$

式中，$f(r)$ 是两个 α 粒子在相距为 r 时的相互作用能，ψE_0 和 ψE_n 分别是 α 粒子在正常态和第 n 个激发态的本征函数，ψE_a 是能量为 $E_{\alpha n} = E_0 - (E_n - E_0)$ 的逃逸 α 粒子的本征函数。

按照这种解释，对应于钍 C 不同精细结构组分的 γ 射线应该被看作是钍 C 的 γ 射线（发射出的电子来自于钍 C'' 原子的 K，L，M 等壳层），而不能被看作是钍 B 的 γ 射线，正如我们在解释长程粒子时预期的那样。图 1 中所示的钍 C'' 原子核能级图画出了能量的精细结构。

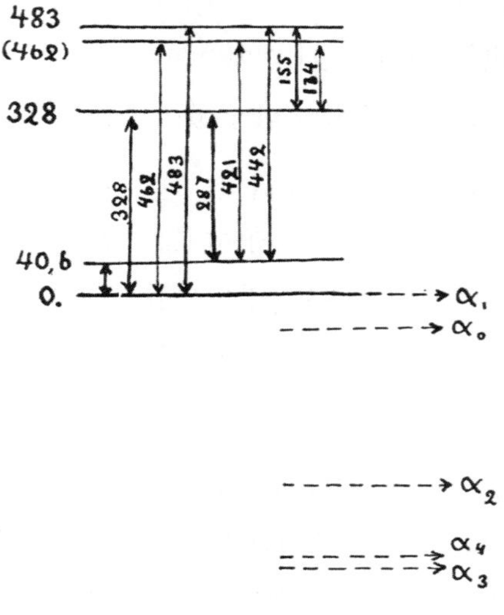

Fig. 1

In the observed γ-ray spectra of thorium-$C + C''$ (Black, *Proc. Roy. Soc.*, pp. 109–166; 1925) we can find lines with the energies: 40.8; 163.3; 279.4; 345.8; 439.0; 478.8; 144.6 kv fitting nicely with the energy differences in Fig. 1.

Thus we see that the fine structure group of highest energy corresponds to the normal level of the nucleus, while the other groups are due to the ordinary α-particles which have lost part of their energy, leaving the nucleus in an excited state.

I am glad to express my thanks to Dr. R. Peierls and Dr. L. Rosenfeld for the opportunity to work here.

(**126**, 397; 1930)

G. Gamow: Piz da Daint, Switzerland, July 25.

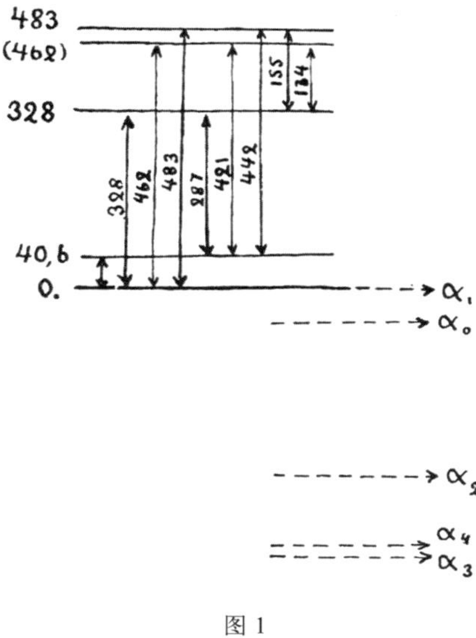

图 1

在实测的钍 C 和钍 C'' 的 γ 射线能谱中（布莱克，《皇家学会学报》，1925 年，第 109~166 页），我们可以找到对应于以下能量的谱线：40.8，163.3，279.4，345.8，439.0，478.8，144.6 千电子伏，这与图 1 中的能级差吻合得很好。

我们可以看到，能量最高的精细结构组分对应于原子核的正常态，而其他组分则是由损失掉部分能量的普通 α 粒子离开处于激发态的原子核造成的。

我非常感谢佩尔斯博士和罗森菲尔德博士给我提供在这里工作的机会。

（王静 翻译；江丕栋 审稿）

Eugenic Sterilisation

J. S. Huxley

Editor's Note

Eugenics—the attempted elimination of "bad genes" in a population—was widely held to be important for maintaining a healthy society for long after Charles Darwin published his evolutionary theory. It was advocated by Darwin's cousin Francis Galton, and Darwin himself assented. So did Julian Huxley, grandson of Darwin's staunch advocate Thomas Henry Huxley, who served in the British Eugenics Society until the 1960s. Here he writes to defend the society's recommendations of enforced sterilization of "mental defectives" against a criticism in a *Nature* editorial. Tellingly, that criticism was of the proposed mechanism, not the principle—*Nature* fully supported eugenic arguments in the 1930s. Only later did they become seen as not just morally but scientifically flawed.

AS a member of the Committee of the Eugenics Society for Legalising Eugenic Sterilisation, I should like to be allowed to say a few words concerning the leading article in *Nature* of Aug. 30 on our proposals. It is stated there: "Is there not a real danger that the advocates of such legislation as here may mistake the assent of the political machine for victory? If assent were gained, would it not be much more accurately determined as the hall-mark of failure? It is not the assent of the State, but the initiative and creative power of the State, that is needed to secure essential progress...."

With the last sentence I entirely agree; but I fail to perceive how a step in the right direction can be regarded as the hall-mark of failure—unless, indeed, the Committee should be so stupid as to believe that the taking of this one step had brought us to our final goal, which is certainly not the case. The article opens with references to the difficulties in the way of progress which are created by timid and ignorant public opinion, and continues, "if, as Sir Walter Fletcher has lately pointed out, a mere ailment, like cancer, has only been made accessible to scientific study through the lifting of foolish and superstitious taboos, how can we expect the direr social maladies to be approached courageously?" I think I can speak for the Committee in saying that we realise to the full the extent of these intangible difficulties, and that it is precisely for that reason that we have concentrated on a small but tangible and urgent beginning. Somehow or other the public has to be made race-conscious, has to be imbued with the eugenic idea as a basic political and ethical ideal. We believe that a campaign of the kind we have launched, directing attention to a gross racial defect, will be the best possible way of turning their thoughts in the desired direction.

优生绝育

赫胥黎

编者按

在查尔斯·达尔文提出进化论很长时间以后，优生学这种试图消除人类“不良基因”的学说被大家公认为是确保社会成员健康的重要方法。优生学是由达尔文的表弟弗朗西斯·高尔顿倡导的，并且得到了达尔文本人的支持。本文的作者朱利安·赫胥黎也是优生学的支持者，他是托马斯·亨利·赫胥黎的孙子。20 世纪 60 年代以前，他一直在英国优生学会工作。他写这篇文章的目的是，为该学会倡导的“精神病患者”应该被强制进行绝育作辩护，以反驳《自然》上持反对意见的一篇评论。事实上，那篇评论批驳的是提案中所说的运作机制，而不是批驳基本原则。早在 20 世纪 30 年代，《自然》就完全赞成优生学的观点。只是后来这些观点变得不仅不人道，而且出现了科学上的谬误。

作为优生学会优生绝育合法化委员会的一名成员，我想就《自然》8 月 30 日发表的那篇针对我们的提议的重要文章说几句。文中说到：“像这里提到的支持如此立法的倡导者们可能会将政治机构的同意误解为自己取得了胜利，这难道不是真正的危险吗？一旦得到同意，那么更准确地说，这更应该被视作失败的标志。因为这不是国家的同意，而只是国家自发的创造性力量的同意，这种力量是用来保障最基本的发展的……”

我完全同意最后一句话，但是我却不理解方向正确的措施怎么会被当作失败的标志呢？除非委员会愚蠢到相信仅仅通过实施这一措施就可以实现我们的终极目标，但事实上他们肯定没有这么认为。那篇文章一开篇就提出，怯懦而又无知的公众意识对优生绝育的实施造成了困难，紧接着写道“正如沃尔特·弗莱彻爵士最近指出的，如果对于一种纯粹的疾病（例如癌症），只有在破除愚蠢迷信的忌讳后才能对其进行科学研究的话，那么我们又怎么能期望研究人员会大胆地去研究更加可怕的社会弊病呢？”我想我可以代表委员会说，我们已经认识到了所有这些无形的困难，也正是因为这个原因，我们才集中精力从一个比较小但很明确很紧迫的问题入手。不管通过什么方法，都必须使公众具有种族意识，必须让他们把优生思想当作一项基本的政治道德理想。我们相信，我们发起的这种引导公众关注整个种族缺陷的运动，是把公众思想扭转到我们预期方向上的最可能有效的途径。

Comment is also made on the fact that the prevention of reproduction by all defectives would only lower the incidence of mental defect by about 17 percent in one generation. The article fails to remind readers that the process is cumulative, and also does not point to any other way in which it could be reduced more rapidly. Finally, the most relevant fact of all is omitted, namely, that one of the greatest obstacles to securing assent to the sterilisation of defectives has been and is the widespread belief that, since two normal persons may have a defective child, therefore preventing defectives from reproducing will have no effect on the proportion of defectives in later generations. Dr. R. A. Fisher has gone carefully into the matter, and has shown that, even when the most unfavourable assumptions are made, prevention of reproduction by all defectives would result in a reduction of some 17 percent—which to me at least seems considerable, as it would mean that there would be above 50,000 less defectives in Great Britain after the lapse of the, biologically speaking, trivial span of one generation.

I am glad that *Nature* has directed attention to the gravity of the problem, and look forward with interest to further discussion of the problem in its columns.

(**126**, 503; 1930)

J. S. Huxley: King's College, London, W. C. 2.

该文章还评论了如下事实：一代人中所有的缺陷个体都不生育，也将只能使下一代中心智缺陷的发生率减少大约 17%。该文章并没有提醒读者注意这一过程是累积的，也没有提出任何别的能使心智缺陷的发生率减少得更加迅速的方法。最后，该文章还遗漏了与减少心智缺陷的发生率关系最密切的事实，即，确保对缺陷个体实施绝育得到人们赞同的最大障碍之一，曾经是并且现在依然是广为流传的一种看法，那就是即使两个正常人也可能会生出一个有缺陷的孩子，因此阻止缺陷个体的生育这一做法对于后代中缺陷个体的出现比例不会有任何影响。费希尔博士已经对这一问题进行了仔细深入的研究，结果表明即使是在最坏的假设下，通过阻止所有缺陷个体的生育也会使下一代中的缺陷个体减少 17% 左右，这个数字至少对我来说算是相当可观了，因为从生物学意义上来说，这意味着仅仅是经过一代人的生育之后，英国心智缺陷者的数量的减少就会超过 50,000。

我很高兴《自然》能够关注这一问题的重要性，我也热切期待其专栏里能够出现对此问题的深入讨论。

（刘皓芳 翻译；刘京国 审稿）

The Proton*

P. A. M. Dirac

Editor's Note

Here English physicist Paul Dirac describes his prediction, reported earlier in another journal, that electrons may have "negative energy" and consequently positive charge. This prediction followed from Dirac's "relativistic" quantum theory, which made quantum mechanics compatible with special relativity. The "positive electrons" appear as "holes" in a pervasive electron sea. Dirac suspects they behave as protons, but there are problems with that: protons have very different masses from electrons, and the electrons and holes were predicted to annihilate. Dirac alludes to an idea of J. Robert Oppenheimer that a positive electron may in fact be a different particle with the mass of an electron. And so it later proved: they were positrons, anti-matter versions of electrons.

MATTER is made up of atoms, each consisting of a number of electrons moving round a central nucleus. It is likely that the nuclei are not simple particles, but are themselves made up of electrons, together with hydrogen nuclei, or protons as they are called, bound very strongly together. There would thus be only two kinds of simple particles out of which all matter is built, the electrons, each carrying a charge $-e$, and the protons, each carrying a charge $+e$.

It should be mentioned here that there is a difficulty in this point of view provided by the nitrogen atom. One can infer from the charge and mass of the nitrogen nucleus that it should consist of 14 protons and 7 electrons, but it appears to have properties inconsistent with its being composed of an odd number of simple particles. However, very little is really known about nuclei, and the opinion is generally held by physicists that some way of evading this difficulty will be found and that all nuclei will ultimately be shown to be made up of electrons and protons.

It has always been the dream of philosophers to have all matter built up from one fundamental kind of particle, so that it is not altogether satisfactory to have two in our theory, the electron and the proton. There are, however, reasons for believing that the electron and proton are really not independent, but are just two manifestations of one elementary kind of particle. This connexion between the electron and proton is, in fact, rather forced upon us by general considerations about the symmetry between positive and negative electric charge, which symmetry prevents us from building up a theory of the negatively charged electrons without bringing in also the positively charged protons. Let us examine how this comes about.

---

* Based on a paper read before Section A (Mathematical and Physical Science) of the British Association at Bristol on Sept. 8.

质　子*

狄拉克

编者按

这篇文章报道了英国物理学家保罗·狄拉克就电子可能具有“负能量”从而带有正电荷这一预测所作的论述，该预测在更早些时候已发表在其他期刊上。狄拉克的“相对论性的”量子理论使量子力学与狭义相对论得以相容，前述的预测正是这一理论的结果。“带正电荷的电子”就像是无处不在的电子海中的“空穴”。狄拉克猜测，这些“带正电荷的电子”的行为方式与质子类似，不过这一猜测存在一些问题：质子与电子在质量上的差别非常大，而且据预测电子和这些空穴相遇会湮灭。在这里，狄拉克也提到了罗伯特·奥本海默的观点，即带正电荷的电子可能就是与电子的质量相同的另一种粒子。后来人们证明确实如此：这些粒子就是正电子，电子的反物质形式。

物质是由原子构成的，每一个原子是由若干个围绕中心原子核转动的电子组成的。原子核很可能不是基本粒子，而是由电子和氢原子核（或者所谓质子）紧密束缚在一起构成的。这样所有的物质都只由这两种基本粒子构成，其中每一个电子带电荷 $-e$，每一个质子带电荷 $+e$。

这里需要指出的是，氮原子的存在给这个观点提出了一个难题。由氮原子核的电荷和质量，我们可以推断出氮原子核是由 14 个质子和 7 个电子组成的，但是氮原子核表现出来的性质似乎与它是由奇数个基本粒子构成这一点不符。然而，关于原子核，我们知之甚少，而且物理学家们普遍认为将来总会有办法克服这个困难，并且最终将会证明所有的原子核都是由电子和质子构成的。

哲学家总是梦想所有的物质都是由一种基本粒子构成的，所以我们的理论——包含两种基本粒子（电子和质子）——并不能使所有人都满意。然而人们有理由相信电子和质子并不是毫无关系的，它们只是一种基本粒子的两种表现形式。而事实上，电子和质子之间的联系在某种程度上是关于正负电荷之间对称性的一般认识强加给我们的，这种对称性使我们不能构建一套只包含带负电的电子，而不包含带正电的质子的理论。下面让我们看看为什么会是这样。

---

* 基于 9 月 8 日在布里斯托尔向英国科学促进会的 A 分部（数学和物理科学）宣读的一篇论文。

The energy W of a particle in free space is determined in terms of its momentum p according to relativity theory by the equation

$$W^2/c^2 - p^2 - m^2c^2 = 0,$$

where m is the rest-mass of the particle and c is the velocity of light. This equation can easily be generalised to apply to a charged particle moving in an electromagnetic field and can be used as a Hamiltonian to give the equations of motion of the particle, and thus its possible tracks in space-time.

Now the above equation is quadratic in W, allowing of both positive and negative values for W. Thus for some of the tracks in space-time the energy W will have positive values and for the others negative values. Of course a particle with negative energy (kinetic energy is referred to throughout) has no physical meaning. Such a particle would have less energy the faster it is moving and one would have to put energy into it to bring it to rest, quite contrary to anything that has ever been observed.

The usual way of getting over this difficulty is to say that the tracks for which W is negative do not correspond to anything real in Nature and are to be simply ignored. This is permissible only provided that for every track W is either always positive or always negative, so that one can tell definitely which tracks are to be ignored. This condition is fulfilled in the classical theory, where W must vary continuously, since W can never be numerically less than mc^2 and is thus precluded from changing from a positive to a negative value. In the quantum theory, however, discontinuous variations in a dynamical variable such as W are permissible, and detailed calculation shows that W certainly will make transitions from positive to negative values. We can now no longer ignore the states corresponding to a negative energy and it becomes imperative to find some physical meaning for them.

We can deal with these states mathematically, in spite of their being physically nonsense. We find that an electron with negative energy moves in an electromagnetic field in the same way as an ordinary electron with positive energy would move if its charge were reversed in sign, so as to be $+e$ instead of $-e$. This immediately suggests a connexion between negative-energy electrons and protons. One might be tempted at first sight to say that a negative-energy electron *is* a proton, but this, of course, will not do, since protons certainly do not have negative kinetic energy. We must therefore establish the connexion on a different basis.

For this purpose we must take into consideration another property of electrons, namely, the fact that they satisfy the exclusion principle of Pauli. According to this principle, it is impossible for two electrons ever to be in the same quantum state. Now the quantum theory allows only a finite number of states for an electron in a given volume (if we put a restriction on the energy), so that if only one electron can go in each state, there is room for only a finite number of electrons in the given volume. We thus get the idea of a *saturated* distribution of electrons.

根据相对论，自由空间中粒子的能量 W 由它的动量 p 决定，即

$$W^2/c^2 - p^2 - m^2c^2 = 0 ,$$

其中 m 是粒子的静止质量，c 是光速。这个方程可以很容易地推广到带电粒子在电磁场中运动的情况，并且可以被用作哈密顿量，给出带电粒子的运动方程，从而得到带电粒子在时空中可能的径迹。

上面的方程中 W 项是二次的，所以 W 既可能是正的，也可能是负的。因此对时空中的一些径迹而言能量 W 是正的，而对其他一些则是负的。当然粒子具有负能量（动能总是会涉及）是没有物理意义的。这样的粒子运动得越快，它的能量就越小，我们不得不给它能量使它静止，然而这与我们观察到的所有现象都是截然不同的。

通常克服这个困难的办法是认为具有负 W 的径迹不对应于任何真实的自然现象，而只需要简单地把它忽略掉。不过这只有在每一条径迹的 W 值恒正或者恒负的前提下才成立，因为只有这样我们才可以明确地判断哪一条径迹应该被忽略。这个条件在 W 连续变化的经典理论中是满足的，因为 W 在数值上不能小于 mc^2，所以排除了 W 从正值变化到负值的可能性。然而，在量子理论中，像 W 这样的动力学量可以不是连续变化的，并且详细的计算表明 W 确实可以从正值变化到负值。因此我们不能再忽略负能量对应的状态，而必须为它们寻找某种物理意义。

我们可以从数学上处理这些状态，而先不去管它们是否具有物理意义。我们发现，如果普通电子的电荷符号发生翻转，即从 $-e$ 变为 $+e$，那么一个具有负能量的电子在电磁场中的运动方式和一个普通的具有正能量的正电子一样。这就意味着负能电子和质子之间存在某种联系。乍一看这种情况，人们可能会说负能电子**就是**质子，但是这无疑是不成立的，因为质子的动能不可能是负的。因此，我们必须在另外的基础上构建它们的联系。

为此我们必须考虑电子的另外一个特性，即它们满足泡利不相容原理。根据这一原理，两个电子永远不可能处于同一个量子态。因为量子理论在给定的空间内只允许有限数目的电子态（如果我们给能量一个限制），所以如果每一个态只允许一个电子占据，那么在给定的空间内只能容纳有限数目的电子。这样我们就会得到电子**饱和**分布的概念。

Let us now make the assumption that almost all the states of negative energy for an electron are occupied, and thus the whole negative-energy domain is almost saturated with electrons. There will be a few unoccupied negative-energy states, which will be like holes in the otherwise saturated distribution. How would one of these holes appear to our observations? In the first place, to make the hole disappear, which we can do by filling it up with a negative-energy electron, we must put into it a negative amount of energy. Thus to the hole itself must be ascribed a positive energy. Again, the motion of the hole in an electromagnetic field will be the same as the motion of the electron that would fill up the hole, and this, as we have seen, is just the motion of an ordinary particle with a charge $+e$. These two facts make it reasonable to assert that *the hole is a proton*.

In this way we see the proper role to be played by the negative-energy states. There is an almost saturated distribution of negative-energy electrons extending over the whole of space, but owing to its uniformity and regularity it is not directly perceptible to us. Only the small departures from perfect uniformity, brought about through some of the negative-energy states being unoccupied, are perceptible, and these appear to us like particles of positive energy and positive charge and are what we call protons.

This theory of the proton involves certain difficulties, which will now be discussed. The theory postulates the existence everywhere of an infinite number of negative-energy electrons per unit volume, and thus an infinite density of electric charge. According to Maxwell's equations, this would give rise to an infinite electric field. We can easily avoid this difficulty by a re-interpretation of Maxwell's equations. A perfect vacuum is now to be considered as a region in which all the states of negative energy and none of those of positive energy are occupied. The electron distribution in such a region must be assumed to produce no field, and only the departures from this vacuum distribution can produce a field according to Maxwell's equations. Thus, in the equation for the electric field E

$$\operatorname{div} E = -4\pi\rho,$$

the electric density ρ must consist of a charge $-e$ for each state of positive energy that is occupied, together with a charge $+e$ for each state of negative energy that is unoccupied. This gives complete agreement with the usual ideas of the production of electric fields by electrons and protons.

A second difficulty is concerned with the possible transitions of an electron from a state of positive energy to one of negative energy, which transitions were the original cause of our having to give a physical meaning to the negative-energy states. These transitions are very much restricted when nearly all the negative-energy states are occupied, since an electron in a positive-energy state can then drop only into one of the unoccupied negative-energy states. Such a transition process would result in the simultaneous disappearance of an ordinary positive-energy electron and a hole, and would thus be interpreted as an electron and proton annihilating one another, their energy being emitted in the form of electromagnetic radiation.

现在我们假设电子的负能态几乎都被占据了，因此整个负能区域电子几乎是饱和的。有一些没被占据的负能态，它们就像饱和分布的负能态电子海中的一些空穴。在我们看来这些空穴是什么样的呢？首先，为了使这些空穴消失，我们需要填充一个负能量的电子，即放入一个负能量。这样空穴本身必须具有正能量。其次，空穴在电磁场中的运动方式和填充空穴的电子的运动方式一样，就像我们之前看到的那样，就是一个带 $+e$ 电荷的普通粒子的运动。这两个事实使我们有理由断言——**空穴就是质子**。

这样我们就看到了负能态起到的作用。整个空间中负能电子几乎处于饱和分布，但由于它们表现出来的均匀性和规律性，因而不能直接被我们觉察到。只有在完美的均匀性上出现一些小的偏离，即有一些负能态没被占据，才能被我们觉察到。这些偏离在我们看来就是些具有正能量和带正电荷的粒子，也就是我们所谓的质子。

这个关于质子的理论存在一些问题，下面我们就来讨论这些问题。这个理论假设空间每处单位体积内都存在无限数目的负能电子，这样电荷密度就是无穷大了。根据麦克斯韦方程，这将会产生一个无限大的电场。不过我们可以很容易地通过重新解释麦克斯韦方程来克服这个困难。理想的真空被认为是所有的负能态都被占据，而所有的正能态都没被占据的空间。我们认为在这样的空间中，电子分布不会产生任何场，只有当电子分布偏离真空分布时才会产生根据麦克斯韦方程得到的场。因此，在电场 E 的方程中

$$\operatorname{div} E = -4\pi\rho\ ,$$

电荷密度 ρ 是由每一个被占据的正能态上的电荷 $-e$ 和每一个没被占据的负能态上的电荷 $+e$ 组成的。这和通常电子和质子产生电场的观点是完全一致的。

第二个困难涉及电子可能存在从正能态向负能态的跃迁，这种跃迁是我们必须赋予负能态以物理意义的最初原因。当几乎所有的负能态都被占据时，这种跃迁是非常受限制的，因为处于正能态的电子只能落入没被占据的负能态。这样的跃迁过程导致一个普通的正能态电子和一个空穴同时消失，所以可以解释成一个电子和一个质子的互相湮灭，它们的能量以电磁辐射的形式发射出来。

There appears to be no reason why such processes should not actually occur somewhere in the world. They would be consistent with all the general laws of Nature, in particular with the law of conservation of electric charge. But they would have to occur only very seldom under ordinary conditions, as they have never been observed in the laboratory. The frequency of occurrence of these processes according to theory has been calculated independently by several investigators, with neglect of the interaction between the electron and proton (that is, the Coulomb force between them). The calculations give a result much too large to be true. In fact, the order of magnitude is altogether wrong. The explanation of this discrepancy is not yet known. Possibly the neglect of the interaction is not justifiable, but it is difficult to see how it could cause such a very big error.

Another unsolved difficulty, perhaps connected with the previous one, is that of the masses. The theory, when one neglects interaction, requires the electron and proton to have the same mass, while experiment shows the mass ratio to be about 1,840. Perhaps when one takes interaction into account the theoretical masses will differ, but it is again difficult to see how one could get the large difference required by experiment.

An idea has recently been put forward by Oppenheimer (*Phys. Rev.*, vol. 35, p. 562) which does get over these difficulties, but only at the expense of the unitary theory of the nature of electrons and protons. Oppenheimer supposes that all, and not merely nearly all, of the states of negative energy are occupied, so that a positive-energy electron can never make a transition to a negative-energy state. There being now no holes which we can call protons, we must assume that protons are really independent particles. The proton will now itself have negative-energy states, which we must again assume to be all occupied. The independence of the electron and proton according to this view allows us to give them any masses we please, and further, there will be no mutual annihilation of electrons and protons.

At present it is too early to decide what the ultimate theory of the proton will be. One would like, if possible, to preserve the connexion between the proton and electron, in spite of the difficulties it leads to, as it accounts in a very satisfactory way for the fact that the electron and proton have charges equal in magnitude and opposite in sign. Further advances in the theory of quantum electrodynamics will have to be made before one can deal accurately with the interaction and see whether it will settle the difficulties, or whether, perhaps, a new idea can be introduced which will answer this purpose.

(**126**, 605-606; 1930)

看起来没有什么理由可以说明为什么这样的过程不能在现实世界的某处发生。它们会遵守自然界所有的一般规律，特别是电荷守恒定律。但是它们在普通条件下必然很少发生，因为即使在实验室中它们也还没有被观察到。一些研究者已经独立地计算出这些过程发生的理论几率，计算中忽略了电子和质子之间的相互作用（即它们之间的库仑力）。计算给出的结果太大了，肯定是不正确的。事实上，结果的数量级都是完全错误的。现在还不知道为什么会出现这样的差异。可能忽略相互作用是不合理的，但是仍然很难理解为什么会导致这么大的错误。

另外一个没有解决的困难就是质量问题，这可能和前一个困难有关。如果忽略相互作用，这个理论就要求电子和质子具有相同的质量，然而实验表明它们的质量比约为 1,840。也许考虑相互作用后理论上的质量会有所不同，但还是很难理解怎样才能得到实验要求的那么大的质量差。

奥本海默最近提出的一个观点（《物理学评论》，第 35 卷，第 562 页）的确可以解决这个困难，但是它却牺牲了关于电子和质子本质的统一理论。奥本海默假定所有的（不仅仅是几乎所有的）负能态都被占据了，因此正能电子不能跃迁到负能态。这里没有我们可以称之为质子的空穴，所以我们必须假定质子是真正独立的粒子。这样，质子本身也有自己的负能态，而且我们必须假设它们也被完全占据了。根据这个观点，电子和质子的独立性允许我们随心所欲地给它们的质量赋值，而且它们之间也不会相互湮灭。

目前断定质子的最终理论还为时尚早。如果可能的话，人们愿意保留电子和质子之间的这种关系，而不管它带来的困难，因为它非常圆满地解释了这个事实——电子和质子携带的电荷大小相等，而符号相反。量子电动力学需要进一步的发展，人们才可以准确地计算相互作用，才可以知道我们的理论是否可以解决这些困难，或者是否会出现新的可以回答这个问题的观点。

（王锋 翻译；李淼 审稿）

The Problem of Epigenesis

E. W. MacBride

Editor's Note

Ernest William MacBride is perhaps in retrospect not the ideal author for this discourse on embryogenesis, being a late supporter of the Lamarckian view of inheritance. Yet he was considered an expert on embryology, and here he anticipates some of the key themes in modern biology: how is the development of an embryo related to its evolutionary heritage (the topic now dubbed "evo-devo"), and how does the organism, "considered as a machine", function? The former question had received much attention in Germany (MacBride's article is a review of three recent German books), especially in Ernst Haeckel's notion that embryogenesis recapitulates evolutionary history. The main issue debated here, however, is the origin of the force that organizes an undifferentiated egg into a structured body, and how that depends on the emerging concept of genes.

(1) *Grundriss der Entwicklungsmechanik*. Von Prof. Dr. Bernhard Dürken. Pp. vii + 208. (Berlin: Gebrüder Borntraeger, 1929.) 12.50 gold marks.

(2) *Die Determination der Primitiventwicklung: eine zusammenfassende Darstellung der Ergebnisse über das Determinationsgeschehen in den ersten Entwicklungsstadien der Tiere*. Von Prof. Dr. Waldemar Schleip. Pp. xii + 914. (Leipzig: Akademische Verlagsgesellschaft m.b.H., 1929.) 85 gold marks.

(3) *Experimentelle Zoologie: eine Zusammenfassung der durch Versuche ermittelten Gesetzmässigkeiten tierischer Formen und Verrichtungen*. Von Prof. Dr. Hans Przibram. Band 6 : *Zoonomie; eine Zusammenfassung der durch Versuche ermittelten Gesetzmässigkeiten tierischer Formbildung* (*Experimentelle, theoretische und literarische Übersicht bis einschliesslich 1928*). Von Prof. Dr. Hans Przibram. Pp. viii + 431 + 16 Tafeln. (Leipzig und Wien: Franz Deuticke, 1929.) 40 gold marks.

THE question of epigenesis may be justly said to constitute one of the two root problems of zoology. For if we think it out there are two main things to be discovered about an animal, namely: (1) How does it fulfil its functions?—in a word, considered as a machine, how does it work? and (2) How does it come into being?—that is, how did it develop and grow? A subsidiary question to the last is: If there be such a thing as evolution, how and why did the powers of growth change from generation to generation? For, as the late Dr. Bateson reminded us so long ago as 1894, the conception of evolution as the remoulding of the adult structures of an animal as we could alter the features of a wax doll by melting the wax and remodelling it, is an entire illusion, since the members of the parent species and of that to which it gives rise both begin as tiny formless germs and what is changed is *the powers of growth*. Now when we begin to analyse growth, we can either directly observe its successive phases—and this is the scope of descriptive embryology; or by operating on the germ by chemical and physical agencies we can seek to discover the

关于渐成论的问题

麦克布赖德

编者按

回想起来，由欧内斯特·威廉·麦克布赖德来作这个胚胎发生学方面的报告可能不是很理想，因为他是拉马克遗传学说的晚期支持者之一。不过，他毕竟是胚胎学方面的专家，而且在这篇报告中他也预见到了现代生物学的一些关键性课题：胚胎的发育过程与其种系的进化过程有什么关系（这一研究课题现在被称为“进化发育生物学”）？“被视为机器”的有机体如何发挥功能？实际上，在此之前德国科学家就对前一个问题给予了极大的关注（麦克布赖德的这篇文章就是对最近相关的3本德国科学家的著作所作的评论），特别是恩斯特·海克尔提出了胚胎发生重演了进化过程的观点。不过，这篇文章讨论的主要问题是，由未分化的受精卵形成有结构的机体的原始动力是什么，以及这一动力与人们最近提出的基因这一概念有何关系。

(1)《发育机制概论》。伯恩哈德·迪肯博士，教授。vii + 208 页。(柏林：施普林格兄弟，1929 年)。12.50 金马克。

(2)《原始发育的决定性：关于动物早期发育中决定因素的结果总结》。瓦尔德马·施莱普博士，教授。xii + 914 页。(莱比锡：学术出版有限责任集团公司，1929 年)。85 金马克。

(3)《实验动物学：关于实验动物挑选的法律规范和步骤概要》。汉斯博士，教授。第 6 卷：动物学；有关实验动物的法律的总结概述（理论和文学概述，以及试验，1928 年）。汉斯·普西布兰博士，教授，xiii + 431 页 + 16 表格。(莱比锡和维也纳：弗朗茨·多伊蒂克，1929 年)。40 金马克。

公正地说，渐成论的问题可能是动物学的两大根本问题之一。因为我们在研究一个动物时会思考两个主要问题：(1) 它是怎样实现它的功能的？简言之，如果把动物看作一台机器的话，它是如何工作的？(2) 它是怎样产生的？即，它是如何发育和生长的？伴随第二个问题又会出现另一个问题：如果真的存在进化的话，那么生长的动力是如何一代代发生变化的，以及为什么会发生变化呢？正如已故的贝特森博士早在 1894 年就提醒我们的，把进化看成是对成体动物结构的重塑，就如同我们通过熔化蜡并进行重建来改变蜡人的造型特点一样，这是完全错误的概念，因为亲代物种的成员以及他们产生的子代都是从无定形的微小生殖细胞开始生长而来的，而**生长的动力**在变化。现在当我们开始分析生长的时候，我们可以直接观察到其连续进行的各个阶段——这属于描述胚胎学的范畴，我们也可以通过我们能够找到的化学和物理手段对生殖细胞进行实验来观察每一个可见因素在成年个体的生长

part which each visible element plays in the upbuilding of the adult individual—and this is the object of experimental embryology.

How this science has grown since its first beginnings with His in 1874 ("Unser Körperform und die physiologische Problem ihrer Entstehung") is witnessed by the three splendid works which are the subject of this review. Each of the three is worthy of unstinted praise: though we may differ from the authors in some of the conclusions reached by them, yet in each case the collection and setting forth of the matter is worthy of our sincere admiration. We hope that too long a time may not elapse before all are translated into English.

As an introduction to the subject Dürken's manual is to be preferred, because it is concise, well illustrated, and includes only typical cases which serve to exemplify the main principles of the subject, so that a beginner can get a good grasp of these principles without being overwhelmed by too much detail. Schleip's large and well-illustrated volume attempts to give a more or less complete account of the present state of our knowledge of the subject, and it will for a long time constitute a classic work of reference. Przibram's work—thorough and excellent as all his work is—is even more ambitious in its scope than that of Schleip, for it includes not only the facts of experimental embryology in the narrower sense, but also a considerable amount of the results of Mendelian experiments. It is, however, extremely condensed and, not being adequately illustrated, somewhat difficult to follow: it seems to us that its chief value will reside in its being a manual in which references to all the important papers on the subject can be easily looked up.

It must be obvious to the reader that, within the limits of the longest review for which space can be found in *Nature*, it would be impossible to refer to a tithe of the new matter contained in these volumes, and so we must limit ourselves to a discussion of the main problems involved and to the attitude of the three authors towards them. In fairness, however, it should be added that this new matter is almost entirely confined to an elaboration of subjects dealt with by the older authors such as Roux, Hertwig, Driesch, Herbst, Boveri, Conklin, and Wilson, and does not consist to consist to any considerable extent of discoveries in newer fields. The number of animals the eggs of which can conveniently be handled and which are tolerant of experiments is limited, and the same familiar figures crop up in successive text-books of experimental embryology. After all, as Driesch has wisely remarked, the biological experimenter cannot produce life at will—he must wait until he finds it, and he is therefore in the same position as a physicist would be if he could only study fire when he found it in the crater of a volcano.

When we approach the analysis of the development of the egg, the first question we encounter is whether the organs of the adult exist in the egg preformed in miniature and development consists essentially in an unfolding and growing bigger of these rudiments, or whether the egg is at first undifferentiated material which from unknown causes afterwards becomes more and more complicated and development is consequently an "epigenesis". This problem is *the* problem of experimental embryology; in varied forms it reappears in every experiment on development which has been made.

过程中都起着什么作用——这是实验胚胎学的目标。

这里即将评论的3部出色的著作描述了这一领域从1874年西斯首次发表著作(《我们的身体形态、构造和生理问题》)以来的发展过程。这3部作品中的任何一部都是值得高度赞扬的，尽管我们可能在某些结论上和作者有不同观点，但是每部作品中对事件的收集及详尽阐述都值得我们致以由衷的敬意。我们希望不久之后，所有这些著作都可以被翻译成英语。

迪肯的这本指南是该学科的首选入门书籍，因为该书简洁明了、插图丰富，只包含解释该学科主要原理的经典事例，所以初学者可以很好地掌握这些原理而不会被铺天盖地的细节吓倒。施莱普的那本插图丰富的大部头试图尽可能全面地向我们阐明该学科的知识现状，该书将在很长时间内成为这一领域的经典参考著作。普西布兰的这部著作和他的其他所有著作一样全面而出色，就其视野来说比施莱普的那本更具远见，因为在这本书里，不仅包括狭义实验胚胎学的内容，也包括相当一部分孟德尔实验的结果。不过，这本书极度浓缩，描述不够详细，因而想要读懂会有些难度。对于我们而言，它的主要价值似乎在于，这是一本可以从中很容易地找到该学科所有重要文章引用的参考文献的手册。

读者们都知道，《自然》上发表的评论是有字数限制的，受此所限，即使只是想谈论这些书中包含的一小部分新内容也是不可能的，因此我们只好仅限于讨论涉及的主要问题以及3位作者对这些问题的看法。不过，应该补充说明的是，公平地讲，这些新内容几乎完全局限于对老一辈作者所讨论主题的细化，这些作者包括鲁、赫特维希、德里施、赫布斯特、博韦里、康克林和威尔逊，而并没有涉及较新领域的任何重要发现。能够方便地对其卵子进行操作并保证卵子可以耐受实验条件的受试动物的数量是有限的,那些熟悉的图片后来连续出现在实验胚胎学的教科书中。毕竟，正如德里施曾经明智地提出过的，生物学实验者不能随意制造生命——他只有在找到受试的生命体后才能对其进行实验，因此从这个角度来看，生物学实验者的处境就如同只有在火山爆发时发现了火之后才能对火进行研究的物理学家的处境。

当我们对卵的发育进行分析时，我们面临的第一个问题就是：是不是卵子中就存在预先成形了的成体器官的缩微版，而发育过程实质上是这些器官雏形逐渐展开显露并变大的过程？又或者卵子最初只是未分化的物质，后来由于未知的原因而变得越来越复杂，因而发育是一个“渐成的”过程？这正是实验胚胎学要解决的问题。在研究发育的实验中，这一问题以各种各样的形式重复出现。

The answer to this question given by the earlier experiments of Driesch was that some eggs, such as those of starfish and sea-urchins, consist of undifferentiated material; but others, like those of Ctenophores, show a specialisation into parts destined to form particular organs of the adult. The experiments of Wilson, Conklin, and Crampton proved that the eggs of Annelida and Mollusca belong also to this latter category. To eggs of the first kind Driesch gave the name of "equipotential systems", since when the egg had divided into eight cells any one of these was capable of forming a tiny larva perfect in all details, and, moreover, when the egg had developed into a hollow sphere or blastula, any considerable piece of this blastula would round itself off and form a perfect blastula of reduced size, which would give rise to a correspondingly reduced larva. On these results, which were a complete surprise to him, Driesch founded his theory of vitalism, arguing that if the organism were to be regarded as a physico-chemical machine, such things could not happen, for no conceivable machine could be divided into parts, each of which would function as a similar machine of reduced size. He inferred that there must be in every egg a non-material force or "entelechy" which was capable of controlling the physical and chemical changes taking place in the germ, so as to direct them towards a definite end. This power of direction was named by Driesch "regulation". This revolutionary idea of Driesch, transcending the bounds of materialistic explanation, evoked the fiercest opposition amongst those biologists by whom life was regarded as nothing more than complicated chemistry. Yet the arguments of Driesch have never been successfully met. The utmost that can be urged against them is the assertion that, although we cannot explain life by physics and chemistry now, some day in the distant future, when we have made further discoveries, we may possibly be able to do so.

Of the authors reviewed in this article, Dürken is inclined to favour Driesch whilst Schleip and Przibram oppose him, but the alternative explanations of the two latter authors when examined in detail resolve themselves into saying the same things that Driesch said, in different phrases. All three authors agree in showing that between equipotential and specialised eggs every conceivable grade of intermediate exists, and that even the eggs of *Echinus* itself are not quite so equipotential as Driesch imagined. Schleip quotes the work of Hörstatius as proving that when the upper half of a blastula is cut off, though it will round itself off so as to form a reduced blastula, yet this will never form endoderm or proceed any further in development. The vegetative half, however, when severed will produce a completely viable gastrula. By a triumph of manipulative skill, Hörstatius succeeded in separating the vegetative pole of a blastula and grafting it in various positions on another blastula in which an appropriate defect had been produced. He thus proved that in all cases development begins in the graft, and that this graft can change cells that would otherwise produce ectoderm into endoderm, in other words, act as an "organiser" of development.

Driesch attributed specialisation in eggs to a "premature stiffening of the cytoplasm" which prevented the "entelechy" from moulding the fragment of the egg into a reduced whole. Przibram in other language comes to exactly the same conclusion. He says that the formation of definite organs is in all cases due to a *solidifying* of a portion of the

德里施早期的实验对于这一问题给出的答案是，像海星和海胆这一类的生物的卵是由未分化的物质组成的；而其他生物，如栉水母类，它们的卵则特化为几个不同的部分，每一个部分都特定发育为成体的特定器官。威尔逊、康克林和克兰普顿的实验证明，环节动物和软体动物的卵也属于第二类。对于第一类卵，德里施称它们是“等潜能系统”，因为当这类卵分裂成 8 个细胞时，其中任何一个都具有成长为所有细节部分都完整的小幼虫的能力。此外，当卵发育成中空的球体或囊胚时，该囊胚的任何一部分有一定大小的片段都可以完善自身，并形成一个体积相对较小的完好的囊胚，相应地这一囊胚可以产生一个体积较小的幼虫。这些结果使德里施感到无比惊讶，基于此，他建立了自己的活力论。该学说认为，如果把生物当作一台物理化学机器的话，这些情况就不会发生，因为想象不出任何机器可以在被分成几部分后，各个部分仍然能够像一台只是尺寸有所减小的类似机器一样正常运转。他推断每个卵中一定都存在一种非物质的驱动力或者“生机”，它具有控制生殖细胞中发生的物理和化学变化的能力，因而可以指导这些变化朝着特定的方向发展。德里施将这种指导能力称为“调控”。德里施这一革命性的想法超越了唯物论解释的范畴，激起了那些认为生命仅仅是一些复杂化学变化的生物学家们最强烈的反对。然而没有任何人在与德里施的观点的交战中取胜。这些极力反对的观点中，分量最重的也只不过是如下的论断：尽管我们现在不能用物理化学变化来解释生命，但是在遥远未来的某一天，当我们取得了进一步发现的时候，我们可能就有能力对其进行解释了。

在本文所评论的这些作者中，迪肯倾向于支持德里施的观点，而施莱普和普西布兰则反对德里施的观点，但是经过仔细推敲后可以看到，后两位作者提出的另外的解释其实与德里施的观点是一样的，只是说法不同而已。所有这 3 位作者都同意，每一种可以想象到的介于等潜能的卵与特化的卵之间的中间状态都是存在的，甚至连海胆本身的卵也并不像德里施想象的那样处于完全等潜能的状态。施莱普引用了赫斯塔提乌斯的工作，以此为证据来证明，当囊胚的上半部分被分离下来时，尽管它可以完善自己而形成一个体积减小了的囊胚，然而却并不能形成内胚层，也不会继续发育。植物极的那一半在被切下来后则可以生成一个完全能存活下去的原肠胚。由于操作技术上的巨大突破，赫斯塔提乌斯成功地分离了囊胚的植物极并将其植入到另一囊胚（此囊胚事先已经进行了相应的切除）的多个位置上。他由此证明了：在所有这些例子中，发育都是从植入物上开始的，这种植入物可以改变细胞，使本来会产生外胚层的细胞转而形成内胚层，换言之，植入物扮演了发育的“组织原”的角色。

德里施将卵的特化归因于“细胞质的过早硬化”，这阻止了“生机”将卵的片段塑造成尺寸有所减小的完整卵。普西布兰用不同的语言书写了几乎完全相同的结论。他说，在所有情况下特定器官的形成都是由部分细胞质的**硬化**引起的，从而形成了

cytoplasm, forming what he calls an "apoplasm" which, if we understand him right, he does not regard as fully alive. In proportion as "apoplasms" are deposited the potentialities of the germ are successively limited, and the reason why the higher animals approximate in their working to mechanisms is the large number of "apoplasms" included in their make-up. Only fluid cytoplasm is completely living and possesses all the potentialities of the race, and Przibram is driven to conclude that these potentialities, so far as embodied at all, must be contained in the molecules of the cytoplasm, and that, therefore, these molecules constitute the real entelechy. Schleip similarly concludes that there must be an ultra-microscopic structure in the cytoplasm which, like a crystal, tends to assume a definite form and to complete itself when a fragment is severed.

In making these admissions, however, it seems to us that both Schleip and Przibram deliver themselves into the hands of Driesch. For in the crystallisation of an inorganic substance from a solution, the crystal assumes a definite form because its molecules have definite corresponding shapes, as Sir William Bragg has taught us. But what kind of structure, whether molecular or super-molecular, are we to envisage in cytoplasm? When the limb of a young newt is cut off and the stump proceeds to regenerate a new limb, are the molecules in the stump in the form of infinitesimal fingers and toes? Moreover, when the stump is cut at different levels and only the missing piece is regenerated, are we to assume that at each level in the limb before amputation the molecules are miniatures of the part distal to them? If we are able to swallow these fantastic assumptions, what are we to say of the experiment recorded by Dürken in which the tail bud of one newt embryo was grafted into the body of another near its forelimb and developed into a new limb? Presumably the cytoplasm of the tail bud was "organised" so as to produce the tissues of an adult tail. How then was this organisation so completely changed as to produce a limb instead? No wonder that Dürken says that in cases like this, physical and chemical explanations leave us completely in the lurch, and we must have recourse to the conception of the "biological field", an influence not in the living matter itself, but in the space, presumably the ether, around it.

Schleip seeks to disprove Driesch's theory by pointing out that the supposititious entelechy sometimes does foolish things, as in the case of the eggs of Nematoda subjected to centrifugal force each of which produces two partial embryos instead of one whole one. But in this objection lurks the childish conception that the entelechy, if it exists, must be the embodiment of Divine Wisdom. the entelechy is not all-seeing—it is a rudimentary "striving" which reacts to its immediate environment, in this case the "apoplasm" or ball of dead matter ejected from the egg by centrifugal force.

The term "organiser" we owe, of course, to Spemann, who wisely abstains from giving any chemical explanation of it. In the course of his marvellous experiments on the newt, Spemann showed that a piece of the dorsal lip of the blastopore of one newt gastrula grafted on the flank of another would change the fate of all the cells in its neighbourhood and force them to develop into a supplementary nerve-cord and underlying notochord. The reviewer might humbly plead that exactly the same conception was reached by him

他称为“质外体”的结构。如果我们没有理解错的话，德里施并不认为“质外体”完全具有生命的结构。根据所形成的“质外体”的多少，生殖细胞的潜能也会成比例地受到相应的限制。高等动物具有相似的生长机理正是由于它们的组成物质中包含大量的“质外体”。只有流动态的细胞质具有完全生命力并拥有该种系的全部潜能。普西布兰认为，所有个体发育中都包含的这些潜能一定存在于细胞质的分子中，因此这些分子构成了真正的生机。与此相似，施莱普认为细胞质中一定存在一种超微观结构，这种结构像晶体一样倾向于形成一种固定的形式，并且当其中一部分被切下来时，它可以再完善自己。

不过，如果认可这些说法，那在我们看来施莱普和普西布兰似乎都成了德里施的支持者。一种无机物从溶液中结晶析出时，正如威廉·布拉格爵士告诉我们的，析出的晶体会呈现出一定的形状，因为这种物质的分子具有相应的确定形状。但是我们应该弄清楚的是，细胞质中的结构到底是什么样子的？构成细胞质的是分子还是超分子？当蝾螈幼体的四肢被切除后，其残肢可以继续再生出新的四肢，那么残肢的分子是不是以极小的手指和脚趾的形式存在？此外，当上述残肢被不同程度地切除时，只有缺失的部分能够再生出来，那么我们是否可以认为，在以不同程度切除四肢之前，四肢末端的微缩版就已经存在于四肢的分子之中？如果我们轻信这些荒谬的设想的话，那么对于迪肯记录的现象，即，在实验中将一只蝾螈胚胎的尾芽移植到另一只蝾螈前肢附近的躯体上，结果发育出一只新的前肢，我们又如何解释呢？大概是尾芽的细胞质被“组织化”而产生了成体尾巴的组织。那么这一组织是如何完全改变发育方向而产生前肢的呢？难怪迪肯说，像这种情况，物理和化学解释根本无能为力，我们必须依靠“生物学领域”的概念，这种影响力不在生命物质上，而可能是在其周围的空间中，有可能是以太。

施莱普不同意德里施的理论。他指出，生机假设有时是很愚蠢的，例如当线虫类的卵受到离心力作用时，每个卵并不是形成一个完整的胚胎，而是产生两个不完整的胚胎。但是这一反对观点中潜藏着一个幼稚的概念，即，如果生机真的存在的话，那么它一定是神性智慧的化身。生机并不是全能的——它是对周围环境作出反应的最基本的“努力”，在上述例子中，它就是“质外体”或者被离心力驱逐出卵的无机球状物。

当然，我们认为“组织原”一词的发明者是施佩曼，他很明智地拒绝给这个词赋予任何化学解释。在其令人称奇的蝾螈实验中，他向我们展示了，将一只蝾螈的原肠胚中的胚孔背唇片断移植到另一只蝾螈的原肠胚的侧面，就会改变附近所有细胞的命运，促使它们发育成一条辅助的神经索和深层脊索。本人在此谦恭地为自己辩护，早在 1918 年我就提出来了同样的概念，并且发表在一篇题为《具有两套水管

and published in a paper which appeared in 1918 entitled "The artificial production of Echinoderm larvae with two water-vascular systems and also of larvae devoid of a water-vascular system" (*Proc. Roy. Soc.*, B, vol. 90). In this paper he showed that when under the stimulus to hypertonic sea-water a second hydrocoele bud was produced in the pluteus, it completely altered the fate of all the tissues near it. It unfortunately did not occur to him to invent the term "organiser."

Of what nature is the influence emitted from the "organiser"? Here again all physical and chemical analogies fail to help us. If the influence were merely a physical or chemical force it would *combine* with the growth-forces of the organised tissue, and what we should observe would be the *resultant* of the two forces. The complete domination of one part by another is not a physical but a vital phenomenon and an instance of Driesch's "regulation".

It would be a fair conclusion to draw from all that has been discovered in the field of embryology to say that in broad outline there are three stages in development, namely: (1) Division of the egg into cells—that is, segmentation; (2) differentiation of these cells so as to form the three primary layers—ectoderm, endoderm, and mesoderm; (3) The action of portions of one layer on the neighbouring parts of other layers so as to form definite organs—that is, the action of organisers.

The ultimate question, however, whence the original organisation of the cytoplasm of the egg is derived, must now be faced. The only answer possible is the nucleus. It is true that, as we have seen, many eggs when ready for fertilisation have an already differentiated cytoplasm. But the cytoplasm of these eggs *when young* is undifferentiated, and during ripening their nuclei are engaged in emissions into the cytoplasm. In particular the nucleolus has been repeatedly observed to become broken into fragments which pass through the nuclear membrane and become dissolved in the cytoplasm. If we take such a specialised egg as that of the Nematode *Ascaris*, Boveri has shown that if it is subjected to centrifugal force *when young*, large portions of the cytoplasm can be shorn away and yet the reduced egg will give rise to a typical embryo. To this conclusion Schleip and Przibram also consent. But it seems to us that a further conclusion follows which they have not clearly envisaged. When differentiation of the cells of the blastula takes place, this must be due to further emissions from the nuclei. But the nuclei in these early stages of development are all alike, and by means of pressure experiments, these nuclei, as Hertwig has put it, may be juggled about like a heap of marbles without altering the result. Moreover, so far as can be judged by the most minute cytological examination, they remain unchanged in their essential make-up throughout the whole of development. So we reach the conception of an *intermittent action of the nuclei on the cytoplasm* giving rise to successive differentiations, that is, stages of development; and as it is by means of these stages that development is directed towards a definite end, if there be an entelechy, we may conclude that the mode of its action is by nuclear emissions. These emissions are the physical correlates of what Uexküll in his "Theoretische Biologie" (1927) calls the "Impulse" to development and the distinguishing of which, he avers, constitutes the utmost limit to which biological analysis can go.

系统的棘皮动物及其缺少水管系统的幼虫的人工培育》(《皇家学会学报》，B 辑，第 90 卷）的论文中。我在该论文中提到，当受到高渗海水的刺激时，长腕幼虫就可以长出第二个水系腔芽体，它的出现会完全改变附近所有组织的命运。但是很遗憾，并不是我发明了“组织原”这个词。

那么“组织原”产生的影响的本质是什么呢？对于这个问题的解释，物理的和化学的推理再一次无能为力。如果这一影响仅仅是一种物理的或化学的力量，那么它就会与有序组织的生长力**结合**起来，我们观察到的就应该是这两种力量作用的**合成**。一部分相对于另一部分来说成为完全主导，这不属于物理现象而是生命现象，这是德里施的“调控”的一个实例。

根据胚胎学领域现已观察到的所有现象，可以很客观地得到如下结论，概括地说，发育可以分为 3 个阶段，即：(1) 卵分裂为细胞——即卵裂；(2) 这些细胞分化形成 3 个主要的胚层——外胚层、内胚层和中胚层；(3) 一个胚层的某些部分作用于邻近的属于其他胚层的部分从而形成特定的器官——即组织原的作用。

然而，最根本的问题是，卵细胞质的最初组成物质来源于何处？这是我们现在必须面对的问题。唯一可能的答案是细胞核。正如我们看到的，事实上许多将要受精的卵子都拥有已分化了的细胞质。但是这些细胞质在卵子**未成熟时**并没有分化，而是在卵子逐渐成熟的过程中，它们的细胞核向细胞质中释放了物质。特别是，在许多研究中都重复观察到核仁分裂成碎片，然后这些碎片穿过核膜，融合到细胞质中。就拿线虫类蛔虫的卵子这样一个特化的卵子来说，博韦里的实验已向我们表明，如果这些卵子**未成熟时**受到离心力的作用，那么大部分细胞质都会丢掉，但是减小了的卵子仍能发育成一个典型的胚胎。施莱普和普西布兰也赞成这一结论。不过我们觉得似乎可以进一步得到他们还没有想清楚的某些结论。囊胚细胞的分化一定是由于细胞核又释放出了某些物质。但是处于这些早期发育阶段的细胞核都是很相似的。赫特维希通过压力实验发现，这些细胞核可能就像被拨弄的一堆大理石一样并没有发生任何变化。此外，从大多数细微的细胞学检查结果可以判断，在整个发育过程中，这些细胞核的基本组成都没有发生变化。所以我们想到，**细胞核对细胞质的间歇性作用**引起了细胞的连续分化，即各个发育阶段。发育过程正是通过这些阶段逐渐走向一个确定的结果，所以如果存在生机的话，我们可以断定它是通过细胞核释放出的物质来发挥作用的。这些释放物就是于克斯屈尔在其《理论生物学》(1927 年）中称为发育“推动力”的相关物质。于克斯屈尔断言，对这些释放物的区分是生物分析所能达到的最大极限。

Comparative embryology, however, can go further, and Schleip rightly insists that experimental embryology ought to be comparative. These embryonic stages are soon discovered to be merely smudged and simplified forms of larval stages which in allied forms lead a free life in the open, seeking their own food and combating their own enemies. These larval forms in turn are seen to be nothing but modified and simplified editions of adult forms in the past history of the race. Therefore, in the last resort, development is found to be due to the successive coming to the surface of a series of racial memories, and the entelechy might be defined as a "bundle" of such memories.

The so-called Mendelian "genes", however, constitute a problem for the embryologist; for the conception of the hereditary make-up which they induce in the minds of geneticists is totally at variance with that which the embryologist draws from the study of development. Schleip and Przibram struggle valiantly to reconcile the two conceptions and fail. Dürken alone boldly questions the validity of the whole conception of the genes and points out how much it is purely arbitrary and theoretical. If the results of a crossing experiment agree with expectation based on the ordinary Mendelian rules, then it proves the reality of genes; if the results do not agree, the geneticist denies that it disproves them, because he immediately postulates the action of an undiscovered "gene" which complicates the result. The real answer to the conundrum was given by Johannsen, when, in his latest publication, deploring the damage and confusion of thought caused by the invention of the word "gene", he states that it represents a mere superficial disturbance of the chromosomes and gives no insight into the real nature of heredity. Even Przibram points out that X-rays will produce "unzählige" mutations, and that there is no correlation between the rays and the nature of the mutation. With these remarks we thoroughly agree.

(**126**, 639-643;1930)

不过，比较胚胎学还可以走得更远，施莱普始终坚持实验胚胎学应该引入比较的方法。人们很快发现这些胚胎发育的阶段仅仅是幼体阶段被混杂和简化后的形式，各个幼体阶段的联合作用最终形成了可以在野外生存的自由生命，它们可以自己寻找食物并与敌人战斗。反过来，这些幼虫形式又仅仅被看作是该种系在过去的发育史中成年形式的修饰简化版本。因此，最终我们会得出发育是一系列种族记忆的逐次苏醒，而生机可以被定义成这些记忆中的“一束”。

但是所谓的孟德尔式“基因”给胚胎学家带来了一个难题，因为遗传学家们推导出的遗传组成的概念与胚胎学家从发育研究中得出的完全不同。施莱普和普西布兰大胆地尝试去调和这两种概念，但最终失败了。迪肯独自勇敢地质疑了基因整个概念的有效性，并且指出这一概念是非常主观和理论性的。如果有交叉实验结果与基于普通孟德尔法则预期的结果一致的话，那么就能证实基因的真实性；如果不一致，那么遗传学家就不会承认该结果能成为他们理论的反证，因为他可以马上假定出一个作用是使该结果变得更加复杂的尚未被发现的“基因”。约翰森给出了这一谜底的真正答案，在他的最后一部著作中，他谴责了“基因”一词的发明给人们造成的思维混乱和困惑，他指出“基因”这个词只不过是给染色体的概念带来了浅薄的干扰，而对于洞察遗传的真正本质并无帮助。甚至普西布兰也指出，X 射线能产生“无数”突变，而射线和突变的本质之间没有任何相关性。我们完全同意这些观点。

（刘皓芳 翻译；刘京国 审稿）

Natural Selection Intensity as a Function of Mortality Rate

J. B. S. Haldane

Editor's Note

J. B. S. Haldane was a prime mover behind the "modern synthesis" in evolutionary biology. This fused Darwin's ideas about selection and Mendel's insights into how traits pass from parents to offspring into a mathematical description of the genetic makeup of populations and how it changes. At the time, some believed these ideas to be antithetical to one another. *Nature* previously published a commentary on Haldane's note by E. W. MacBride, who concluded that he was "convinced that Mendelism has nothing to do with evolution." Here Haldane gives a specific example of his advocacy of mathematical analysis in evolutionary biology, and its superiority over verbal arguments, disproving the notion that selection is limited to the stage of highest mortality.

IN *Nature* of May 31, Prof. Salisbury points out that most of the mortality among higher plants occurs at the seedling stage, and concludes that natural selection is mainly confined to this stage. I believe, however, that this apparently obvious conclusion is fallacious, for the following reason:

Consider two pure lines A and B originally present in equal numbers, and with a common measurable character, normally distributed according to Gauss's law in each group. Let the standard deviations of the character be equal in each group, but its mean value in group A slightly larger than that in group B. Johanssen's beans furnish examples of this type of distribution. Now let selection act so as to kill off all individuals in which the character falls below a certain value. I think that this type of artificial selection furnishes a fair parallel to natural selection, in which chance commonly plays a larger part than heritable differences. Let x be the proportion of individuals eliminated to survivors, and $1+y$ the proportion of A to B among the survivors, so that x measures the intensity of competition, y that of selection.

Then when x is small y is roughly proportional to it. Thus when x increases from 10^{-4} to 10^{-1}, y increases 200 times. But when x is large y becomes proportional to $\sqrt{\log x}$. In consequence y only increases 9 times when x increases from 1 to 10^{12}, and is only doubled when x increases from 1 to 1,800. In other words, when more than 50 percent of the population is eliminated by natural selection, the additional number eliminated makes little difference to the intensity of selection. The theory, which I hope to publish shortly, has been extended to cover cases where the standard deviations differ, and also where populations consist of many genotypes. In general y changes its sign with x, but when x is large y never increases more rapidly than $\log x$.

自然选择强度与死亡率的关系

霍尔丹

编者按

在进化生物学领域，霍尔丹是“现代综合论”的先驱。现代综合论融合了达尔文的自然选择学说和孟德尔关于性状如何从亲代传递到子代的观点，并用数学方法描述了人类的基因构成和基因如何变化。当时，一些人认为达尔文的理论和孟德尔的观点是相互对立的。《自然》早先曾发表了麦克布赖德对霍尔丹的一篇论文的评论，麦克布赖德在评论中声称，他“对孟德尔的遗传学说对于进化毫无意义这一点深信不疑”。在这篇文章中，霍尔丹为了反驳自然选择只局限于最高死亡率阶段的观点，他给出了一个具体的例子。这个例子表明，他支持在进化生物学中采用数学分析的方法，并认为这种方法比文字说明更好。

在 5 月 31 日的《自然》中，索尔兹伯里教授指出高等植物中绝大多数的死亡发生在幼苗时期，由此他断定自然选择主要发生在这个阶段。我认为，这个看似很显然的结论是不正确的，主要原因如下：

假定纯系 A 和纯系 B 在初始状态时具有相同的个体数并都具有某一相同的可测性状，并且这一可测性状在每组群体中都服从高斯分布。假定每组群体中性状分布具有相同的标准差，而 A 系群体的均值比 B 系群体的均值略高。约翰森豆正是能够满足这种分布类型的实例。现在让选择起作用，杀死那些性状低于某一给定值的所有个体。我认为这样的人为选择是非常类似于自然选择的，在自然选择中通常是偶然事件所起的作用比遗传差异更大。用 x 表示死亡个体与存活个体的比例，$1+y$ 表示纯系 A 的存活个体与纯系 B 的存活个体的比例。那么 x 表示的就是竞争的强度，而 y 则表示选择的强度。

这样，当 x 较小时 y 与 x 大致是成比例的。当 x 从 10^{-4} 增加到 10^{-1} 时，y 增加了 200 倍。但是当 x 较大时，y 则变成与 $\sqrt{\log x}$ 成比例。其结果是当 x 从 1 增加到 10^{12} 时，y 仅增加了 9 倍，当 x 从 1 增加到 1,800 时，y 仅翻了一番。换言之，当超过 50% 的个体在自然选择中被淘汰时，再淘汰更多个体对应的自然选择强度的改变是很小的。这一理论已经扩展到纯系间性状分布的标准差不同的情况，以及群体中包含多种基因型的情况。我希望能很快发表这一理论。总体上讲，y 的数值随 x 的改变而改变，但是当 x 较大时 y 增加的速度不会比 $\log x$ 更快。

Careful mathematical analysis seems to disclose the extraordinary subtlety of the natural selection principle, and merely verbal arguments concerning it are likely to conceal serious fallacies.

(**126**, 883; 1930)

J. B. S. Haldane: John Innes Horticultural Institution, Merton Park, London, S.W. 19, Nov. 1.

看起来，细致的数学分析才能够解析自然选择原理的精细之处，而文字说明中很可能隐藏着严重的谬误。

（刘晓辉 翻译；刘京国 审稿）

The Ether and Relativity

J. H. Jeans

Editor's Note

Here the English physicist James Hopwood Jeans responds to a letter from Oliver Lodge criticizing Jeans' recent claim that the laws of the universe would only be penetrated by the use of mathematics. Jeans affirms his belief that "No one except a mathematician need ever hope fully to understand those branches of science which try to unravel the fundamental nature of the universe—the theory of relativity, the theory of quanta and the wave mechanics." Lodge suggested that the universe might ultimately turn out to have been created or designed on aesthetic, rather than mathematical lines. If so, one might expect artists, not mathematicians, to be best suited to fundamental science. But Jeans notices no such aptitude in his artist friends.

I obviously must not ask for space to discuss all the points raised in Sir Oliver Lodge's interesting letter in *Nature* of Nov. 22, and so will attempt no reply to those parts of it which run counter to the ordinarily accepted theory of relativity. For I am sure nothing I could say would change his views here. But I am naturally distressed at his thinking I have quoted him with a "kind of unfairness", and should be much more so, had I not an absolutely clear conscience and, as I think, the facts on my side.

In the part of my book to which Sir Oliver objects most, I explained how the hard facts of experiment left no room for the old material ether of the nineteenth century. (Sir Oliver explains in *Nature* that he, too, has abandoned this old material ether.) I then quoted Sir Oliver's own words to the effect that many people prefer to call the ether "space", and his sentence, "The term used does not matter much."

I took these last words to mean, not merely that the ether by any other name would smell as sweet to Sir Oliver, but also that he thought that "space" was really a very suitable name for the new ether. He now explains he was willing to call the ether "space", "for the sake of peace and agreement". If I had thought it was only *qua* pacifist and not *qua* scientist that he was willing to call the ether "space", I naturally would not have quoted him as I did, and will, of course, if he wishes, delete the quotation from future editions of my book. But I did not know his reasons at the time, and so cannot feel that I acted unfairly in quoting his own words verbatim from an Encyclopaedia article.

Against this, I seem to find Sir Oliver attributing things to me that, to the best of my belief, I did not say at all, as, for example, that a mathematician alone can hope to understand the universe. My own words were (p. 128):

以太与相对论

金斯

编者按

英国物理学家詹姆斯·霍普伍德·金斯写这篇文章是为了回应奥利弗·洛奇的一篇快报，奥利弗·洛奇在该快报文章中批评了金斯最近提出的观点，即宇宙中的定律只有用数学方法才能解释清楚。金斯坚持认为“除了数学家以外，没有人能完全理解那些试图揭示宇宙基本性质的科学分支——相对论、量子理论和波动力学。”洛奇认为宇宙的创造和设计最终有可能是按照美学原则而非数学原理进行的。如果事实真的如此，那么艺术家应该比数学家更适合研究基础科学。但金斯在他的艺术界朋友中没有发现这种特别的倾向。

显然，我没有必要要求一个很大的版面来讨论奥利弗·洛奇爵士发表在 11 月 22 日《自然》上的那篇引人注意的快报中提到的全部要点，因而也不会试图回复该快报文章中那些与人们普遍接受的相对论不相符合的内容。因为我知道无论我在这里说什么也不能改变他的观点。但是，如果他认为我是以“一种不公正的方式”引述他的话，我自然会感到不安，而且，如果我不曾拥有一个绝对清晰的意识并认为事实一定站在我这一边的信念，我将会感到更加不安。

在我的书中，奥利弗爵士最不赞同的部分是我关于铁一般的实验事实如何使得 19 世纪的旧的物质性以太再无容身之地的解释。（奥利弗爵士在《自然》上曾解释说他也已经抛弃了这种旧的物质性以太。）接着我引述了奥利弗爵士本人针对很多人更愿意称以太为“空间”这一现象所说的话，他的原话是，“用什么样的术语关系不大。”

我引用最后这句话是想说明，不仅以任何其他方式命名的以太一样合乎奥利弗爵士的胃口，而且他认为“空间”对于新的以太来说确实是一个很合适的名称。他现在解释说自己乐于称以太为“空间”，“为的是息事宁人和意见统一”。如果我以前认为他只是作为和平主义者而不是作为科学家才乐于称以太为“空间”，自然就不会像我之前那样引述他的话，当然，如果他希望的话，我将在我那本书的新版中删去那段引文。但当时我并不知道他是这样想的，所以我不知道我从一篇百科全书的文章中一字不差地引述他的原话是不公正的做法。

与此相反，我可以非常负责任地说，我发现奥利弗爵士把我根本没有说过的话强加在我身上，比如，单凭数学家就能理解宇宙。我的原话是这样的（第 128 页）：

"No one except a mathematician need ever hope fully to understand those branches of science which try to unravel the fundamental nature of the universe—the theory of relativity, the theory of quanta and the wave-mechanics."

This I stick to, having had much experience of trying to explain these branches of science to non-mathematicians. In the same way, if the material universe had been created or designed on aesthetic lines—a possibility which others have contemplated besides Sir Oliver Lodge—then artists ought to be specially apt at these fundamental branches of science. I have noticed no such special aptitude on the part of my artist friends. Incidentally, I think this answers the question propounded in the News and Views columns of *Nature* of Nov. 8, which was, in brief:—If the universe were fundamentally aesthetic, how could an aesthetic description of it possibly be given by the methods of physics? Surely the answer is that if the objective universe were fundamentally aesthetic in its design, physics (defined as the science which explores the fundamental nature of the objective universe) would be very different from what it actually is; it would be a *milieu* for artistic emotion and not for mathematical symbols. Of course, we may come to this yet, but if so, modern physics would seem rather to have lost the scent.

However, I am glad to be able to agree with much that Sir Oliver writes, including the quotations from Einstein which he seems to bring up as heavy artillery to give me the final *coup de grace*:—"In this sense, therefore, there exists an ether", and so on. On this I would comment that nothing in science seems to exist any more in the good old-fashioned sense—that is, without qualifications; and modern physics always answers the question, "To be or not to be?" by some hesitation compromise, ambiguity, or evasion. All this, to my mind, gives strong support to my main thesis.

(**126**, 877; 1930)

J. H. Jeans: Cleveland Lodge, Dorking, Nov. 23.

“除了数学家以外，没有人能完全理解那些试图揭示宇宙基本性质的科学分支——相对论、量子理论和波动力学。”

我仍然坚持这一点，因为在试图向非数学家解释上述科学分支方面我已经积累了许多经验。同样地，如果物质性宇宙的创造和设计是按美学原则进行的——除了奥利弗·洛奇爵士之外还有其他人也曾考虑过这种可能性，那么艺术家应该特别容易理解这些基础的科学分支。迄今为止，在我的艺术家朋友中我并没有发现这种特别的倾向。顺便提一句，我认为这解答了 11 月 8 日《自然》的“新闻与视点”栏目中提出的问题，简单地说就是：如果宇宙基本上是美学的，那么物理学方法又怎么能给出一个关于宇宙的美学描述呢？答案只能是这样的，如果客观的宇宙基本上是依美学观点设计的，那么物理学（定义为研究客观宇宙基本性质的科学）将会与它实际的样子大为不同；这种依美学观点设计的**宇宙环境**将适合于艺术情感而非数学符号。当然，我们可以这样做，但是果真如此的话，现代物理学似乎会迷失方向。

不过，我很高兴在很多方面与奥利弗爵士意见一致，包括他引用的爱因斯坦的话——“所以就这个意义而言，是有某种以太存在”等，看起来他要把这些当作重炮给我最终的**致命一击**。对此我的评论是，在科学中没有什么东西能以旧有的形式安然存在，这是绝对的；现代物理学经常会以某种迟疑不决的折中、模棱两可或者遁词来回答“存在还是不存在？”的问题。在我看来，所有这些都强有力地支持了我的主要观点。

（王耀杨 翻译；张元仲 审稿）

The X-Ray Interpretation of the Structure and Elastic Properties of Hair Keratin

W. T. Astbury and H. J. Woods

Editor's Note

William Astbury was an X-ray crystallographer based in a university department supported by funds from the wool and leather industries of the county of Yorkshire in northern England. He was one of the first to use X-ray analysis for the study of the structure of complicated polymer molecules; during his career in Leeds, he also studied the structure of deoxyribonucleic acid (DNA) and even arrived at the correct spacing between successive units in the polymer. His biggest handicap was that he made enemies easily. Here he describes the X-ray analysis of a fibrous "structural" protein, keratin, the main component of hair. Such proteins were Astbury's forte.

RECENT experiments,[1] carried out for the most part on human hair and various types of sheep's wool, have shown that animal hairs can give rise to two X-ray "fibre photographs" according as the hairs are unstretched or stretched, and that the change from one photograph to the other corresponds to a reversible transformation between two forms of the keratin complex. Hair rapidly recovers its original length on wetting after removal of the stretching force, and either of the two possible photographs may be produced at will an indefinite number of times. Both are typical "fibre photographs" in the sense that they arise from crystallites or pseudo-crystallites of which the average length along the fibre axis is much larger than the average thickness, and which are almost certainly built up in a rather imperfect manner of molecular chains—what Meyer and Mark[2] have called *Hauptvalenzketten*—running roughly parallel to the fibre axis.

Hair photographs are much poorer in reflections than are those of vegetable fibres, but it is clear that the α-keratin, that is, the unstretched form, is characterised by a very marked periodicity of 5.15 Å along the fibre axis and two chief side-spacings of 9.8 Å and 27 Å (? mean value), respectively; while the β-keratin, the stretched form, shows a strong periodicity of 3.4 Å along the fibre axis in combination with side-spacings of 9.8 Å and 4.65 Å, of which the latter is at least a second-order reflection. The β-form becomes apparent in the photographs at extensions of about 25 percent and continues to increase, while the α-form fades, up to the breaking extension in cold water, which is rarely above 70 percent. Under the action of steam, hair may be stretched perhaps still another 30 percent, but no other fundamentally new X-ray photograph is produced. The question is thus immediately raised as to what is the significance of a crystallographically measurable transformation interpolated between two regions of similar extent where no change of a comparable order, so far as X-ray photographs show, can be detected.

X 射线衍射法解析毛发角蛋白的结构与弹性

阿斯特伯里，伍兹

编者按

威廉·阿斯特伯里是一位在大学工作的X射线晶体学家，他所在的系受到了北英格兰约克郡羊毛和皮革业多家企业提供的研究基金的支持。他是利用X射线解析复杂多聚体分子结构的先驱之一。在利兹工作期间，他还研究了脱氧核糖核酸(DNA)的结构，甚至得出了这一聚合物中相邻单体分子之间的正确距离。对他来说最不利的是他太容易树敌。在这篇文章中，他描述了对角蛋白这种纤维状“构造的”蛋白质的X射线衍射分析。角蛋白是毛发的主要组成部分。阿斯特伯里非常擅长于研究这种蛋白。

最近的一些主要针对人类毛发和各种类型羊毛的实验[1]表明，动物毛发在拉伸状态和非拉伸状态下可以产生两种不同的X射线“纤维衍射图”，而且从一种衍射图向另一种衍射图的转变对应于角蛋白复合体的两种形式之间的可逆转变。在撤去外界拉力后，浸湿的毛发会很快恢复到原来的长度，因此可以无限次地重复得到这两种衍射图中的任意一种。这两种衍射图都来自于晶体或伪晶体，从这一意义上来说，这两种衍射图都是典型的“纤维衍射图”。在相应的晶体和伪晶体中，沿中心轴方向的平均长度比平均厚度大得多，这些晶体和伪晶体极有可能是由一种非常不完整的、几乎平行于中心轴的分子链（被迈尔和马克[2]称为**主分子链**）搭建起来的。

与植物纤维的衍射图相比，毛发衍射图中的反射线非常少，但很清楚的是，非拉伸状态的 α 角蛋白明显表现出了沿中心轴 5.15 Å 的周期性以及宽度分别为 9.8 Å 和 27 Å（均值？）的两个主要的侧向间隔。处于伸展状态的 β 角蛋白则表现出沿中心轴 3.4 Å 的很强的周期性以及宽度分别为 9.8 Å 和 4.65 Å 的两个侧向间隔，其中后一个侧向间隔至少是二次反射。当毛发在冷水中被拉伸 25% 时，衍射图中开始出现 β 类型，随着拉伸幅度的增大 β 类型越来越多，与此同时，α 类型逐渐消失，直到毛发被拉断（在冷水中毛发的拉伸幅度很少超过 70%）。在蒸汽的作用下，毛发也许还能再被拉伸 30%，但即便这样也不会出现本质上全新的X射线衍射图。这样立刻就提出了一个问题，就X射线图而言，在拉伸程度接近、没有发生相对次序改变的两个区域之间，晶体学上可测量的相互转变的意义是什么呢？

The elastic properties of hair present a complex problem in molecular mechanics which up to the present has resisted all efforts at a satisfactory explanation, either qualitative or quantitative. Space forbids a detailed discussion here of the almost bewildering series of changes that have been observed, and we shall merely state what now, after a close examination of the X-ray and general physical and chemical data, appear to be the most fundamental.

(1) Hair in cold water may be stretched about twice as far, and hair in steam about three times as far, as hair which is perfectly dry. (2) On the average, hair may be stretched (in steam) to about twice its original length without rupture. (3) By suitable treatment with steam the discontinuities in the load/extension curve may be permanently smoothed out, the original zero is lost, so that the hair may be even contracted by as much as one-third of its original length, and elasticity of form may be demonstrated *in cold water* over a range of extensions from −30 percent to +100 percent. (4) The elastic behaviour in steam is complicated by "temporary setting" of the elastic chain and ultimately by a "permanent setting" of that part which gives rise to the fibre photograph. (5) That part of the elastic chain which is revealed by X-rays acts *in series* with the preceding and subsequent changes.

On the basis of these properties and the X-ray data, it is now possible to put forward a "skeleton" of the keratin complex which gives a quantitative interpretation of the fundamentals, and may later lead to a correct solution of the details. The skeleton model is shown in Fig. 1. It is simply a peptide chain folded into a series of hexagons, with the precise nature of the side links as yet undetermined. Its most important features may be summarised as follows:—(1) It explains why the main periodicity (5.15 Å) in unstretched hair corresponds so closely with that which has already been observed in cellulose, chitin, etc., in which the hexagonal glucose residues are linked together by oxygens. (2) When once the side links are freed, it permits an extension from 5.15 Å to a simple zigzag chain

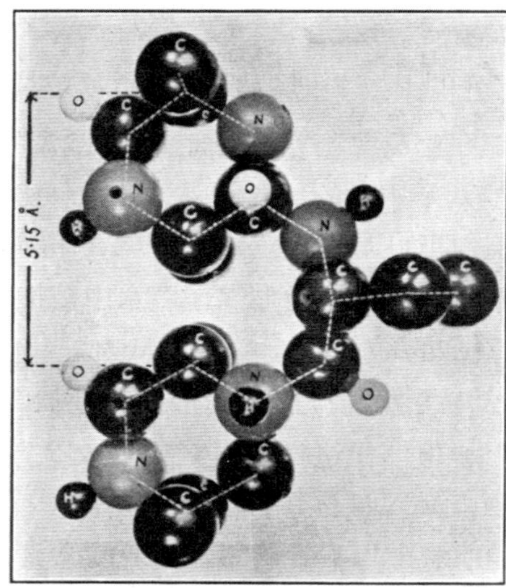

Fig. 1

毛发弹性的分子机制是个复杂的问题，无论是定性的还是定量的，至今为止还没有得到任何令人满意的结论。篇幅所限，我们这里将不再详细讨论曾观测到的纷繁复杂的一系列变化，只介绍对 X 射线结果和基本的物理和化学数据进行详细考察后得到的一些看起来最重要的结论。

（1）和完全干燥的毛发相比，在冷水中毛发可以被拉伸到大约原长的两倍，而在蒸汽中可以被拉伸到大约原长的三倍。（2）平均而言，毛发（在蒸汽中）可以被拉伸到原长的两倍而不断裂。（3）适当的蒸汽处理可以永久地消除毛发载荷–伸长曲线的不连续性，原来的零点就没有了，这样毛发甚至可以收缩到其原长的 2/3，而**在冷水中**其所表现的弹性介于收缩 30% 到伸长 100% 之间。（4）在蒸汽中毛发的弹性行为由于存在弹性链的“临时形态”而变得复杂，但最后变成“固定形态”，从而产生纤维衍射图。（5）X 射线衍射图显示，部分弹性链**连续地**进行先前和随后的变化。

在这些特性以及 X 射线衍射数据的基础上，我们现在可以提出一种角蛋白复合体的“骨架”，这可以定量地解释一些基本问题，将来或许还能引导对细节问题的正确解答。骨架模型如图 1 所示。该骨架模型只是简单地展示了折叠成几个六边形的一条多肽链，其中侧链的准确性质目前还没有被测定。该模型最重要的特征可以总结如下：（1）它解释了为什么非拉伸状态下毛发中的主要周期（5.15 Å）与在纤维素、几丁质等物质中观察到的周期十分相近，这是因为在这些物质中六边形

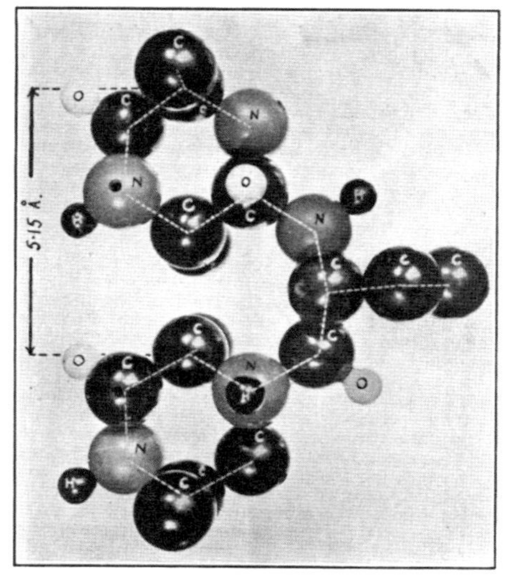

图 1

of length 3 × 3.4 Å, that is, 98 percent, and also allows for possible contraction below the original length, without altering the inter-atomic distances and the angles between the bonds. (3) It explains why natural silk does not show the long-range elasticity of hair, since it is for the most part already in the extended state,[3] with a chief periodicity of 3.5 Å. We may now hope to understand why it is that the photographs of β-hair and silk are so much alike. (4) It gives a first picture of the "lubricating action" of water and steam on the chain, since X-rays show that the direction of attack is perpendicular to the hexagons and that this spacing remains unchanged on stretching. Furthermore, it now seems clear that the new spacing, 4.65 Å, is related to the old by the equation $27/(3\times4.65) = 3\times3.4/5.15$ (very nearly), that is, the transformation elongation takes place directly at the expense of the larger of the two side-spacings. In the particular arrangement of the hexagons shown in the model, the side chains occur in pairs on each face, and it may well be that the action of water is the opening-up of an internal anhydride between such adjacent side chains. (5) The chain being built up of a succession of ring systems stabilised and linked together in some way by side chains of the various amino-acids, we have here an explanation of the well-known resistance of the keratins to solvents and enzyme action. In addition, each hexagon is effectively a diketo-piperazine ring, an interesting point in view of the evidence which has been brought forward by Abderhalden and Komm[4] that such groups pre-exist in the protein molecule. It may also throw light on the stimulating researches of Troensegaard.[5] (6) There are three principal ways of constructing the model, according to which group lies at the apex of a hexagon. It thus affords an explanation of the apportioning of a transformation involving a 100 percent elongation into three approximately equal regions which may be opened up in turn under the influence of water and temperature and other reagents. The modification shown in the model must be ascribed to the crystalline phase, since it would, alone of the three, be expected to give rise to a strong reflection at 5.15 Å, as in the α-photograph.

A detailed account of the above work will be published shortly.

(**126**, 913-914; 1930)

W. T. Astbury, H. J. Woods: Textile Physics Laboratory, The University, Leeds, Nov. 15.

References:

1. W. T. Astbury, *J. Soc. Chem. Ind.*, **49**, 441; 1930.
2. Meyer and Mark, "Der Aufbau der hochpolymeren organischen Naturstoffe".
3. Meyer and Mark, *Berichte*, **61**, 1932; 1928.
4. Abderhalden and Komm, *Z. Physiol. Chem.*, **139**, 181; 1924.
5. Troensegaard, *Z. physiol. Chem.*, **127**, 137; 1923.

的葡萄糖残基可以通过氧原子相互连接起来。(2) 当侧向的连接解开后，模型中长为 5.15 Å 的主链单元就会伸展成 3 × 3.4 Å 的简单“之”字形长链，也就是伸展了 98%，同样也可能会因为收缩而使长链短于初始长度，但不论伸展还是收缩，原子间距和键–键之间的角度都不会发生改变。(3) 它解释了为什么自然丝的伸缩性没有毛发那么大，因为大多数情况下自然状态的丝已经处于主周期为 3.5 Å 的伸展状态。[3] 这样我们就能够理解为什么 β 型毛发的衍射图与丝蛋白的衍射图十分相似。(4) 这一模型首次给出了水和蒸汽对主链的“润滑作用”的图片。X 射线衍射结果显示，水分子攻击主链的方向是垂直于六边形平面的，而且在主链拉伸过程中六边形中各原子之间的空间间隔没有发生改变。另外，比较明确的是，拉伸后新的侧向间隔（4.65 Å）与原来的侧向间距之间的关系符合方程 $27/(3 \times 4.65) = 3 \times 3.4/5.15$（非常近似），这就是说，主链的伸长是以两个侧向间隔的减小为直接代价的。在模型中显示的六边形的某些特殊排列中，在六边形平面任何一边的侧链总是成对出现，水分子的作用很可能就是打开这些相邻侧链内部的酐键。(5) 由连续的环形系统构成的主链是通过各种不同的氨基酸的侧链以某种方式保持稳定并连接在一起的，这样我们就可以解释众所周知的角蛋白对溶剂作用和酶作用的抗性。另外，每个六边形都是一个有效的二酮哌嗪环，这一点很有趣，考虑到阿布德哈尔登和科姆 [4] 曾提出证据说这种基团结构预先就存在于蛋白分子中，那么这个有趣的特点也许会对特森加德的研究工作有所启示。[5] (6) 根据位于六边形顶点上的基团的不同，有 3 种主要的模型构建方法。这样就可以把长链 100%的伸长分配到 3 个近似相等的区域上。在水、温度或其他试剂的影响下，这些区域会依次伸展开。对模型的这些修正要归因于晶相结构的特点，因为可以预期在 3 种结构中只有晶相结构会给出如 α 衍射图所示的在 5.15 Å 处出现的强反射。

关于以上工作的详细说明将在不久之后发表。

（高如丽 翻译；刘京国 审稿）

Embryology and Evolution

Editor's Note

In 1930 Irish-born zoologist Ernest William MacBride had exposed his Lamarckian leanings in a review of several books on experimental embryology, arguing that the genes-eye view of the world was "totally at variance with that which the embryologist draws from the study of development." Here, embryologist G. L. Purser conjures up a nice analogy that he thinks illuminates the role that genes do (and do not) play in embryonic development or "epigenesis". This letter contains some prescient reflections about the mutual interdependence of genes and their immediate environment. However, Purser goes too far in claiming a new proof for the inheritance of acquired characteristics when he suggests "that the environment is in some way responsible for the appearance of the gene".

I have read with much interest Prof. MacBride's review entitled "The Problem of Epigenesis", and I should like to make a few remarks upon what he says at the end. First of all, I wonder if the following analogy will help him, as it has helped me, to reconcile the conceptions of the geneticist with those of the embryologist. In a modern motor works the cars, so I understand, move along a track past a series of workmen, each of whom has one particular job to do, which is related to what has already been done and also to what is going to be done afterwards. Now if we imagine that all the parts and materials which are going to make up the finished car represent the substances in the developing embryo and that the workmen are the genes, we have an analogy which can be carried surprisingly far. Not only will it give us a picture of normal development, but we can see, by altering one of the parts, how a variation may occur; by altering a workman, how "sports" may arise; and, by adding a new workman with a new job, how progressive evolution may take place.

There is no need for me to occupy space in working the analogy out, for anyone can do it for himself: what is more important is to point out where the analogy fails. A motor-car is adapted for life on the road, and, until it is completed, it has, for all practical purposes, no environment at all comparable with that which bears upon an embryo throughout its development. So whereas a feature of a car is simply due to the action of the workman on the materials, a feature of an animal is the result of the combined action of the genes and of the environment upon the materials of the embryo. Genes without the appropriate materials can produce nothing; genes with the appropriate materials can only produce a partially developed structure; but genes with the appropriate materials and environment can produce the fully developed functional character. Hence it is that in the development of the frog, for example, the gill-clefts, etc., are full developed, whereas in the Amniota, with the radical change in the environment of the early stages, such structures are only

胚胎学与进化

编者按

1930 年，出生于爱尔兰的动物学家欧内斯特·威廉·麦克布赖德在评论几本关于实验胚胎学的著作时，表示了自己对拉马克学说的认同。他认为从基因的角度看到的世界“与胚胎学家从发育研究中得出的完全不同。”在这里，胚胎学家珀泽想出了一个很好的类比，他认为这样就可以解释基因在胚胎发育或“渐成论”中是否起作用。这篇文章在基因与周围环境的相互依赖关系这个问题上提出了一些很有远见的观点。但是珀泽在阐述获得性遗传假说的新证据时说，“环境在某种程度上决定了基因的表现形式”，这就过于偏激了。

我怀着极大的兴趣读完了麦克布赖德教授那篇名为《关于渐成论的问题》的书评，看完后我想对他的书评结尾处的内容发表一下我的看法。首先，我想知道如下的类比是否会像其帮助过我一样，也能够帮助他化解遗传学家的观念与胚胎学家的观念之间的分歧。我是这样理解的，现代汽车工业中，小汽车是沿着一条由一系列工人组成的生产线移动的，生产线上的每个人都有自己特定的工作要做，每个人的工作都与已经完成的工作以及之后将要进行的工作有关。这时如果我们想象组成成品车辆的所有零件和材料代表正在发育的胚胎里的物质，工人代表基因，那么我们就有了一个蕴含着深远意义的类比。这一类比不仅可以给我们描绘出一幅正常发育的图像，而且从中我们可以看到：变异是如何通过改变众多零件中的一个而实现的；“变种”是如何通过更换一名工人而产生的；以及渐进演化是如何通过增加一名从事新工作的新工人而发生的。

我没有必要浪费版面在这里推演这一类比，因为任何人都可以独自完成，但更重要的是指出这种类比在哪些地方不适用。一辆汽车是适于在路面上奔跑的，并且在生产出来后它就已经适于各种实用目的，而在胚胎生长发育的整个过程中都没有任何环境可以与此相类比。因此汽车的特征仅仅取决于工人对原材料进行的处理，而动物的特征则是基因和环境对胚胎物质共同作用的结果。如果没有适当的胚胎物质，那么基因什么也产生不了；如果只有适当的胚胎物质，那么基因只能产生部分发育的结构；只有既有适当的胚胎物质，又有环境时，基因才能产生充分发育的有功能的结构。因此，如腮裂等结构，在青蛙的发育过程中是充分发育的，在羊膜动物中则因为在发育早期遭遇到剧烈的环境变化而只是部分发育的。引用麦克布赖德

partially developed and the stages, to quote Prof. MacBride, are smudged.

Looked at from this point of view, two other conclusions of great importance are unavoidable. The first is that the recapitulation of an ancestral stage of the evolution of an animal, as distinct from the repetition of an ancestral character, will only occur when the early stage of development is passed in the same environment as that of the ancestor, which environment is different from that of the present-day adult. Only under such conditions will the genes responsible for the adult ancestral characters give rise to them all together without any great admixture of other features; though it must always be borne in mind that such stages in the life history, being larvae, may evolve on their own account and, therefore, may have features which the ancestor never had. In parenthesis, I should just like to add here that, so far as I know, a larva has never been properly defined: such a definition would be "A free-living stage in an animal's life history which fends for itself and possesses certain characters which it has to lose before it can become a young adult": the possession of *positive* characters distinguishes a larva, not its lack of adult ones.

The other conclusion is reached thus. The appearance of a functional feature is dependent, as we have seen, upon the interaction of three things: the materials of the embryo, the genes, and the environment. Now the facts of Mendelian inheritance give clear evidence that there need be no change in the materials of an embryo for a new gene to modify the form, so, in discussing the origin of a new feature, there is no need to consider a change in the materials as one of the essential factors. The fortuitous appearance of a gene without the appropriate environment would produce a partially developed character, but, in actual experience, we do not find features in a partially developed condition which *have never been functional* at any period in the history of the race. So the genes must, in actual fact, only arise after the suitable environment is present; and the only conclusion to be drawn from that is that there is a causal relation between the two; that is, that the environment is in some way responsible for the appearance of the gene, which is surely nothing more or less than the basis of a new proof of the inheritance of acquired characters.

G. L. Purser

* * *

I have read with interest Mr. Purser's thoughtful letter on the subject of my review. If he will substitute the term "race-memory" for "gene", we shall not be far apart. But the gene of the Mendelian stands out as something that is never functional. "No one," said the late Sir Archdall Reid, "ever heard of a useful gene." When one takes into consideration the fact that the Mendelian genes in *Drosophila* have been shown to increase in their damaging effect on the viability of the organism in proportion to the structural change which they involve, and when further it is discovered that genes can be artificially produced by irradiating insect eggs with X-rays—a process which kills most of the eggs—one is driven

教授的话说，这些阶段都是遗留下的痕迹。

从这个观点来看，将会不可避免地得出另外两个非常重要的结论。第一个是，只有当胚胎在发育早期阶段所处的环境与该种生物的祖先所处环境一样时，才会出现对其祖先进化过程的重演。这种环境与现在的成体所处的环境是不同的，这种重演与重现祖先特征是完全不同的概念。只有在这种情况下，那些决定祖先成体特征的基因才完全产生效果，而不与其他特征发生大规模的混合；但是我们必须时刻牢记，生命过程中的这些阶段（即幼虫）都是自行进化的，因此它们可能具有祖先从不具有的特征。另外，我想在此补充说明的是，据我所知，幼虫这一概念从未被恰当地定义过。如下描述可能是一个比较恰当的定义，“在动物生命史中的一个自由生活的阶段，在此阶段中它们自己照料自己并具有一定的特征，但一旦它们成长为年轻的成体，这些特征就会消失。”幼虫拥有**初级的**特征而成虫没有，根据这一点能够将它们区别开来。

第二个结论是按照如下所述得出的。正如我们看到的一样，功能性特征的出现依赖于 3 个条件的相互作用：胚胎物质、基因和环境。现在，孟德尔的遗传结果已经给出了明确的证据，表明一个新基因形式发生改变时胚胎物质并没有发生改变，所以在讨论新特征的起源时，就没有必要将胚胎物质的变化作为基本因素之一来考虑。偶尔会出现某个没有适当环境的基因，这时该基因会产生部分发育的结构，然而实际上在种系史的任何阶段我们都没发现部分发育的**无功能**结构的出现。因此，事实上基因肯定是在适当环境出现之后才产生的。从这一点，我们可以得到的唯一结论是：基因与环境之间存在因果关系，即，从某种意义来说，环境在某种程度上决定了基因的表现形式，这正好可以作为获得性性状遗传的新证据。

珀泽

* * *

我已经饶有兴趣地读完了珀泽先生就我的评论所写的颇具思想性的快报。如果他用“种族记忆”一词来代替“基因”，那么我们的分歧并不大。但是，孟德尔式基因正是因其根本不具有功能而格外引人注目。已故的阿奇德尔 · 里德爵士说：“没有人曾听说过一个有用的基因。”考虑到果蝇的孟德尔式基因对该物种生存能力的损伤有所增加，且损伤增加的程度与相关基因发生的结构变化成比例这一事实，再加上人们又发现可以通过 X 射线辐射昆虫卵子来人为地产生基因（这一辐射过程能够杀

to the conclusion that a gene is germ damage of which the outward manifestation is a mutation. The only effect that natural selection would have on such aberrations would be to wipe them out. In my opinion, mutations and adaptations have nothing to do with one another and only adaptations are recapitulated in ontogeny.

E. W. MacBride

(**126**, 918-919; 1930)

G.L. Purser: The University, Aberdeen, Oct. 29.

死大部分卵子），我们就会被引向如下结论：基因是外在表现为突变的配子损伤。自然选择对这种异常的唯一反应就是清除它们。在我看来，突变和适应没有任何关系，只有适应性能够在个体发育中重演。

麦克布赖德

（刘皓芳 翻译；刘京国审稿）

Unit of Atomic Weight

F. W. Aston

Editor's Note

Francis Aston, working at the Cavendish Laboratory of the University of Cambridge, had devised an instrument for measuring the atomic masses of individual atoms, now called the mass spectrometer. This led him earlier to postulate the notion of isotopes, which have identical chemical properties but different masses. Having used the device to identify the isotopes of more than 80 different chemical elements, Aston here advocates the need for a new standard of atomic mass, to replace the practice then current of referring all masses to that of oxygen—for this element, having several isotopes, is not an appropriate reference point.

THE discovery of the complexity of oxygen clearly necessitates a reconsideration of the scale on which we express the weights of atoms. Owing to the occurrence of O^{17} and O^{18}, now generally accepted, it follows that the mean atomic weight of this element, the present chemical standard, is slightly greater than the weight of its main constituent O^{16}. The most recent estimate of the divergence is 1.25 parts per 10,000.

This quantity, even apart from its smallness, is not of much significance to chemists, for the experience of the last twelve years has shown that complex elements do not vary appreciably in their isotopic constitution in natural processes or in ordinary chemical operations. Physics, on the other hand, is concerned with the weights of the individual atoms, and by the methods of the mass-spectrograph and the analysis of band spectra it is already possible to compare some of these with an accuracy of 1 in 10,000. Furthermore, the theoretical considerations of the structure of nuclei demand an accuracy of 1 in 100,000, which there is reasonable hope of attaining in the near future. The chemical unit is clearly unsuitable, and it seems highly desirable that a proper unit for expressing these quantities should be decided upon.

The proton, the neutral hydrogen atom, one-quarter of the neutral helium atom, one-sixteenth of the neutral oxygen atom 16, and several other possible units have been suggested. None of these is quite free from objection. It is desirable that this matter should be given attention, so that when a suitable opportunity occurs for a general discussion of the subject, each point of view may be afforded its proper weight in arriving at a conclusion.

(**126**, 953; 1930)

F. W. Aston: Trinity College, Cambridge, Dec. 4.

原子量的单位

阿斯顿

编者按

在剑桥大学卡文迪什实验室工作的弗朗西斯·阿斯顿设计了一种用来测量单个原子的原子质量的仪器，现在我们称之为质谱仪。这使得他更早地提出了同位素（化学性质完全相同但质量不同）的概念。利用该仪器鉴定了超过 80 种化学元素的同位素以后，在这篇文章中阿斯顿主张，有必要采用一种新的原子质量的标准，以取代当时测量所有其他原子质量时利用氧作为标准的做法——因为氧有多种同位素，不适合作为原子质量的参考标准。

氧元素具有多种同位素，这一发现无疑使我们必须重新考虑用以表述原子量的标度。目前在化学上是以氧元素的平均原子质量作为原子量的标准，但由于现在同位素 O^{17} 和 O^{18} 的发现已经得到了普遍的认可，所以氧元素的平均原子质量略大于氧元素中的主要组成部分 O^{16} 的原子质量。最新的估计表明这一差别是 0.125‰。

这个差值对于化学家来说没有太大的意义，更别说它还非常微小，因为最近 12 年的研究经验已经表明，对于同位素形式复杂多样的元素来说，在自然过程或者化学操作中其各种同位素的相对丰度并不发生明显改变。但是，物理学研究要考虑单个原子的质量，通过质谱仪和谱带分析的方法目前已经可以以万分之一的精度来分辨某些原子质量的差别。此外，对原子核结构的理论研究需要十万分之一的精度，在不久的将来对原子质量的测定有望能够达到这一精度。因此，目前化学上采用的原子量的单位显然是不合适的，对于物理学研究来说，似乎迫切需要确定一个能够描述这些量的合适的原子量单位。

质子的质量，中性氢原子的质量，氦原子质量的 1/4，中性氧同位素 O^{16} 原子质量的 1/16，以及其他几种可能的原子量单位都被提出来了。但所有这些无一例外都遭到了一些反对。这个问题应该受到关注，一旦出现一个对这一主题进行广泛讨论的合适时机，那么为了得出结论每一种观点都可以适当地发挥作用。

（王锋 翻译；李芝芬 审稿）

Embryology and Evolution

J. B. S. Haldane

Editor's Note

Renowned biologist J. B. S. Haldane was one of the key figures in the development of population genetics, a field underpinned by a Mendelian view of inheritance. Here Haldane has no time for the views of the neo-Lamarckian zoologist Ernest William MacBride, one of a dwindling number of scientists still prepared to dismiss the notion of a gene. With characteristic rhetorical flair, Haldane ridicules MacBride for his outdated views. Publicly aired disagreements like these played an important part in turning the scientific community towards a modern evolutionary synthesis, the idea that a Mendelian mechanism of inheritance could result in the sort of gradual natural selection that Charles Darwin envisaged.

FOUR of Prof. MacBride's statements, in *Nature* of Dec. 6, call for comment. "...no one has ever seen 'genes' in a chromosome." Genes cannot generally be seen, because in most organisms they are too small. In *Drosophila* more than 100, probably more than 1,000, are contained in a chromosome about 1 μ in length. They are therefore invisible for exactly the same reasons as molecules. But the evidence for their existence is, to many minds, as cogent. Where the chromosomes are larger, as in monocotyledons, competent microscopists—for example, Belling, in *Nature* of Jan. 11, 1930—claim to have seen genes. In a case where I (among others) postulated the absence of a gene in certain races of *Matthiola*, my friend Mr. Philp has since detected the absence of a trabant, which is normally present, from a certain chromosome. I shall be glad to show this visible gene to Prof. MacBride.

"...if Prof. Gates were a zoologist instead of being a botanist, he would know that the assumption that 'genes' have anything to do with evolution leads to results...that can only be described as farcical." I should like to direct Prof. MacBride's attention to the droll fact that in a good many interspecific crosses various characters behave in a Mendelian manner, that is, are due to genes. This is so, for example, with the coat colour of *Cavia rufescens*, which, on crossing with the domestic guinea-pig, behaves as a recessive to the normal coat colour, but a dominant to the black. Hence there has been a change in a gene concerned in its production during the course of evolution. Scores of similar cases could be cited.

"All known chemical actions are inhibited by the accumulation of the products of the reaction. An 'autocatalytic' reaction, in which the products of the reaction accelerated it, must surely be a vitalistic one!" Autocatalytic reactions are common both in ordinary physical chemistry and in that of enzymes. Thus the acid produced by the hydrolysis of an

胚胎学与进化

霍尔丹

编者按

在以孟德尔遗传学说为基础的群体遗传学的发展中，著名生物学家霍尔丹算是一个重要的人物。在这篇文章中，霍尔丹没有理会支持新拉马克主义的动物学家欧内斯特·威廉·麦克布赖德（他仍然拒绝接受基因的概念，持这种观点的科学家已经越来越少了。）的观点，而是以他特有的文字才能嘲弄了麦克布赖德的过时思想。在公开场合进行这样的争论，对于学术界最终转向现代进化综合论起到了重要的作用。在现代进化综合论中，可以由孟德尔的遗传机制推出查尔斯·达尔文设想的自然选择学说。

我要对麦克布赖德教授发表于12月6日的《自然》上的4项陈述稍作评论。他说，"……没有人看见过染色体上的'基因'。"确实，基因一般是看不见的，因为大多数生物的基因都太小了。果蝇的一条长约1μ的染色体上可能就有一百多个甚至一千多个基因。因此，就像分子是不可见的一样，基因也是不可见的。不过许多人都认为用来证明基因确实存在的证据是令人信服的。另外，一些有能力的显微镜专家曾公布过他们在具有较大染色体的单子叶植物中看到了基因，例如贝林就在1930年1月11日的《自然》上宣称看到了基因。我（和其他人一起）曾经在一项研究中推测紫罗兰属的某些种是没有基因的，后来我的朋友菲尔普先生检测到了特定染色体上随体的缺失，而在正常情况下这一染色体上是存在随体的。我很乐意向麦克布赖德教授展示这个可见的基因。

"……如果盖茨教授不是植物学家而是动物学家的话，他将了解'基因'与进化具有某些关联，这样的假设产生的结果……只能用滑稽可笑来形容。"我期望麦克布赖德教授能注意到下面这个很有趣的事情：即在许多种间杂交的实例中，很多性状都表现出孟德尔式遗传的特点，也就是说它们是由基因决定的。例如，有一种野生巴西豚鼠，当它与家养豚鼠杂交时，其毛色对于正常毛色表现为隐性，但对于黑色却表现为显性。因此，在进化过程中，与产生该性状相关的基因一定发生了某种变化。与此类似的例子还有很多。

"所有已知的化学反应都会被反应产物的积累所抑制。而在'自催化'反应中，反应产物却可以加快反应的进行，因而这种反应显然就是活力论的一个实例！"其实，自催化反应在普通物理化学和酶化学中都是很常见的。酯水解产生的酸可以促

ester may accelerate its further hydrolysis. As an example of an enzyme action, which for quite simple physico-chemical reasons proceeds with increasing velocity up to 75 percent completion, I would refer Prof. MacBride to Table 7 of Bamann and Schmeller's[1] paper on liver lipase.

In view of such facts, Prof. MacBride's statement that "The term 'autocatalysis' is a piece of bluff invented by the late Prof. Loeb to cover up a hole in the argument in his book" would seem to be a wholly unfounded attack on a great man who can no longer defend himself. If Prof. MacBride would acquaint himself with the facts of chemistry and genetics, he might be somewhat more careful in his criticism of those who attempt to analyse the phenomena of life. He might also cease to ask the question propounded by him in *Nature* of Oct. 25, "whether the organs of the adult exist in the egg preformed in miniature and development consists essentially in an unfolding and growing bigger of these rudiments, or whether the egg is at first undifferentiated material which from unknown causes afterwards becomes more and more complicated and development is consequently an 'epigenesis'." The formation of bone in the embryo chick was shown by Fell and Robison[2] to be due to the action of the enzyme phosphatase, which is neither a miniature bone nor an unknown cause. But so long as he does not take cognisance of recent developments in science, Prof. MacBride will no doubt remain a convinced vitalist.

(**126**, 956; 1930)

J. B. S. Haldane: Biochemical Laboratory, Cambridge University, Dec. 8.

References:

1. *Zeit. Physiol. Chem.*, **188**, p. 167.

2. *Biochem. Jour.*, **23**, p. 766.

进酯的进一步水解。酶反应中也有因为非常简单的物理化学原因而使得反应速率提高 75% 的例子，我想请麦克布赖德教授看一看巴曼和施梅勒关于肝脂酶的那篇文章[1]中的表 7。

麦克布赖德教授还认为，"'自催化'是已故的洛布教授为掩盖其书中观点的漏洞而发明的欺骗性词语。"而从前述的事实来看，麦克布赖德教授的这一观点似乎完全是对一位不可能再为自己辩解的伟人的毫无根据的攻击。如果麦克布赖德教授了解化学和遗传学事实的话，那他在批评那些试图分析生命现象的人时可能会更加谨慎些。他可能也就不会提出自己在 10 月 25 日的《自然》上提到的问题："是不是卵子中就存在预先成形了的成体器官的缩微版，而发育过程实质上是这些器官雏形逐渐展开显露并变大的过程？又或者卵子起初只是未分化的物质，后来由于未知的原因而变得越来越复杂，因而发育是一个'渐成的'过程？"费尔和罗比森的研究[2]显示，鸡胚中骨骼的形成是由于磷酸酶的作用，因而既不存在微型骨骼，也不是由于未知的原因。但是，在麦克布赖德教授认识到科学领域的最新进展之前，他无疑将依旧是一个固执的活力论者。

（刘皓芳 翻译；刘京国 审稿）

Appendix: Index by Subject
附录：学科分类目录

Physics
物理学

Chemistry
化学

Biology
生物学

Astronomy
天文学

Geoscience
地球科学

Engineering & Technology
工程技术

Others
其他